Bioinspired Materials Surfaces

Bioinspired Materials Surfaces

Nature-Inspired Micro-/ Nanostructures

Yongmei Zheng

JENNY STANFORD
PUBLISHING

Published by

Jenny Stanford Publishing Pte. Ltd.
101 Thomson Road
#06-01, United Square
Singapore 307591

Email: editorial@jennystanford.com
Web: www.jennystanford.com

British Library Cataloguing-in-Publication Data
A catalogue record for this book is available from the British Library.

Bioinspired Materials Surfaces: Nature-Inspired Micro-/Nanostructures

ISBN 978-981-5129-34-2 (Hardcover)
ISBN 978-1-003-51388-9 (eBook)

Contents

Preface

In this book, we will focus on the functions and models of biological surfaces with unique wettability and elucidate methods for realizing bioinspired surfaces. We introduce the theories and mechanisms of fabrication and functions to help researchers understand the nature of these surfaces as well as how to design improved functional surfaces for applications. The models of structures are introduced based on lessons from biological surfaces, including lotus leaves, spider silk, butterfly wings, beetle backs, etc. By establishing an understanding of natural biological features, we introduce examples for developing new functional materials related to research areas of physics, chemistry, biology, and materials science that have attracted increasing attention in recent years. Our purpose is to explore concepts of surface design to develop promising applications such as microfluidic devices and functional textiles, as well as corrosion-resistant, liquid transport, antifogging, and water-collecting engineering systems, which offer researchers and students some innovative ideas for the design of novel materials for future scientific works.

Inspired by nature, an exciting confluence of research areas of physics, chemistry, biology, and materials science has arisen in recent years. As is known, the designs of biological surfaces in terms of micro- and nanostructures have received increasing attention. In biological design, we discover that a key aspect consists in using organic or inorganic materials, including those with high/low surface energy or regular/irregular, ordered/disordered, or rough/smooth surfaces, in addition to being endlessly arranged and combined with micro- and nanostructures of various forms. After natural evolution over millions of years, these biological materials exhibit great environmental adaptability and display ideal biological functions. To study biological materials, novel methods and technologies have been developed, such as soft lithography, nanocrystalline growth, dip coating, electrodynamics, fluid coating, self-assembly, and nanocomposite techniques. By these methods and techniques, novel materials and surfaces can be designed. It is interesting that some designs are focused on micro- and nanostructures due to their robust properties. In particular, gradients existing in micro- and nanostructures can effectively achieve functional surfaces. For example, smart wetting control surfaces can be achieved by learning from biological surfaces, including the ultrahydrophobic water repellency on lotus leaves, directional water collection on wetted spider silk, direction adhesion of superhydrophobic butterfly wings, and fog-collecting hydrophobic/hydrophilic patterns on beetle backs, in addition to robust mechanical properties, smart antiadhesion/adhesion, environmental self-adaption, etc.

The book offers novel theories, methods, and techniques to fabricate bioinspired material surfaces, including water-repellent surfaces in Chapter 2, water-collecting surfaces in Chapter 3, anti-icing/ice-phobic surfaces in Chapter 4, surfaces for droplet transport in Chapter 5, and antiadhesive/adhesive surfaces in Chapter 6, which have considerably promising future applications. This book is highly suitable for readers such as scientific PhD students and researchers to learn about the design of smart and advanced materials.

I would like to thank the authors for their publications and my students for collecting literature. I also thank Stanford Chong for offering me the opportunity to publish a book on my favorite subject.

I would also like to acknowledge the support of the National Natural Science Foundation of China (grant nos. 21234001, 21473007, 21771015, and 22275008) and the International School of Beihang University for enabling the preparation of this book.

Yongmei Zheng
March 2024

Chapter 1

Introduction

In this chapter, biological surfaces in nature are introduced, such as lotus leaves, spider silk, cactus spines, butterfly wings, beetle backs, water strider legs, Nepenthes petals, feathers, gecko feet, fish skin, and compound eyes. The physical and chemical theories and foundations are introduced, including the Young equation, Wenzel's law, Cassie's law, two superhydrophobic states, anisotropic wettability, dynamic contact angle and contact angle hysteresis (CAH), liquid spreading and transport, and nonequilibrium interface forces. Based on biological surfaces, biologically structured models have been proposed, such as isotropic micro- and nanostructures, anisotropic structures, heterogeneous wettable patterns, gradient roughness and conical geometry, spindle-knots and humped structures, and liquid-infused interfaces. Some bioinspired methods and techniques have been developed to fabricate surfaces, including electrochemistry, soft lithography, dip coating, microfluidics, electrospinning, imprinting, etc., which have achieved the preparation of bioinspired materials and surfaces, such as spindle-knotted structured fibers, micro- and nanostructures of lotus leaves, and compound-eye structured surfaces.

Bioinspired Materials Surfaces: Nature-Inspired Micro-/Nanostructures
Yongmei Zheng

ISBN 978-981-5129-34-2 (Hardcover), 978-1-003-51388-9 (eBook)
www.jennystanford.com

1.1 Biological Surfaces in Nature

Over billions of years, the surfaces of creatures in nature have evolved to generate almost perfect functions for their activities and survival against harsh environments. The functional properties of biological surfaces are valuable for inspiring us to design novel materials for potential applications in our daily lives, such as materials with water repellency, self-cleaning, anticorrosion, anti-icing, low hydrodynamic friction, water harvesting, and fluid-transport properties. The following sections will present various natural functional surfaces with micro- and nanoscale structures, including lotus leaves, spider silk, cactus spines, butterfly wings, beetle backs, water strider legs, Nepenthes petals, feathers, gecko feet, fish skin, compound eyes of insects, etc., where the wettability functions are related to the types of micro- and nanostructures that are generated by natural creatures. Learning the mechanism of functions from biological surfaces is significant to improve our understanding.

1.1.1 Lotus Leaves

The lotus leaf has been known as a symbol of sacred purity in Asian religions for over 2000 years due to the self-cleaning and superhydrophobic properties of its surfaces. The lotus is called "He Hua" in Chinese, is botanically known as *Nelumbo nucifera*, and is a variety of aquatic plants. The lotus is native to China and is now widely cultivated in China, Japan, India, North America, West Asia, and Middle Asia. Lotuses were planted over 3000 years ago (Bhushan, 2009; Nishimotoab and Bhushan, 2013) and have been regarded as food (seeds and rhizomes), decoration, and medicine. Lotuses also play a significant role in Chinese cultural activities, with a poem written by the famous Chinese poet Tao Yuanming, who lived during the Dongjin Dynasty (approximately 400 AD), in which he says "I admire [that] she [lotus] grows and blooms in the mud, yet neatly and cleanly out of it. I admire [that] she flows in the drops of water the yet graceful seen." For an understanding based on scientific research, Dettre and Johnson found that the surface of the lotus leaf was superhydrophobic (Bhushan, 2012), making the lotus flow "in the drops of water the yet graceful seen". Barthlott and Ehler found that the lotus leaf had a self-cleaning property due to its microstructure, helping the lotus bloom "in the mud, yet neatly and cleanly out of it" (Barthlott and Neinhuis, 1997; Nosonovsky and Bhushan, 2008).

Figure 1.1 Photograph of a lotus leaf with dew droplets in the morning.

In nature, we often observe the phenomenon that water or rain can wash off dust and contaminants on the surfaces of lotus leaves, thereby resulting in the cleaning of the surface. Due to the superhydrophobicity of the surface, a water droplet on it has a spherical shape, as shown in Fig. 1.1, which presents a photograph of a lotus leaf with dew droplets appearing as spheres from water condensation in the morning. According to reports (Koch et al., 2008, 2009), the contact angle of water on lotus leaves is ~164°, and the CAH is ~3°. Thus, water droplets easily roll off the surface at a tilt angle of less than 5°. We, therefore, call such a property superhydrophobicity. Based on biology, it has been revealed that the leaf surface consists of papillae epidermal cells that form microscale papillae or asperities, and the surface of the papillae consists of nanoscale asperities composed of epicuticular hydrophobic hydrocarbon wax tubules (Koch et al., 2008). We can observe the microstructure of the surface on the lotus leaf in situ via environmental scanning electron microscopy (ESEM). Figure 1.2 shows the micropapillae (Fig. 1.2a) and nanohairs of micropapillae (Fig. 1.2b) on the surface of the lotus leaf. It has been demonstrated that the superhydrophobicity of the surface is attributed to the combination of hierarchical structures that trap air beneath a water droplet and the low-surface-energy waxy covering. In particular, the waxy covering is waxy crystalline on the nanoscale, for example nanohairs observed by ESEM (Fig. 1.2b).

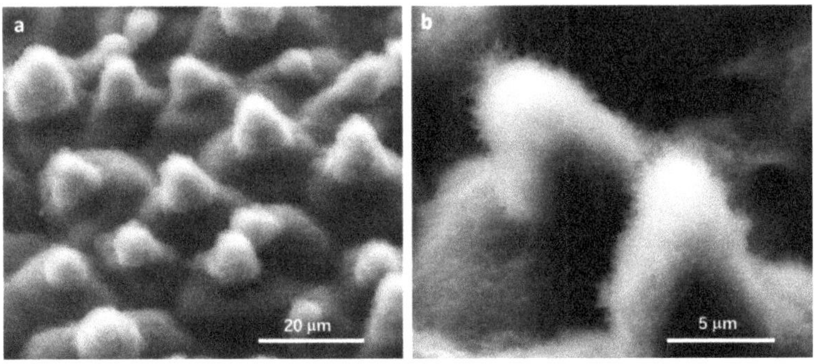

Figure 1.2 Lotus surface with micropapillae (a) and nanohairs of micropapillae (b) observed by ESEM.

The superhydrophobicity of the surface would result in a low adhesiveness as well as a low coefficient of friction due to the roughness and trapped air in the micro- and nanostructures. If the surface is contaminated by dust particles, the contact area between the contaminating particles and the surface is very small, resulting in decreased adhesion. Accordingly, water droplets from rainfall can pick up the contaminating particles and carry them away when the droplets roll off the surface (which protects leaves from pathogens), called self-cleaning or the "lotus effect" (Barthlott and Neinhuis, 1997; Koch et al., 2008). This effect generally occurs on the surfaces of most plant leaves that have rough structures and waxy coverings (Barthlott and Neinhuis, 1997).

In particular, the superhydrophobicity of the lotus leaf surface can be observed in situ by ESEM, where different sizes of condensed droplets are formed, all exhibiting spherical shapes. As shown in Fig. 1.3a, a larger droplet is ~370 μm in diameter. Figure 1.3b shows a droplet (with a diameter of ~30 μm) on the surface of a lotus leaf with micropapillae (diameter of ~10 μm), indicating the water repellency of the lotus leaf at the micro- and nanolevel. The sphere-shaped droplets are the result of dynamic formation on lotus leaves. During water condensation, a wettability gradient is formed along micropapillae from the bottom to the top. There is relatively high surface energy at the top location and relatively low surface energy at the bottom location. Such a wettability gradient effectively generates an asymmetric force that acts on two sides of the as-grown condensed droplet. Therefore, along with droplet growth, the droplet would be moved from the bottom to the top. Finally, the droplet is suspended on the top of micropapillae. The dynamic process of condensed droplets moving from the bottom to the top of micropapillae sweeps or washes the structure of the surface to clean at the microlevel and nanolevel (Zheng et al., 2008). This phenomenon fully demonstrates the robust self-cleaning property of lotus leaves.

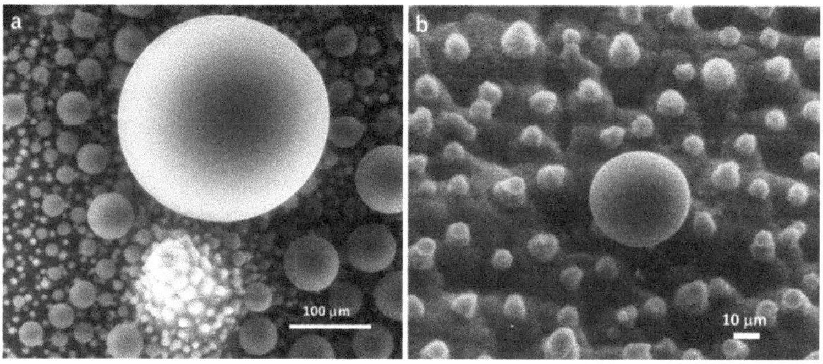

Figure 1.3 Condensed droplets on the surface of the lotus leaf under ESEM. Both large droplets (a) and small droplets have spherical shapes (b) on the surface of the lotus leaf.

1.1.2 Spider Silks

Spider silk is a well-known biological silk that is thin and lightweight, slightly visible under sunlight, and composed of humidity-sensitive (Vollrath and Edmonds, 1989; Vollrath et al., 1990; Edmonds and Vollrath, 1992; Vollrath, 2000; Vollrath and Porter, 2006) and hydrophilic flagelliform proteins (Peter, 1984; Backer et al., 2003), enjoying a high reputation as a functional fiber with excellent mechanical properties (Vollrath, 2000; Emile et al., 2006; Porter and Vollrath, 2007; Liu et al., 2005). Spider silks are usually classified into various types that serve different functions, including drag silk to help the spider move from one place to another, radial silk to fix spider web, capture silk to catch prey insects, etc. For example, capture silks possess distinctly different structures that are made by different species of spiders (e.g., cribellate or ecribellate) (Zheng et al., 2010; Huang et al., 2011). Ecribellate capture silk includes periodic glue droplets along a main-axis fiber. Cribellate capture silk includes periodic nanofibril puffs along both main-axis fibers. Figure 1.4 presents a photograph of a cribellate spider on its silk web (the visually highlighted lines are the capture silks), which is built by a cribellate spider using a cribellum—a comb-like organ—to separate silk fibers drawn from its spinnerets into many exceedingly fine fibers (i.e., nanofibrils).

Figure 1.4 Photograph of a cribellate spider and its silk web.

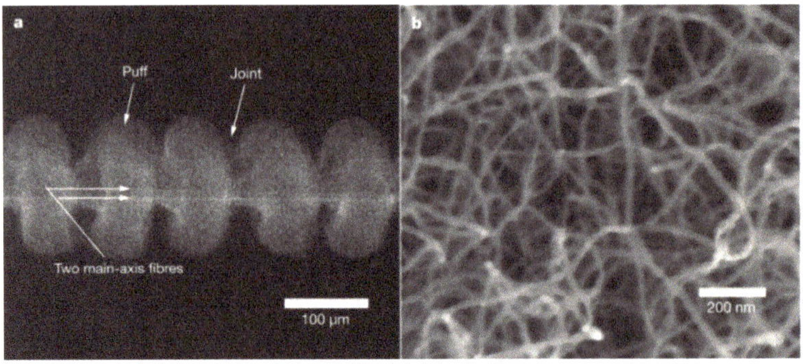

Figure 1.5 In situ observed ESEM images of cribellate spider capture silk in the web. The capture silk is composed of puffs and joints and is supported by two main-axis fibers (a), where the puffs are composed of nanofibrils (b) (Zheng et al., 2010).

As shown in Fig. 1.5, ESEM is used to observe the structures of the capture silk of the cribellate spider. The capture silk of the cribellate spider is composed of periodic puffs separated by joints along two main-axis fibers (Fig. 1.5a) (Zheng et al., 2010). The zoomed-in image in Fig. 1.5b shows the puffs to be composed of random nanofibrils (20–30 nm in diameter). These highly hydrophilic nanofibrils enhance the adhesion of spider silk (Peter, 1984; Backer et al., 2003), which is greatly favorable for prey capture.

An intriguing ability of spider silk is to collect water from humid air (Fig. 1.6). After rain or in the morning, we often see that pearly droplets (black balls in Fig. 1.6) hang on the web silk, like a beautiful necklace, exhibiting a bead-on-string appearance due to coverage by water droplets. This hanging-droplet phenomenon implies the water-collecting ability of spider silk. Recently, this phenomenon has been revealed as the result of the unique, wet-built gradient micro-/nanostructure of cribellate spider capture silk after being wetted by mist (Zheng et al., 2010; Huang et al., 2011).

Figure 1.6 Pearly water droplets hanging on spider silk by microscopy observation.

In particular, the highly hydrophilic nanofibrils in the puffs absorb water molecules in air; thus, the puffs (Fig. 1.7a) shrink into spindle-knots separated by joints (called wetted spider silk) (Fig. 1.7b). In a humid environment, the spindle-knots with gradient micro- and nanostructures on wetted spider silk capture the as-condensed droplets and drive the condensed droplets to gather together and finally form larger hanging droplets in an ultrashort time (Fig. 1.7c). The details are shown in Fig. 1.7d–f. Once the tiny as-condensed droplets occur on the surface of wetted spider silk, they move from the joints toward the spindle-knots in a short time, between 0.156 s and 0.796 s (Fig. 1.7f). This phenomenon is attributed to the property of the surface, which is exactly related to the unique micro- and nanostructures on the surface. Figure 1.7g shows a scanning electron microscopy (SEM) image of wetted spider silk with a spindle-knot and joint. A gradient structure forms between random nanofibrils on the spindle-knot (Fig. 1.7h) and aligned nanofibrils on the joint (Fig. 1.7i). Based on the interface physics, both gradients (i.e., Laplace pressure difference, surface-energy gradient between spindle-knots and joints) would form an interface force on the surface of wetted spider silk. Therefore, in a humid environment, once the tiny droplets are condensed on the surface of wetted spider silk, both gradient forces would generate a cooperative force to drive efficient transport of these tiny condensed droplets from joints to spindle-knots, allowing cribellate spider capture silk to exhibit an excellent water-collection ability.

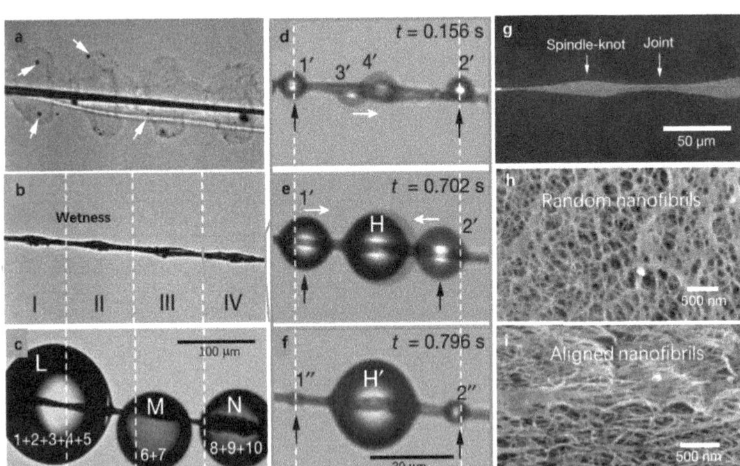

Figure 1.7 The cribellate spider capture silk changes its surface structure for water collection (a–f). A change in the spider silk forms the spindle-knot and joint (g). The spindle-knots are composed of random nanofibrils (h), and the joints are composed of aligned nanofibrils (i) (Zheng et al., 2010).

1.1.3 Cactus Spines

Cactus plants are well known for their strong abilities to live in the desert even though drought is common and water evaporates very quickly. One feature of cactus is that it can retain water via its body, and another feature is that its surface is covered with clusters of oriented spines.

Figure 1.8 Optical images of a representative *Opuntia microdasys* stem (a) and the magnified optical image of trichomes including spines (b).

Recently, cactus spines were reported to be able to harvest fog due to the microstructure of their surfaces. The spine surface has many gradient grooves to drive droplets in certain directions (Ju et al., 2012). For instance, the cactus *Opuntia microdasys* in the Chihuahua Desert has an integrated multifunctional system for efficient fog collection. This unique fog-collection system is attributed to the integration of its multilevel cone structures. Figure 1.8 shows optical images of a representative *O. microdasys* stem. Clusters of needle-like spines and trichomes grow as an array on the stem's surface, with distances ranging from ~7 to 23 mm. A single cluster contains ~100 spines. These spines grow in arbitrary directions, with an average angle of 18.1° between the nearest two spines, forming a hemispherical structure. Trichomes at the base of the spines also form a hemispherical structure. The trichomes are superhydrophilic, which aids cactus spines in efficient water collection.

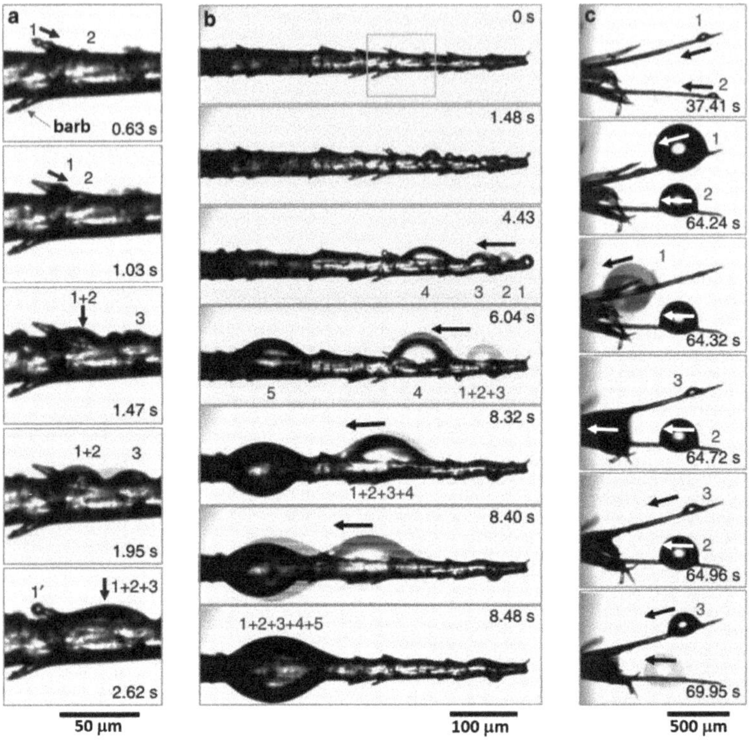

Figure 1.9 In situ optical microscopic observation of water collection on a cactus spine. (a) Tiny condensed droplets transport and coalescence on barbs of a spine. Droplets 1, 2, and 3 coalesce together at 2.62 s. (b) The as-coalesced droplets (droplets 1, 2, 3, 4, and 5) are transported in one direction (arrows). (c) The droplets transport and coalescence on two adjacent spines with trichomes for water collection in 69.95 s (Ju et al., 2012).

Figure 1.9 shows the in situ observation of water collection on a horizontal spine. By focusing on the behavior of the water drops on the tip side of a spine covered with several oriented barbs, the initial, simultaneous deposition of tiny water drops on the barbs (drop 1) and the spine (drop 2) can be observed (Fig. 1.9a). As the deposition proceeded, the growing drop 1 moved toward the base of the barb and coalesced with drop 2, forming the larger drop 1 + 2 (arrows in Fig. 1.9a). With continuous deposition, drop 1 + 2 further coalesced with drop 3, which had been deposited on the adjacent barb. After water drop 1 moved away from the barb, a new cycle of water deposition and directional collection began. During this process, the gradient of the Laplace pressure between the two sides caused the water drops on the barbs to continuously move toward the base of the barbs. After the initial stage, drops that were collected from the barbs or deposited directly from the fog could be observed on the tip side of the spine (identified by numbers 1–5 in Fig. 1.9b). As the water deposition continued, the drops increased in size; the tip-side drops coalesced and moved directionally along the spine (black arrows in Fig. 1.9b), ultimately forming a large drop (1 + 2 + 3 + 4 + 5). These observations demonstrated that the water drops could move from the tip to the base of the spine by driving forces from the gradient of the surface free energy and the gradient of the Laplace pressure. In addition, the oriented barbs on the tip side adopted an asymmetrical structure, enabling the unidirectional rolling and spreading of the water drops (Chu et al., 2010; Malvadkar et al., 2010) and further aiding the directional movement of the water drops toward the base of the spine. Figure 1.9c shows this directional movement in detail. The individual drops that were collected on adjacent spines both moved to the base side that was covered with trichomes. The trichomes and spines formed conical internal surfaces, similar to shorebird beaks (Prakash et al., 2008), which can generate a strong capillary force. When the water drops contacted the trichomes, the drops were rapidly absorbed (Fig. 1.9c). The rapid absorption process (that is, the rapid removal of the water drops) guaranteed the rapid regeneration of the fog-collection process. As described in previous studies on various species, water drops can be absorbed into the stem (Mauseth, 2006; Loik, 2008; Koch and Barthlott, 2009; Mosco, 2009), especially in foggy areas; from there, the drops can be reserved in mucilage cells containing polysaccharides with a high affinity for water, which can reduce evaporative loss (Ogburn and Edwards, 2009). The integration of multiple functions within the spines and trichomes in the cactus, including water deposition, collection, transport, and absorption, collectively facilitates an efficient fog-collection system.

Recently, it was also found that the scale-covered spines of the cactus *Gymnocalycium baldianum* (which lives in the desert area of Argentina), the structure of which is totally different from that of other cacti, could also propel fog drops directionally with the assistance of tilted scales and splayed capillary tubes (Liu et al., 2015). In this system, the subsequent droplets could move more quickly than the first droplet because the surface adhesion was greatly lowered by the thin liquid film that became trapped in the splayed capillary tubes after the procession of the first droplet.

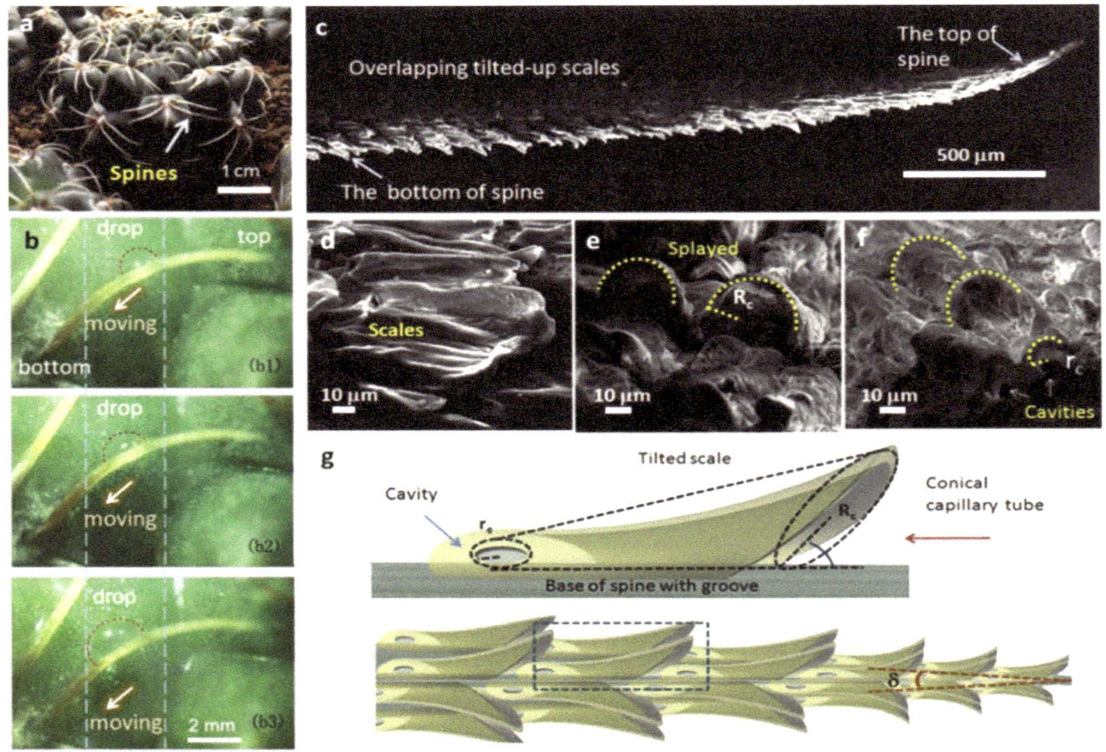

Figure 1.10 Photograph of *Gymnocalycium baldianum* cactus with clusters of 5–6 spines (a) and droplet movement on the spine and splayed microscale structure (b). (c–f) SEM images of spines with overlapping titled-up scales (splayed scales) and cavities under the splayed scales. (g) Illustration of the spine, including the tilted-up scales and cavities that form conical capillary tubes (Liu et al., 2015).

Figure 1.10 shows a photograph of spines of the cactus *G. baldianum* in mist (Fig. 1.10a) and a directional drop self-transport for efficient water collection, where the drop moves from the tip to the base of the spine (arrows, Fig. 1.10b), as observed by optimal microscopy. ESEM was used to observe the microstructure of the spine of *G. baldianum* (Fig. 1.10c). It is found that there are tilted scales overlapping along the length of the cone spine (~5 cm). The cone-apex angle δ of the spine is ~15°, and the tilted angle of the scale (α) is ~20° from the baseline of the axis of the spine (Fig. 1.10c). The scales are ~100–130 μm in length and ~30–50 μm in width (Fig. 1.10d). Under the tilted-up scales to the base of the spine, there are some cavities that fan out, with a maximum radius (R_C) of ~20 μm (Fig. 1.10e) at the openings of the scales and a minimum radius (r_C) of ~8 μm at the ends of the scales (Fig. 1.10f). The overlapping scales and the surface of the spine construct the gradient splayed capillary microtube array along the cone-shaped spine of the cactus (Fig. 1.10g). Such gradient splayed capillary structures would generate the driving force, along with the conical geometry of the spine and the slippery of the water thin film after the process of first droplet movement on the spine, which together propels droplets for efficient self-transport in direction.

1.1.4 Morpho Butterflies

The brilliant Morpho butterflies that live in high-humidity rainforests have wings with various colors in metallic, shimmering shades of blue and green. Usually, male butterflies have much brighter colors on their wings than female butterflies. The male butterfly wings reflect their bright, iridescent color over an extremely wide angle to maximize visibility in the rainforest. The colors of a male butterfly are used to intimidate any rivals that might fly into his territory. The colors vary with viewing angle due to nanoscale pores or layers in the structures of wings, defined as structural color (Zheng et al., 2007; Liu et al., 2014). The iridescent colors are the result of structural arrangements instead of pigmentation. Such structures exhibiting iridescent color are considered natural "photonic crystals", which have attracted ever-increasing interest and have inspired the development of optimal devices and productions.

Figure 1.11 Photograph of *Morpho aega* butterfly wings with directional water repellency.

In particular, Morpho butterflies (e.g., *Morpho aega*) have also been demonstrated to present superhydrophobic and water-repellent properties (Fig. 1.11), i.e., self-cleaning properties, similar to lotus leaves, the phenomenon of which is also called the "lotus effect". In contrast, the surface of the butterfly wing exhibits anisotropic micro- and nanostructures due to the overlapping microscales and oriented nanolamellar arrays on the microscales. The arrangement induces directional adhesion on the superhydrophobic wings; accordingly, the droplet easily rolls off the surface of the wing along the outside radial direction from the center of the body (arrows in Fig. 1.11) (Zheng et al., 2007).

In addition, the wings of butterflies (e.g., *Morpho deidamia*) exhibit static and vibrating modes for water repellency (Fig. 1.12) (Srygley and Thomas, 2002; Tanaka et al., 2008; Tanaka et al., 2005) because butterflies need their wings to be lightly loaded to ensure their flying flexibility (Zheng et al., 2007; Mei et al., 2011; Wanasekara and Chalivendra, 2011). In addition, Morpho butterflies live in rainforests, which have high humidity all year. Therefore, driving fog drops away from the wings becomes necessary to keep the wings dry and clean and lighten the weight.

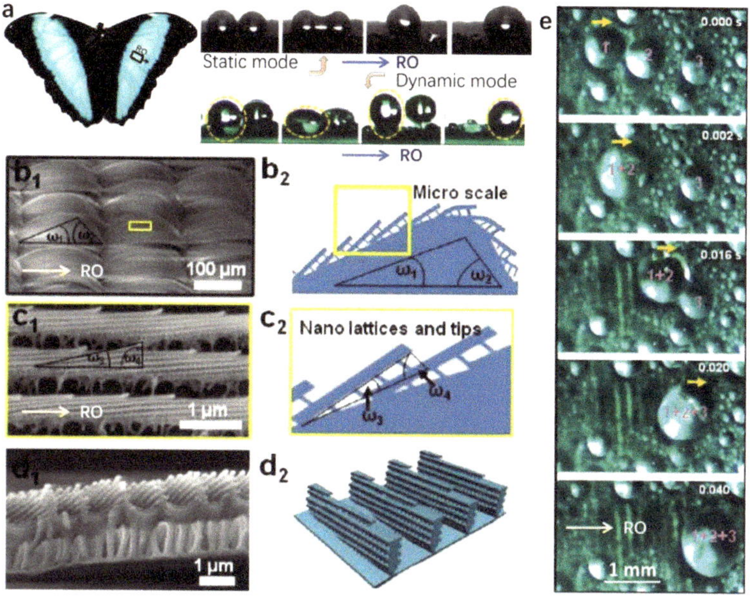

Figure 1.12 (a) Photograph of a *Morpho deidamia* butterfly. The RO direction indicates the radial outside direction of the wing (arrow). Droplet behaviors can be in static and dynamic modes along the RO direction. (b–d) SEM image and illustration of directional micro- and nanostructures of wings. (e) The tiny fog droplets grow and coalesce and are transported along the direction of the micro- and nanostructures (i.e., RO direction) in static and dynamic modes (e) (Liu et al. 2014).

It is known that high humidity condenses water droplets onto the wings of butterflies. The as-condensed droplets grow and coalesce. Interestingly, the wings of butterflies (e.g., *Morpho deidamia*) can drive droplets away from the surfaces in both static and dynamic modes without the assistance of gravity. In static mode, the wings transport droplets directionally by means of asymmetric drop growth and coalescence on the micro/nano ratchet-like structure of the wings. In dynamic mode, the wings transport droplets directionally by means of asymmetric droplet dewetting from the micro/nano ratchet-like structure triggered by vibration (Fig. 1.12b–d). An optimal vibration manner could propel droplets in a large-diameter range (Fig. 1.12e), where the droplets coalesce and move away from the surface of the wing. Such features inspire us to fabricate two-dimensional surfaces that promote directional droplet transport via both static and dynamic modes, which have prospects in the design of materials and surfaces in some fluid-controlling realms, including heat transference, antifogging, water harvesting, etc.

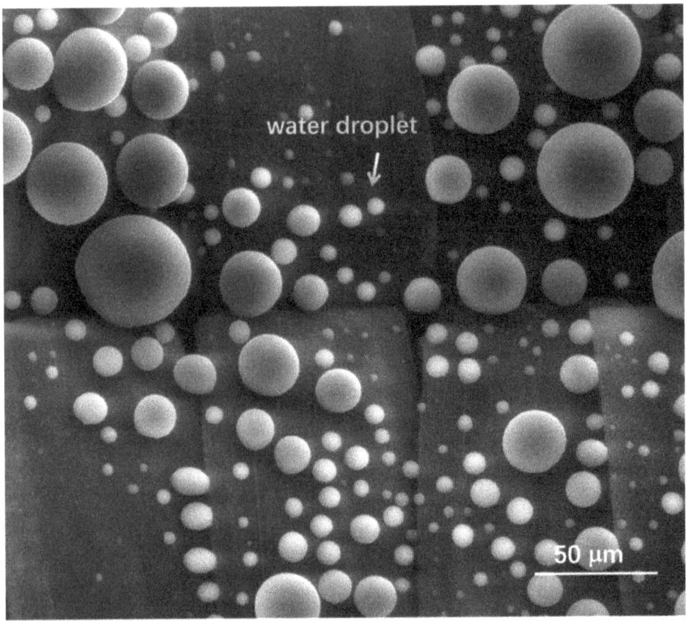

Figure 1.13 As-condensed droplets on the *Morpho aega* butterfly wing in ESEM observations.

At the microlevel, interestingly, the superhydrophobicity of the butterfly wing can be observed in situ by ESEM. Figure 1.13 shows the sphere-like condensed droplets on the *Morpho aega* butterfly wing as observed by ESEM (at a relative humidity of ~70% and a temperature of ~ −2°C), where all the droplets appear as spheres (white balls in Fig. 1.13) and range in size from 10 to 30 μm. These observations fully demonstrate the robust superhydrophobicity of butterfly wings.

1.1.5 Beetle Backs

Beetles can survive in semiarid desert regions, particularly in coastal areas where morning fog is abundant and rainfall is scarce, where fog harvesting would be a crucial source of water for beetle survival. Interestingly, beetles have a tailor-made covering for collecting water from early-morning fog and survive by collecting water from fog by using their unique back structure (Wagner et al., 1996). In particular, the back of a beetle has an array of hydrophilic bumps across a waxy superhydrophobic surface, which collect fog droplets and then route the droplets into the beetle's mouth as drinking water when its back faces the fog-laden wind (Andrew and Lawrence, 2001). Such a structure can inspire scientists and researchers to design surfaces for fog-water harvesting that can be applied to fog engineering.

Figure 1.14 Photograph of the fog-collecting manner of desert beetles (Wagner et al., 1996).

An example is the Namib Desert beetle; the Namib Desert in southern Africa has unique sand-dune fauna (Hamilton and Seely, 1976), high winds, and extreme daytime temperatures and densities, along with early-morning fog (Koch, 1962; Crawford, 1981), but almost no rainfall (Roberts et al., 1991). The beetle harvests water in a unique way. More specifically, the tenebrionid beetle *Stenocara* sp. (Fig. 1.14) tilts its body forward into the wind to harvest water (Koch, 1962; Wagner et al., 1996). Droplets form on the top (front) of the back wing (i.e., elytra) and roll down the beetle's surface to its mouthparts, similar to the rolling of water droplets from lotus leaves due to the superhydrophobic surface (Louw, 1972). However, the fog-water droplets encountered by the Stenocara beetle are much finer (1–40 μm diameter) than raindrops. Without a special controlling mechanism, the droplets would quickly be lost due to the heat and winds of the desert because water droplets in fog are so light that they travel almost horizontally in a breeze (Machin, 1970; McGill and King, 1989; Watanabe, 2000). Interestingly, beetles use their backs to capture fog-water droplets in a highly effective manner.

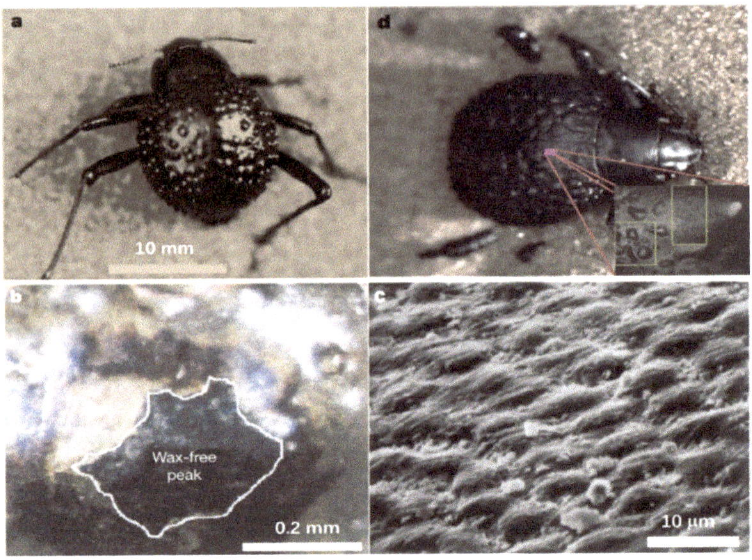

Figure 1.15 (a) Photograph of Stenocara beetle with the rough surface of the elytra. (b) Wax-free peak on the rough surface of the elytra. (c) The microstructure consists of flattened hemispheres in a regular hexagonal array. (d) Water-capturing surface of the fused overwings (elytra) of the desert beetle *Stenocara* sp., along with hydrophilic and hydrophobic regions (inset) (Andrew and Lawrence, 2001; Guadarrama-Cetina et al., 2014).

It has been revealed that the fog-harvesting mechanism is due to the unique microstructures on the surface of the beetle back. On a macroscopic scale, the elytra of Stenocara are covered in a near-random array of bumps 0.5–1.5 mm apart, each approximately 0.5 mm in diameter (Fig. 1.15a) (Andrew and Lawrence, 2001). At the microscopic level, the peaks of these bumps are smooth, with no wax covering (i.e., hydrophilic regions) (Fig. 1.15b). However, the troughs, including their sloping sides, are covered by a wax-coated microstructure and are hydrophobic (Fig. 1.15b) (Andrew and Lawrence, 2001). The microstructure consists of flattened hemispheres that are 10 μm in diameter and arranged in a regular hexagonal array (Fig. 1.15c) (Andrew and Lawrence, 2001), creating a superhydrophobic system similar to that of lotus leaves (Guadarrama-Cetina et al., 2014). The droplet is rapidly formed on the hydrophilic region of the peaks (Fig. 1.15d) (Guadarrama-Cetina et al., 2014) and gathers fast-growing droplets there. In addition, the water striking the hydrophobic slopes can be collected as it may bounce or be blown to a hydrophilic region. When each attached droplet eventually reaches or exceeds a size that covers the entire hydrophilic island, gravity detaches the droplet, which rolls down the tilted back of the beetle. This mechanism provides inspiration for designing fog-harvesting systems. The design of this fog-collecting structure can guide inexpensive reproduction on a commercial scale and may find applications in water-trapping tent and building coverings, for example, or in water condensers and engines.

1.1.6 Water Strider Legs

The water strider (*Gerris remigis*) is a kind of insect that has remarkable nonwetting legs that enable it to stand effortlessly and move quickly on water (Fig. 1.16a). Such a feature is believed to be due to a surface-tension effect caused by secreted wax or microstructure effects (Caponigro and Erilsen, 1976; Dickinson, 2003; Hu et al., 2003). To reveal the effects, a high-sensitivity balance system has been used to construct a force–displacement curve for a water strider's leg as it is pressed on the surface of water. Surprisingly, the leg does not pierce the water surface until a dimple of ~4.38 mm depth is formed (Fig. 1.16b). The maximal supporting force of a single leg is 152 dynes, or approximately 15 times the total body weight of the water strider. The corresponding volume of water ejected is approximately 300 times that of the leg itself, indicating that the surface of the leg is strikingly water repellent. It can be inferred that there is a superhydrophobic effect on a water strider's legs. It is revealed that the contact angle with water is greater than 150°, indicating the superhydrophobicity. However, the contact angle of water on the smooth cuticle wax that is secreted reaches ~105° (Holdgate, 1955). Based on Cassie's law, the high contact angle of water (e.g., 150°) is the result of microstructures on the surface that exhibit low surface energy, which enhances the hydrophobicity (Feng et al., 2002).

Therefore, the superhydrophobic force is the result of the special hierarchical structure of the legs, which are covered by large numbers of oriented tiny hairs (microsetae) with fine nanogrooves, a feature that is more important in inducing this water resistance. There are numerous oriented hairs (or setae) on the legs, and the hairs have a needle shape, with diameters ranging from 3 μm to 500 nm (Fig. 1.16c). Most hairs are approximately 50 μm in length and arranged at an inclined angle of approximately 20° from the surface of the leg. There are many elaborate nanoscale grooves on each microseta, indicating a unique hierarchical structure (Fig. 1.16d). Therefore, the available air is trapped in spaces in the microsetae and nanogrooves to form a cushion at the leg–water interface that prevents the legs from being wetted. This unique hierarchical micro- and nanostructuring of the leg's surface is responsible for its water resistance and strong supporting force (Cassie and Baxter, 1944).

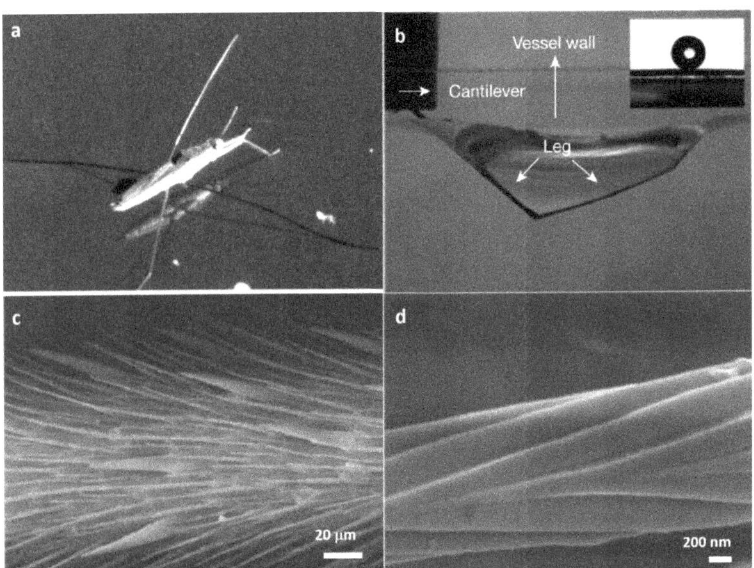

Figure 1.16 (a) Photograph of a water strider on the surface of water. (b) Superhydrophobicity of the water strider's leg to repel water. The inset shows the high contact angle of water on the leg (more than 150°). (c) An SEM image of oriented hairs. (d) Nanogrooves on a hair (Gao and Jiang, 2004).

ESEM is used to observe the water repellency of the water strider's leg when the leg is pressed in water (Fig. 1.17). The microseta can form air-trapped gasbags. This clever arrangement allows water striders to survive on water even if they are bombarded by raindrops, as they can bounce to avoid being drowned. This discovery may be helpful in the design of miniature aquatic devices and nonwetting materials.

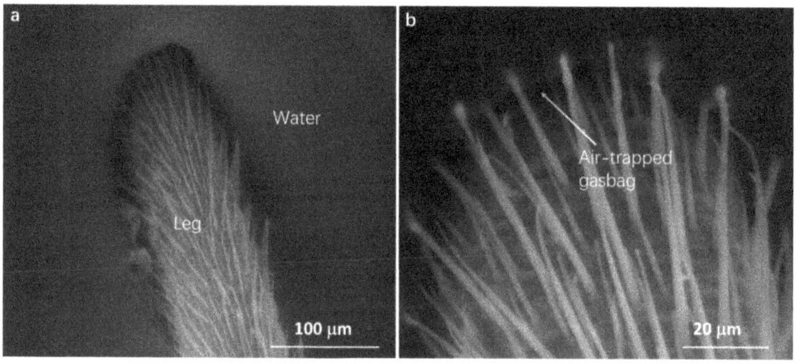

Figure 1.17 The air-trapped gasbag when the leg is pressed in water via in situ observation with ESEM: (a) low-resolution observation and (b) high-resolution observation.

Recently, it was found that water striders live at the water surface in a highly humid environment. Interestingly, without any external force, tiny condensed droplets in the range of femtoliters (fL) to microliters (μL) are removed from the water strider's legs owing to the presence of oriented conical hairs (Wang et al., 2015). As shown in Fig. 1.18, when droplets (at a picoliter or nanoliter scale) merge at the leg surface, the center of gravity of the resulting drop is found to be shifted (Fig. 1.18a) in the direction of "easy" motion—that is, in the direction favored by the anisotropic CAH at the surface of the hairs. The tilted hair forms a kind of ratchet that promotes directional movement in the direction of smaller adhesion (Chu et al., 2010; Malvadkar et al., 2010; Prakash and Bush, 2011; Hancock et al., 2012) as activated by vibrations or coalescence. The amplitude of the jump is one drop size, as expected for anisotropic coalescence. Drops can sometimes even jump off the leg (Fig. 1.18b), a direct expulsion event similar to that encountered on cicada wings or lotus leaves (Boreyko and Chen, 2009; Wisdom et al., 2013).

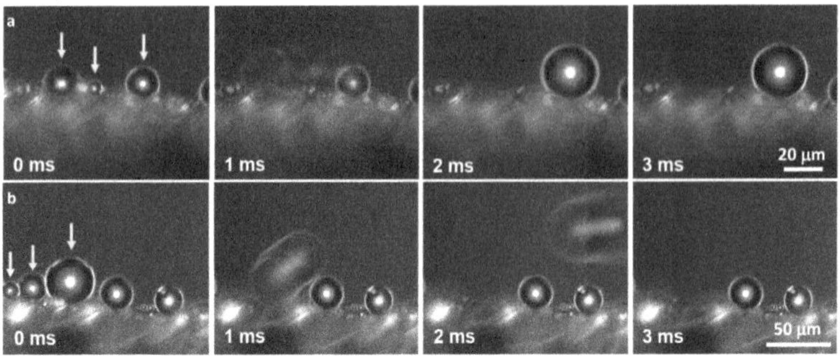

Figure 1.18 (a) The condensed droplet directionally grows and coalesces. (b) Directional removal of the drop on the anisotropic hair array. Droplets involved in coalescence are indicated by yellow arrows (Wang et al., 2015).

The sophisticated textures of the legs provide a unique ensemble of functions. Robust superhydrophobicity is first necessary to deform the water surface in such a way that the creature's weight can be opposed by Laplace pressure. The anisotropy of textures is also exploited to establish gliding and resisting directions, allowing the water strider to control its movements (Extrand, 2007; Prakash and Bush, 2011). The texture also enables the expulsion of condensed droplets, which prevents a major source of risk for the creature, whose hydrophobicity would reverse (in a fatal way) if hairs became impregnated by water as humid air condenses.

1.1.7 Nepenthes Petals

Carnivory in plants is a relatively rare phenomenon that can be an adaptation in nutrient-poor habitats; it is considered an additional pathway for the acquisition of supplemental nutrients (Adamec, 1997; Miguel et al., 2018). According to a report by Juniper et al. (1989), a plant is defined as carnivorous if it has developed an ability to attract, catch, retain, and digest prey into easily assimilated compounds in its apparatus. Subsequently, such a plant absorbs nutrients for its growth (Ellison, 2006).

The following example is focused on the prey-trapping pitcher organs of the carnivorous plant *Nepenthes alata*. There is continuous, directional water transport on the surface of the "peristome" (Chen et al., 2016)—the rim of the pitcher—because of its unique multiscale structures and the enhancement of capillary

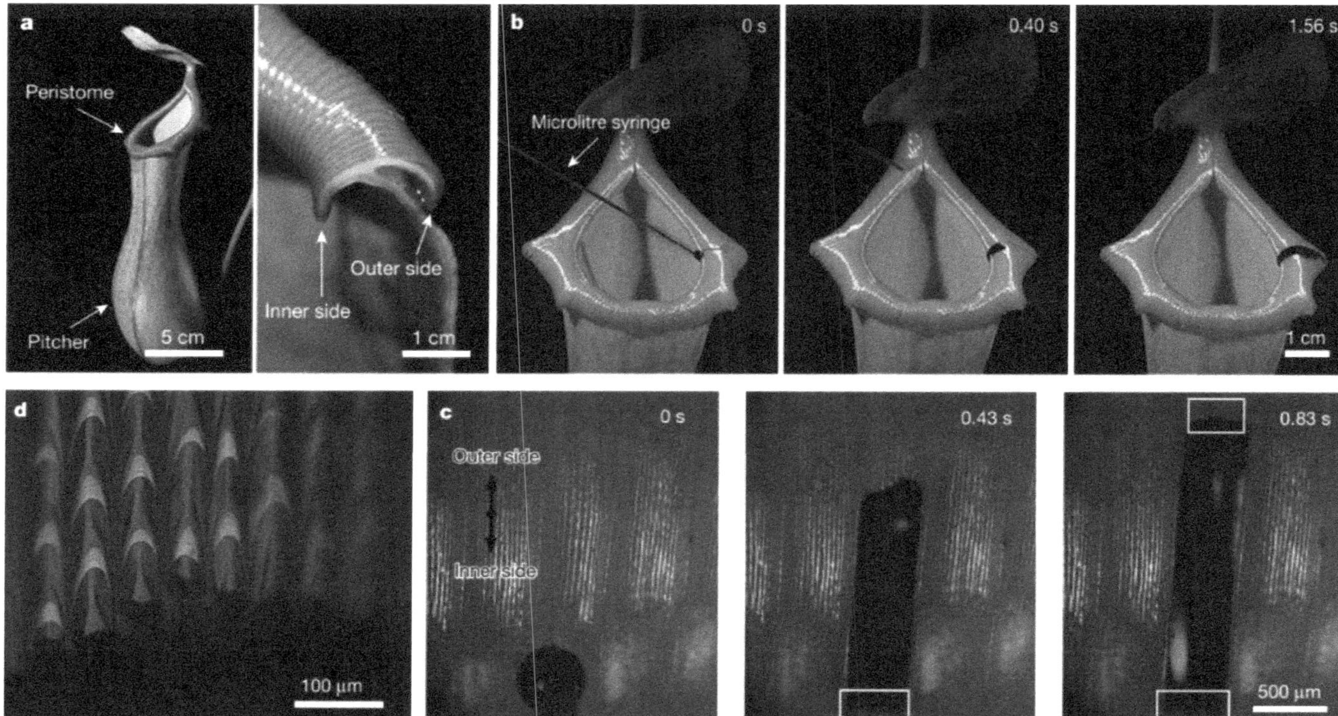

Figure 1.19 (a) Photograph of the plant *Nepenthes alata* with its pitcher and the upper rim as the peristome. (b) In situ optical observation of directional water transport on the peristome surface. (c–d) Optical images of water transport on the peristome surface with arch-shaped microcavities at low resolution (c) and high resolution (d) (Chen et al., 2016).

rise (Hauksbee, 1712; Taylor, 1712) in the transport direction, which prevents backflow by pinning in place any water that is moving in the reverse direction. The tropical plant *N. alata* has a "pitcher"—a highly modified, prey-trapping organ, filled with digestive fluid—that is 10–15 cm in length; at the upper rim is an arch-shaped ring of tissue (i.e., peristome), which is approximately 1–2 cm in width from the inner to the outer side (Fig. 1.19a). The pitcher can digest insects to meet fundamental nutrient needs (Juniper and Burras, 1962; Eillson, 2006). Because the pitcher captures the insects when they aquaplane on the peristome surface (Bohn and Federle, 2004), it is completely wettable by water from rain, dew, or high humidity. The nectar is secreted from extrafloral nectaries at the pitcher's inner margin (Bauer et al., 2008; Bauer and Federle, 2009). Slippery surfaces have been designed to mimic the structure and behavior of the natural peristome surface (Lafuma and Quéré, 2011; Wong et al., 2011), but questions remain about the mechanism(s) underlying its function. Figure 1.19b shows the directional transport from the inner to the outer side: A water droplet deposited on the inner peristome margin moves to the outer margin within several seconds, while a water droplet placed on the outer margin cannot move inward. In situ optical microscopic observations (Fig. 1.19c) show that water transport is confined within a single large channel running perpendicular to the peristome rim, where the water does not spread laterally beyond the width of the initial deposition area (marked by white rectangles in Fig. 1.19c). The higher-resolution image in Fig. 1.19d shows that water transport is, in fact, confined to approximately 10 microchannels distributed within the large channel and that each microchannel consists of arch-shaped microcavities with a spacing of approximately 100 μm. These structural features ensure that water evaporating from the pitcher and condensing on the inner margin will spread over the entire peristome surface to make it more slippery, and such features also likely promote the directionality of the water transport.

Detailed structural characterization of the peristome surface with SEM reveals a two-order hierarchical arrangement of regular radial ridges aligned so that they form a two-order hierarchy of parallel microgrooves (Fig. 1.20a). Each first-order microgroove with a width L_1 (~461.72 μm) (Fig. 1.20a) contains approximately 10 second-order microgrooves whose width L_1 (~50.18 μm at the valley)

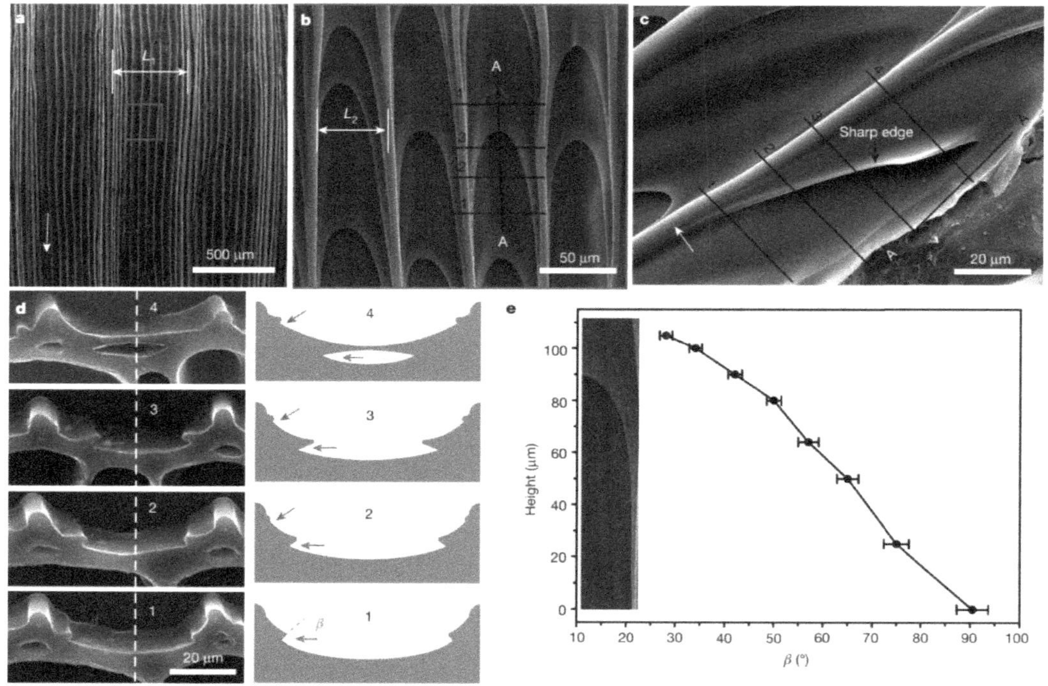

Figure 1.20 Surface features of the peristome in the pitchers of *N. alata*. (a) The hierarchy of parallel microgrooves, where L_1 is the width of the first-order microgroove. (b) The second-order microgroove with arch-shaped (or duck-billed) microcavities along the width of L_1. The four positions are marked with numbers 1, 2, 3, and 4. (c) There is a sharp edge in the arch-shaped (or duck-billed) microcavities. (d) The side-view microgroove structures of four positions 1, 2, 3, and 4. Movement occurs toward the top of the microcavity (red arrows). The sides of each cavity form a wedge (green arrows), with the opening angle β. (e) The relationship between the opening angle β and the height of the wedge along four positions. The small β corresponds to position 4 (Chen et al., 2016).

increases from the ridge to the valley of the first-order microgroove. Arch-shaped (or duck-billed) microcavities are regularly distributed along the second-order microgrooves (Fig. 1.20b), with an overall upward slant and with the tops of the arches pointing toward the outer side. A vertical section through the peristome shows that the top sections of the microcavities are closed (Fig. 1.20c), and that the enclosing surface is slightly slanted and has a sharp edge. The periodic microcavities overlap along the ridge (the starting point of which is marked by a white arrow in Fig. 1.20c), with the overlap between a given microcavity and its overlying neighbor becoming more evident when moving toward the top of the microcavity (red arrows in Fig. 1.20d). The sides of each cavity form a wedge (green arrows in Fig. 1.20d), with the opening angle β decreasing from an initial value of approximately 90° at the bottom of the microcavity to approximately 28° at the top (Fig. 1.20e). As a result, not only unidirectional flow happens (despite the absence of any surface-energy gradient) but also the transport speed is much higher than previously thought. The biological design principles can inspire us to develop artificial fluid-transport systems for practical applications.

1.1.8 Feathers

Biological feathers (from birds, geese, etc.) usually exhibit an array of natural fibers, which consist of well-ordered hair-like structures (barbs and barbules); accordingly, these feathers provide hydrophobicity and thermal insulation (Rijke, 1968; Dawson et al., 1999; Rijke and Jesser, 2010). Small amounts of oil disrupt this arrangement by clumping adjacent barbules, affecting their water repellency and insulating properties and thus reducing the survival rate of oiled birds (Hurtung, 1967; O'Hara and Morandin, 2010).

Interestingly, as recently reported, there are three possible final states experimentally observed when a polydisperse aerosol of oil is sprayed on goose feathers (Fig. 1.21a–e). It is found that a volume of oil less than the critical volume (V_c, along with a drop radius less than 20 mm) would totally spread, thus clumping adjacent barbules and making the cleaning process difficult. Drops larger than the critical volume (V_S, along with drop radius ~140 mm) do not spread and may be dislodged from the bird's plumage. Figure 1.21f shows the map of the three spreading states, namely, no spreading, partial spreading, and total spreading (Dupart et al., 2012). The blue region indicates the no-spreading state as $V > V_S$ and $L > z_S$. The pink region indicates the partial-spreading state as $V > V_C$. The white region indicates the total-spreading state as $V < V_C$. When $V < V_C$, total spreading happens in the length $L > L_{dry}$, where z_S is the location of the drop that starts spreading and the fibers are drawn together, V_S is the critical volume of droplet spreading, and L and L_{dry} are the fiber length and fiber dry length, respectively. This finding offers insight into the design of materials by controlling fiber arrays for oil-repellent or transport functions in some surfaces for promising applications.

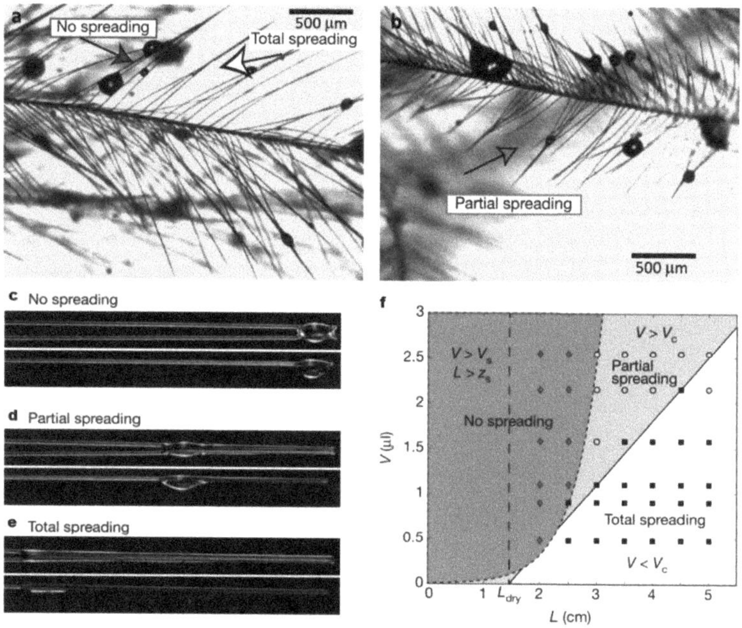

Figure 1.21 The three different final states of a drop between two flexible fibers on a goose feather. (a) Observation of no-spreading and total-spreading phenomena on fiber. (b) The partial spreading between no spreading and total spreading. (c–e) Optical images of three states on fibers: no spreading (c), partial spreading (d), and total spreading (e). (f) Map of three spreading states: no spreading as $V > V_S$ and $L > z_S$, partial spreading as $V > V_C$, and total spreading as $V < V_C$ (Dupart et al., 2012).

Another example is penguins, which live in the world's coldest environments and have a thick layer of feathers that prevent the penetration of cold seawater to the skin and play a role in insulation. The morphology of the contoured, downy after feathers allows them to function as thermal insulators (Dawson et al., 1999), and their material composition affects the mechanical performance of penguins (Taylor, 1986; Bonser and Dawson, 2000). Penguins are the fastest swimming birds. Frost and ice are not observed on penguin feathers. Figure 1.22a shows a photograph of a penguin and penguin superhydrophobic feathers (bottom, inset, contact angle of water: 147°). The feathers are composed of two types of barbules: with hamules (Fig. 1.22b) and without hamules (Fig. 1.22c). As observed under a low temperature of −5°C (Frames I and II, Fig. 1.22b) and by the supercooled water microdroplets (Frames I and II, Fig. 1.22c), no frozen ice appears on the barbules. This result indicates that their feathers exhibit excellent antifrosting and anti-icing properties (Wang et al., 2016a).

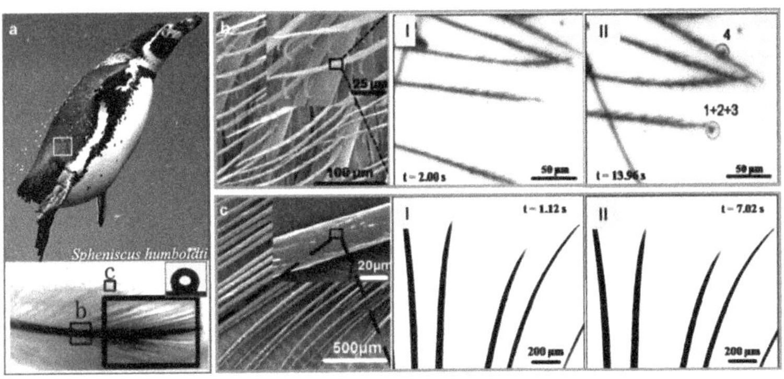

Figure 1.22 (a) Photograph of a penguin. The inset indicates the superhydrophobic feather with a water contact angle of 147°. (b–c) SEM images of hierarchical micro- and nanostructures on body feathers of the penguin *Spheniscus humboldti* and antifogging (b) and anti-icing properties (c) (Wang et al., 2016a).

1.1.9 Gecko Feet

A tokay gecko can cling to virtually any surface and support its body mass with a single toe by using millions of keratinous setae on its toe pads. Each seta branches into hundreds of 200 nm spatulae that make intimate contact with a variety of surface profiles (Fig. 1.23a,b). The combined surface area of billions of spatulae maximizes the van der Waals interactions to generate large adhesive and shear forces. Setae occur in uniform arrays on overlapping lamellar pads at a density of 14,400 per mm^2 (Fig. 1.23c) (Schleich and Kästle, 1986). Each seta (Fig. 1.23d) is 110 μm in length and 5 μm in diameter (Ruibal and Ernst, 1965) and branches at the tip into 100–1000 spatulae, which are 0.2 μm in length and maximal width (Williams and Peterson, 1982) (Fig. 1.23e).

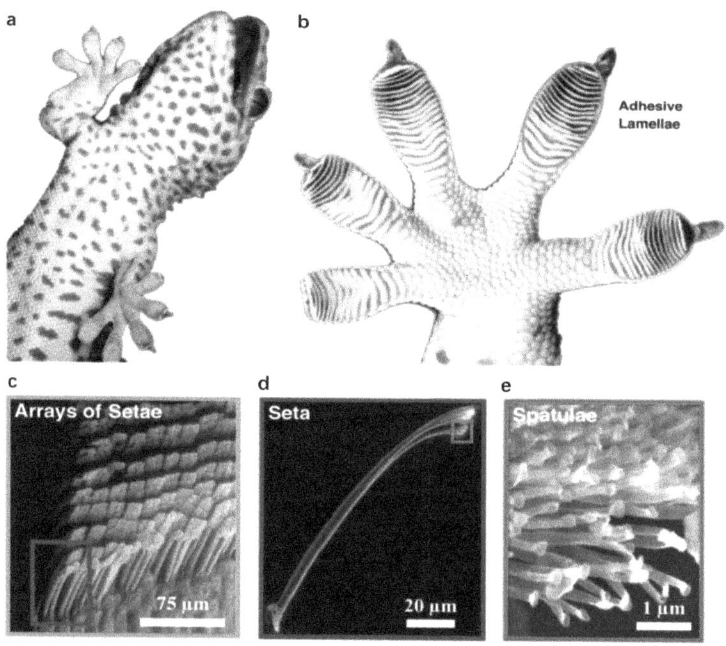

Figure 1.23 Structure of gecko foot (Hansen and Autumn, 2005). (a) Photograph of a tokay gecko. (b) The foot with adhesive lamellae. (c–e) Structural hierarchy of the gecko adhesive system with arrays of setae (c) and individual seta (d) composed of spatulae (e).

The adhesive foot hairs, or setae, on the tokay gecko (*Gekko gecko*) are remarkably sticky: A single seta can generate up to 200 μN of force (Autumn et al., 2000). Geckos use this adhesive to rapidly and repeatedly navigate a wide variety of surfaces (Autumn et al.,

2002; Autumn and Peattie, 2002), and they can even climb vertically on smooth surfaces at a speed >1 m/s (Russell, 1975). Geckos neither groom their footpads, as do some beetles (Stork, 1983), nor secrete fluids that can remove adhering particles, as do certain ants (Federle et al., 2002), crickets (Jiao et al., 2000), and tree frogs (Hanna and Barnes, 1991). Geckos with dirty feet recover their ability to cling to vertical surfaces after only a few steps. Self-cleaning also occurs in arrays of setae isolated from the gecko. Contact mechanical models suggest that self-cleaning occurs by an energetic disequilibrium between the adhesive forces attracting a dirt particle to the substrate and those attracting the same particle to one or more spatulae. The property of self-cleaning is intrinsic to the setal nanostructure and, therefore, should be replicable in synthetic adhesive materials in the future (Hansen and Autumn, 2005).

1.1.10 Shark Skin

Shark skin has a unique microstructure and distribution of denticles, which has attracted considerable interest from biologists for possible effects on the pattern of water flow over the body. For instance, engineers have focused on adjusting the surface roughness to reduce drag forces during locomotion by using scaled-up models of denticles to study how surface roughness affects drag forces (Reif, 1982; Bixler and Bhushan, 2013; Wen et al., 2014), including biomimetic models of shark skin to reduce locomotor drag, for example human-designed structures to mimic the effects of denticle-like structures. Figure 1.24 shows ESEM images of the bonnethead shark (*Sphyrna tiburo*, as illustrated in Fig. 1.24a) skin surface at different body locations, i.e., skin pieces extracted at the positions of the head (Fig. 1.24b), the leading edge dorsal fin (Fig. 1.24c), and the anal fin (Fig. 1.24d). As observed from Fig. 1.24b–d, the denticles have either three or five top ridges, with an oriented direction, resembling the shape of a tree leaf. Thus, this structure ensures low-friction drag reduction when the shark is swimming; the natural flow direction across the denticle surface is from the lower left to the upper right and from the denticle base to the tip.

Figure 1.24 Structures of the bonnethead shark (*Sphyrna tiburo*) skin surface at different body locations (Bixler and Bhushan, 2013). (a) Illustration of a bonnethead shark. There are three regions marked with b, c, and d. (b–d) ESEM images of the bonnethead shark skin surface at different body locations: the head (b), the leading edge dorsal fin (c), and the anal fin (d).

1.1.11 Compound Eyes

As is known, insects (butterflies, mosquitoes, green bottle flies, etc.) have faceted compound eyes consisting of numerous anatomically identical units, the ommatidia. The eyes are classified according to the optical system that is used to efficiently focus light onto the light-sensitive parts of the photoreceptors. In apposition eyes, employed by butterflies, a facet lens together with its crystalline cone channels light into a fused rhabdom, a long, cylindrical structure, which contains the visual pigment molecules of photoreceptors. In optical superposition eyes, used by moths, light reaches the photoreceptive rhabdom via several facet lenses. Moths, thus, exhibit a much higher light sensitivity than butterflies, allowing a nocturnal instead of diurnal lifestyle (Stavenga et al., 2006).

In other insects, their compound eyes were revealed to have antifogging properties, such as the mosquito *Culex pipiens* (Gao et al., 2007) or the green bottle fly, *Lucilia sericata* (Sun et al., 2014). The compound eyes of *C. pipiens* possess ideal superhydrophobic properties that provide an effective protective mechanism to maintain clear vision in a humid habitat. Figure 1.25a shows the eyes of *C. pipiens*. Figure 1.25b shows an SEM image of a single eye. The mosquito eye is a compound structure composed of hundreds of microscale hemispheres, which act as individual sensory units called ommatidia. These ommatidia are uniform, with a diameter of ~26 μm, and organize in a hexagonal close-packed (HCP) arrangement (Fig. 1.25c). The surface of each microhemisphere is covered with numerous, fine, nanoscale nipples as observed by using high-resolution field emission SEM (Fig. 1.25d). These nipples are very uniform, with average diameters of ~101.1 nm and interparticle spacings of ~47.6 nm, which organize in an approximately hexagonal non-close-packed (NCP) array (Fig. 1.25e).

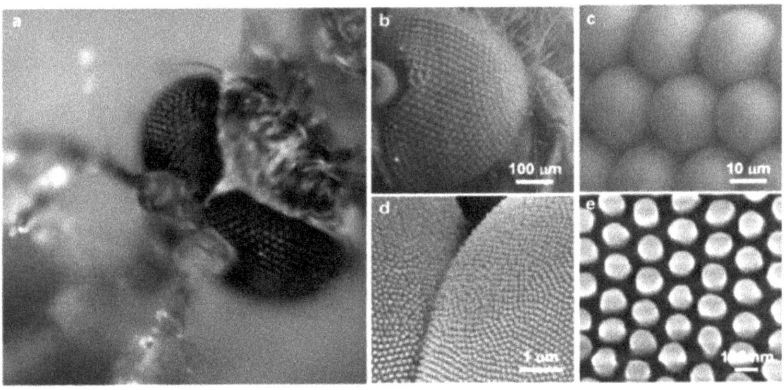

Figure 1.25 Nanostructures on the eyes of the mosquito *Culex pipiens* (Gao et al., 2007). (a) Photograph of the eyes of the mosquito *C. pipiens* and (b–e) micro-/nanostructures of the compound eye.

Another example is the common green bottle fly, *L. sericata* (Beijing), which plays a significant role in human life and medicine (Rueda et al., 2010). The eyes of the fly can remain functional and uncontaminated in some extremely dusty, miry, and moist environments. Figure 1.26a presents an optical image of a common green bottle fly after exposure to a fogging testing chamber, in which an ultrasonic humidifier was used to regulate the relative humidity of the atmosphere and mimic a mist composed of numerous tiny water droplets with diameters less than 10 μm. It is interesting that the compound eyes of the fly present ideal superhydrophobic and antifogging properties so that the fog droplets can only condense on the body while none of them can condense on the compound

Figure 1.26 Structures of a common green bottle fly (Sun et al., 2014). (a) Photograph of a green bottle fly, *Lucilia sericata*. (b) Illustration of the microstructures of the fly's compound eye. (c–e) SEM images of nano-/microstructures of the compound eye.

eyes. The dry-style antifogging properties of such superhydrophobic compound eyes can provide clear vision for insects in highly humid environments. This special wettability of the compound eyes is attributed to the combination of surface chemistry and roughness on multiple scales. Figure 1.26b shows the anatomic structure of the fly's eyes. It was found that each ommatidium consists of a cornea lens (the front surface of which makes up a single facet), a transparent crystalline cone, a rhabdom or channel to guide light transmission, pigment cells that separate the ommatidium from its neighbors, and a center zone containing the optic nerves (Nilsson, 1989). Figure 1.26c–e shows the microstructures of the fly's compound eye on different scales. The compound eyes of the green bottle fly are approximately 5 mm in size and made up of thousands of repeating hexagonal units (Fig. 1.26d) called ommatidia, each with a diameter on the order of ~20 μm, in a hexagonal arrangement. Upon further magnifying the surface of an ommatidium, as shown in Fig. 1.26e, each ommatidium is observed to be covered with numerous, near-hexagonal, bubble-like protuberances with diameters of ~100 nm. These well-ordered, close-packed, hierarchical hexagonal nanostructures are one of the origins of the superior superhydrophobicity and antifogging properties of the green bottle fly's eyes. In addition, the superhydrophobicity of the compound eyes is attributed to a thin layer of biopolymer or wax on the surface of the ommatidia, which can lower the surface energy of the compound eyes. These unique surface structures of the compound eyes provide inspiration for how to design inorganic nanostructures to produce inexpensive functional nanostructures and coatings to provide environmentally durable antifogging performance (Sun et al., 2014).

1.2 Theory and Foundation Based on Physics and Chemistry

When we explain some phenomena in nature—for instance, a pearly water droplet rolling off surfaces of lotus leaves, while a water droplet spreading on paper—the theory and foundation of such processes can be explained and elucidated based on concepts of physics and chemistry. Generally, we observe the wetting phenomenon ubiquitously around us, which can be introduced with the concept of wettability. If a surface causes droplets to roll off it, it is defined as a superhydrophobic surface, and if the surface causes droplets to be pinned or spread, it is defined as a superhydrophilic surface. These wetting properties can be the result of the surface chemistry (high/low surface tension, chemical components), physical structure (micro-/nanoscale geometry), types of structures (anisotropic, ratchet, etc.), and the interactions at solid–liquid–gas interfaces.

1.2.1 Young's Equation

When a droplet is placed on a surface, the droplet takes on a static state with a shape, and the droplet is considered in equilibrium by interactions among interfaces of the solid, liquid, and vapor phases; thus, the droplet state can be described by the contact angle (θ) (Fig. 1.27) with the following relationship:

$$\cos\theta = \frac{\gamma_{SV} - \gamma_{SL}}{\gamma_{LV}} \quad (1.1)$$

Equation (1.1) indicates Young's law (Bico et al., 2001), where γ_{LV}, γ_{SL}, and γ_{SV} are the liquid–gas, solid–liquid, and solid–gas interface tensions, respectively. Young's law explains why droplets on a surface would exhibit a contact angle, which is characterized by the equilibrium among the solid–liquid, liquid–air, and solid–air surface tension at solid–liquid–air three-phase interfaces on a smooth surface. Generally, θ can be considered the intrinsic contact angle of the material surface.

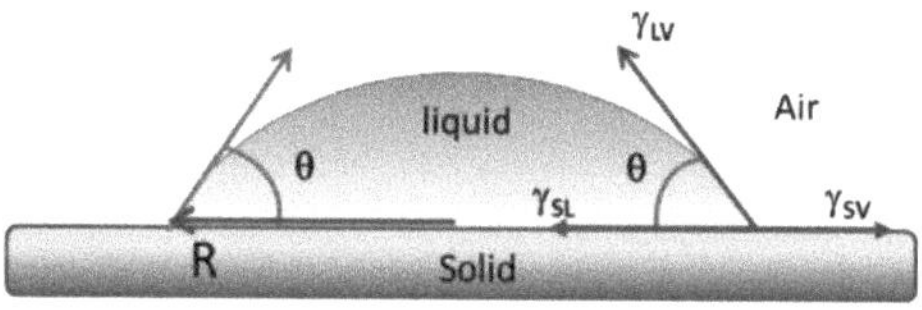

Figure 1.27 A liquid drop on a flat surface (*R* and θ are the radius and contact angle, respectively, of the liquid drop, and γ_{LV}, γ_{SL}, and γ_{SV} are the liquid–gas, solid–liquid, solid–gas interface tensions, respectively).

1.2.2 Wenzel's Law

Realistic solid surfaces are usually complex and rough. When a droplet contacts a rough surface, the droplet shape is different from that on a smooth surface. Wenzel's law can basically explain the droplet behavior on a rough surface and was developed based on Young's law. Figure 1.28a shows a simple model of water on a rough surface. θ and θ* denote the microscopic contact angle (i.e., Young's angle) and the macroscopic contact angle (i.e., apparent contact angle), respectively. It can be easily understood that the states of the two contact angles (θ and θ*) have distinct differences: θ is measured using the tangent of each microscopic plane (the red dashes) based on Young's law, while θ* is measured using the horizontal line (the blue dash) (De Gennes, 1985). The relationship, established by Wenzel (Wenzel, 1936), is defined as Wenzel's law. In Wenzel's law, the roughness (r) is proposed and defined as the ratio of the effective (A_{eff}) and apparent surface areas (A_{app}), i.e., $r = A_{eff}/A_{app}$.

Based on energy, as shown in Fig. 1.28b, when the contact line makes a small displacement (dx), the change in the surface energy (dE) would be $dE = r\,(\gamma_{SL}-\gamma_{SV})\,dx + \gamma_{LV}\,dx\,\cos\theta^*$, and given $dE/dx = 0$, the following relationship can be obtained:

$$\cos\theta^* = r\cos\theta \quad (1.2)$$

Equation (1.2) indicates Wenzel's law (Wenzel, 1936), where θ and θ^* represent Young's angle and apparent angle, respectively (Carroll, 1986).

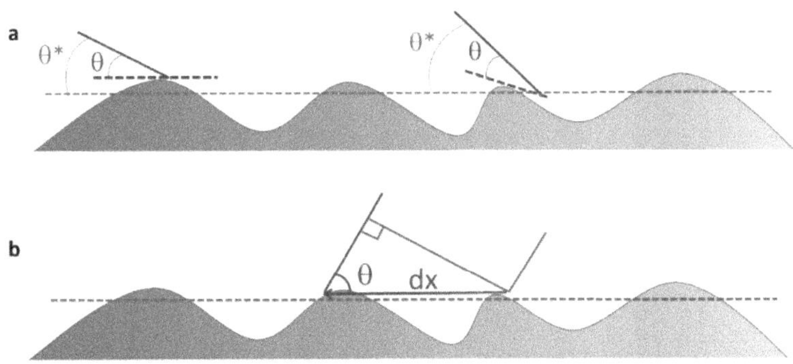

Figure 1.28 Apparent contact angles θ^* on a rough surface (a) and the drop movement on the rough surface (b).

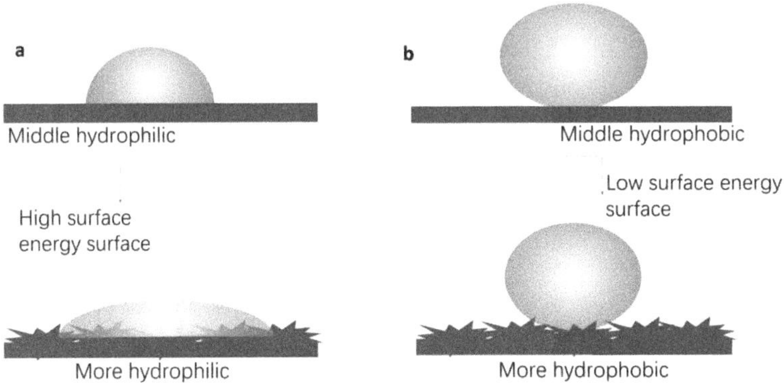

Figure 1.29 Different surface energies and roughness change the shapes of droplets.

According to Eqs. (1.1–1.2), the surface always has $r > 1$ (there is no absolute flat solid surface with $r = 1$), and the sequence of θ and

θ^* can be qualitatively described as follows: with $0 < \theta^* < \theta < 90°$, the hydrophilic surface would be more hydrophilic (Fig. 1.29a); with $90° < \theta < \theta^* < 180°$, the hydrophobic surface would be more hydrophobic (Fig. 1.29b). However, Wenzel's law is limited to a wet contact mode, i.e., the droplet closely contacts the solid, which usually pins the droplet on this surface or makes the surface highly sticky. This law cannot be used to explain why droplets roll off the surface of lotus leaves.

1.2.3 Cassie's Law

Based on the fact that droplets roll off lotus leaves, it is possible that the droplet does not closely contact the solid surface due to the rough structures, where air would be trapped in the surface structures. Simply put, based on the hemiwicking model built by Quéré (Bico et al., 2002), the rough surface can be considered a pore array with trapped air and may have a "two-stage" architecture (Fig. 1.30, side view): Stage 1 is at the top location and stage 2 is at the bottom location. f_1 and f_2 are the fractions of the surface area of stage 1 and stage 2 normalized by the total surface area, respectively. The droplet contact angle can be expressed as:

$$\cos\theta^* = f_1\cos\theta_1 + f_2\cos\theta_2 \tag{1.3}$$

where θ_1 and θ_2 are Young's contact angles of stage 1 and stage 2, respectively; θ^* represents the apparent contact angle; and $f_1, f_2 < 1$ and $f_1 + f_2 = 1$ are given.

With the impregnated mode, i.e., completely hydrophilic surface (Fig. 1.30a), $\theta_1 = \theta$ (with θ the Young's angle of the liquid on the surface) and $\theta_2 = 0$. Thus, θ^* can be:

$$\cos\theta^* = (1 - f_1) + f_1\cos\theta \tag{1.4}$$

When $0 < f_1 < 1$ and $0 < \cos\theta < 1$, $0 < \theta^* < \theta < 90°$ is true, which means that hemiwicking improves the solid wettability; it indicates the wet contact mode between liquid and solid, where the droplet is in Wenzel states. Equations (1.1–1.4) also predict the impossibility of inducing complete wetting ($\theta^* = 0$) by texturing a solid.

With the nonimpregnated mode, i.e., hydrophobic surfaces including air in their structures (Fig. 1.30b), $\theta_1 = \theta$ and $\theta_2 = \pi$. Thus, θ^* can be:

$$\cos\theta^* = -1 + f_1(\cos\theta + 1) \tag{1.5}$$

When $0 < f_1 < 1$ but $-1 < \cos\theta < 0$, $90^\circ < \theta < \theta^* < 180^\circ$ is true. Equation (1.5) presents Cassie's law (Cassie and Baxter, 1944), which indicates the composite contact mode between liquid and solid, where the droplet is in the Cassie state. The solid–liquid interface is replaced with a solid–liquid–air interface on the superhydrophobic texture surface. Cassie's law can be generally used to explain the extent of hydrophobicity on a low-surface-energy surface, for example the effect of the lotus leaf, and determine the relationship between the fraction of solid and contact angles. A larger fraction of solid induces a high contact angle on a rough textured surface, where the droplet is in composite contact mode. Note that the value $\theta = 90^\circ$ (e.g., water) represents the critical angle threshold between hydrophilicity and hydrophobicity based on a smooth surface (i.e., flat surface). Recently, one study noted that the threshold can reach 65° on a rough surface (Su et al., 2016). Otherwise, it is hard for a surface to produce a liquid contact angle above 150° if the surface is not rough.

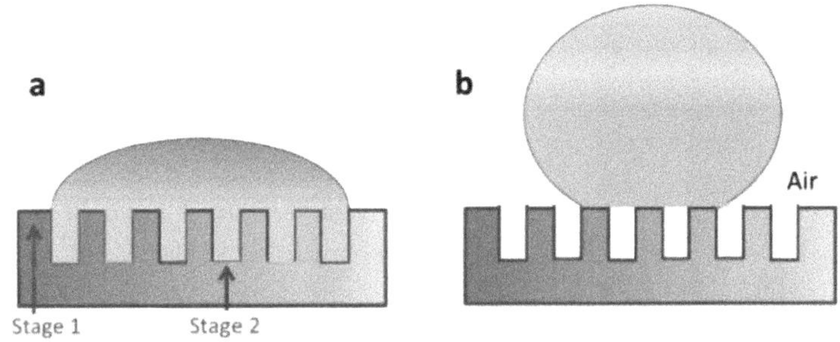

Figure 1.30 Drops with different states on surfaces: (a) hydrophilic, complete wetting solid–liquid contact and (b) hydrophobic, composite solid–liquid contact.

1.2.4 Special Wettability States

There are also special states when droplets contact a surface with complex physical chemistry. According to the wetting features of surfaces, the following special states can be defined: When the surface is entirely wetting, i.e., the liquid spreads over the solid surface, and the surface is ultra-superhydrophilic (Fig. 1.31a), $\theta = 0$.

In fact, when $0 < \theta < 5°$, the surface is defined as superhydrophilic (Fig. 1.31b). When the liquid is suspended on the surface, i.e., $150° < \theta < 180°$, the surface is defined as superhydrophobic (Fig. 1.31c). When the liquid is completely nonwetted on a surface, i.e., $\theta = 180°$, the surface can be considered ultra-superhydrophobic (Fig. 1.31d); such a surface would be extremely ideal and is expected to be achieved by combining both surface chemistry and surface roughness in optimal design.

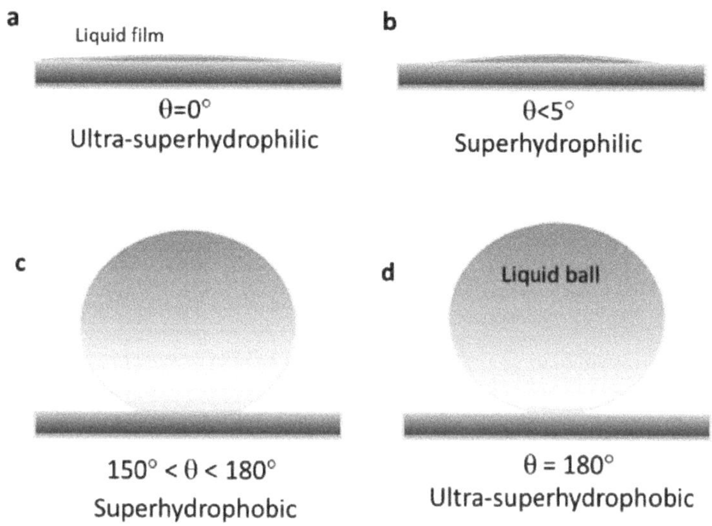

Figure 1.31 Special wettability states on surfaces.

1.2.5 Two Superhydrophobic States

Natural surfaces generate two typical superhydrophobic properties; for example lotus leaves cause droplets to roll off, whereas rose petals cause droplets to be pinned. According to the properties of solid–liquid interfaces, a droplet on a surface can be in one of two different superhydrophobic states (Fig. 1.32), which are further defined as the Cassie state and Wenzel state of the droplet. In the Cassie state, the droplet is in composite contact mode, along with discontinuous solid–liquid–gas three-phase contact lines (TCLs). In the Wenzel state, the droplet is in wet contact mode, along with a continuous TCL.

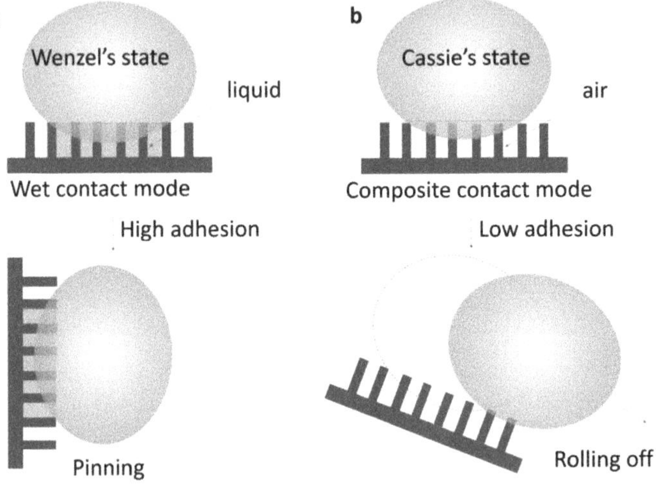

Figure 1.32 Two superhydrophobic states with wet contact mode and composite contact mode.

1.2.6 Dynamic Contact Angle and CAH

The dynamic contact angle can be used to characterize the CAH and adhesion of the surface. CAH occurs when the movement of the liquid is blocked due to the pinning of the TCL (Aussillous and Quéré, 2001; ElSherbini and Jacobi, 2004 a,b; McHale et al., 2009; Eral and Oh, 2013). The example can be seen on car windscreens or windows; some smaller drops are pinned to the glass, while some larger drops slide down. The dynamic contact angle can be characterized by the following methods:

The general characterization method is used for a horizontal surface (moderate hydrophilicity or hydrophobicity) (Fig. 1.33a), and the dynamic contact angle can be estimated by the change in the contact angle when the droplet volume is increased; i.e., by injecting liquid into the droplet on the solid surface, the droplet extends its volume to make the edge of the droplet move. When the contact angle reaches a critical value, it is defined as the advancing angle (θ_A). In contrast, when some liquid is pumped away, the edge of the droplet will shrink, and then the contact angle reaches another threshold, which is defined as the receding angle (θ_R) (Fig. 1.33a).

Another characterization method is used for hard fiber materials; the CAH of a surface can usually be tested (Brabdon et al., 1997) when a solid sample moves into liquid, generating θ_A, while when it moves away from liquid, it generates θ_R, as illustrated in Fig. 1.33b.

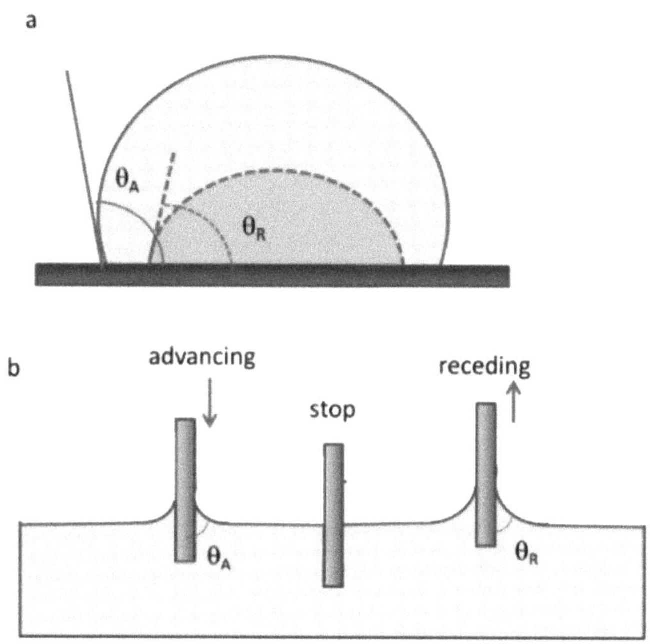

Figure 1.33 Characterization of dynamic contact angles.

Based on the dynamic contact angle, i.e., the advancing angle θ_A and receding angle θ_R, the difference between both can characterize the CAH (Brabdon et al., 1997; ElSherbini and Jacobi, 2004a,b): $\Delta\theta_{(CAH)} \approx \theta_A - \theta_R$. If $\Delta\theta_{(CAH)} > 10°$, the surface is in a high adhesion state. In general, for a superhydrophobic surface, $\Delta\theta_{(CAH)} < 10°$, and the surface is in a low adhesion state; for example, a droplet rolls off the surface of a lotus leaf, along with a self-cleaning effect.

1.2.7 Wettability Induced by Anisotropic Structures

The anisotropic structures of a surface change the shape of a droplet due to differences in the solid–liquid–air TCLs. A moderately hydrophobic or hydrophilic surface induces a distinct anisotropy in wettability when the surface structures have distinct differences in direction. Figure 1.34 indicates the typical features of a droplet changing shape when it is placed on a surface with a grooved structure. Observed in the Y-direction, the droplet has a contact angle θ_X, and the continuous TCL enlarges the droplet in length along the X-direction. Observed in the X-direction, the droplet has a contact angle θ_Y, and the discontinuous TCL decreases the droplet contact length along the Y-direction.

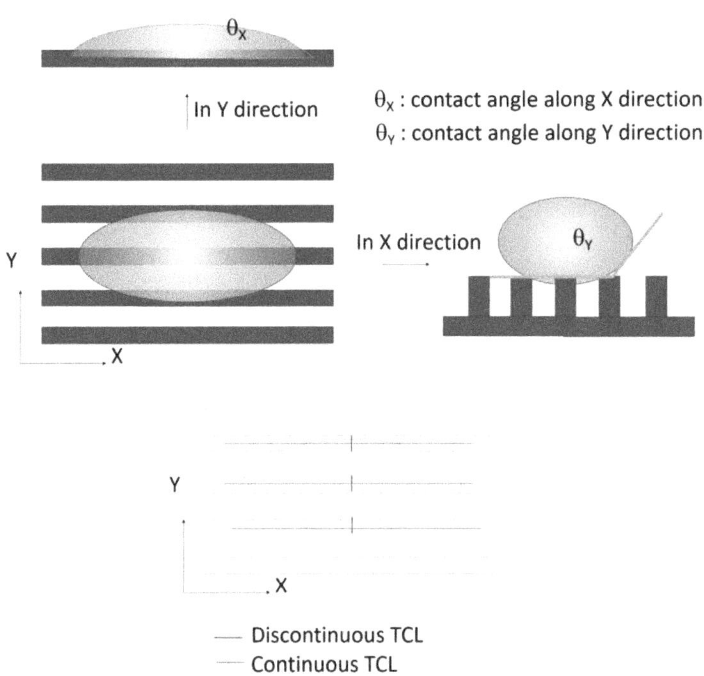

Figure 1.34 Anisotropic-structure-induced wettability based on directional structures. The directional groove generates two oriented (X, Y) wetting differences, along with contact angle differences; θ_x and θ_y indicate the contact angle along the X-direction and Y-direction, respectively, corresponding to the continuous TCL and discontinuous TCL.

1.2.8 Nonequilibrium Interface Forces for Liquid Spreading and Transport

When a surface is designed with a gradually changing chemical composition that causes a surface-energy gradient, i.e., one side has a high surface energy (γ_H) and another side has a low surface energy (γ_L), the droplet on such a gradient surface would have a different contact angle (θ_L at location γ_L and θ_H at location γ_H) and can be acted on by the unbalanced force (F) pointing in the high surface-energy direction (Fig. 1.35). Given the unbalanced force dF, considering Young's law and Wenzel's law, dF can be given by the following relation:

$$dF = r\gamma_{LV}(\cos\theta_H - \cos\theta_L)dx \quad (1.6)$$

where r is the roughness of the surface and γ_{LV} is the tension of the liquid.

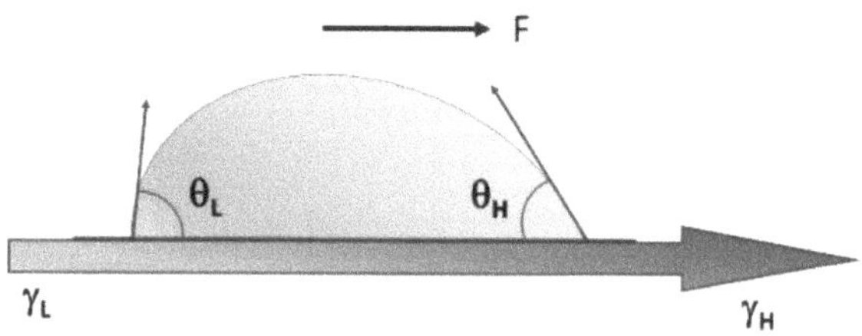

Figure 1.35 A droplet on a composition gradient surface. The high or low contact angle is induced by the high or low surface energy.

Conical features can be observed on biological materials such as cactus spines and spindle-knots of wetted spider silk. It is known that multiple droplet shapes can be present on a fiber (Loreceau et al., 2004): If the volume is large or the fiber is thin, the drop will form around the fiber, adopting a barrel shape (Fig. 1.36a, left). In contrast, small droplets would have a clamshell shape (Fig. 1.36a, right). There is a pressure difference (ΔP) between the inside and outside of a droplet, which is defined as the Laplace pressure (Loreceau et al., 2004; Revoise et al., 2009). For the barrel-shaped droplets on a fiber, $\Delta P = 2\gamma/(R + R_0)$, where R is the radius of the fiber, R_0 is the radius of the droplet, and γ is the surface tension of the liquid. For the clamshell-shaped droplet, $R \approx R_0$; thus, $\Delta P = \gamma/R$. If the fiber is conical (apex angle is 2α), as shown in Fig. 1.36b, the Laplace pressure ΔP will decrease as R increases along the z-axis of the conical fiber. The gradient ΔP would be formed along the length of the conical fiber; thus, the droplet can be driven by the force (F_P) of the Laplace pressure difference (Fig. 1.36c). The F_P can be described as (Song and Zheng, 2014):

$$F_P \sim -\int_{R_S}^{R_L} \frac{2\gamma}{(R+R_0)^2} \sin\alpha dz \tag{1.7}$$

where R is the local radius of the conical region, R_S and R_L are the local radii of the region at the two opposite sides of the droplet, and dz is the displacement increment along the conical axis. Such force will drive the droplets to move directionally along the conical fiber, achieving liquid transport.

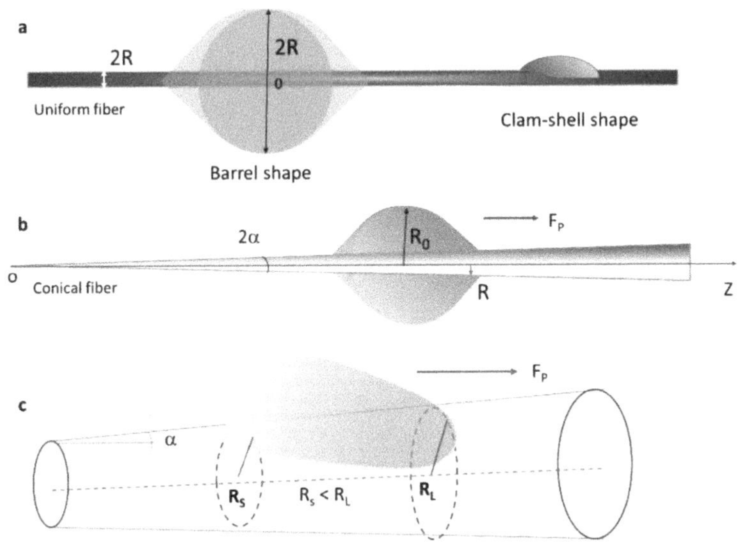

Figure 1.36 (a) The barrel shape (droplet height more than the fiber radius) and clamshell shape (droplet height far less than the fiber radius) of droplet formation on a uniform fiber. (b) Droplet on a conical fiber with apex angle 2α, where the directional gradient force (F_p) of the Laplace pressure on the conical fiber acts on the droplet. (c) Illustration of F_p, which is related to R_s and R_L ($R_s < R_L$) on conical fibers with apex angle 2α.

1.3 Biological Structure Models

Based on the features of biological wettable surfaces, biological surface structures can be modeled to understand the surface functions and further design artificial or bioinspired surfaces. Such structures can be classified in terms of isotropic micro- and nanostructures, anisotropic structures, heterogeneous wettable patterns, gradient toughness, conical geometry, spindle-knot and humped structures, and liquid-infused interfaces for different functions.

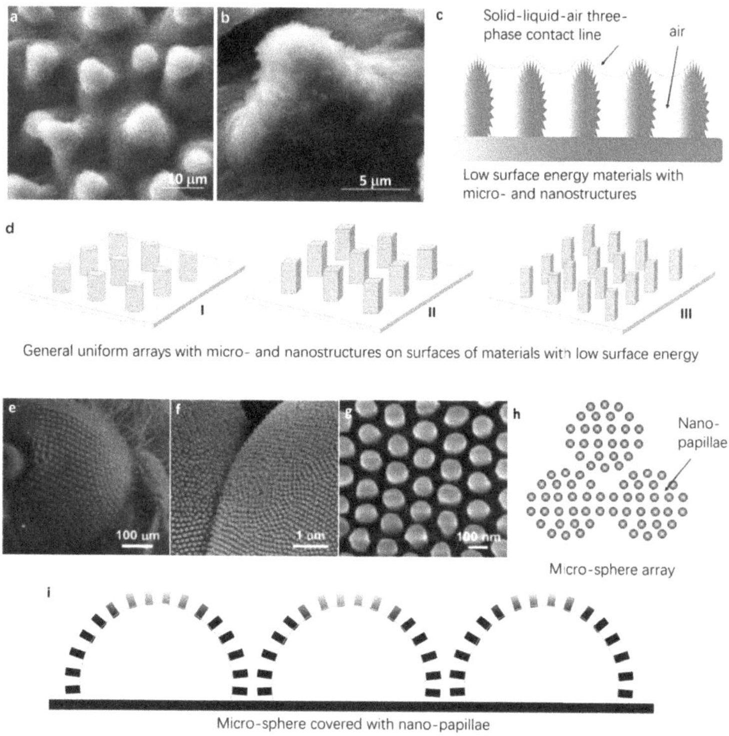

Figure 1.37 Isotropic micro- and nanostructures of biological surfaces. (a–b) Micropapillae (a) and nanohairs (b) on the surface of a lotus leaf observed by ESEM. (c) Illustrations of micropapillae and nanohairs with trapped air. (d) Illustration of general uniform arrays with micro- and nanostructures on an as-designed low-surface-energy surface. (e–g) The compound eyes (e) of mosquitoes composed of microspheres (f) and nanopapillae arrays (g). (h) Illustration of microsphere and nanopapillae arrays. (i) Illustration of the general design for microspheres covered with nanopapillae.

1.3.1 Isotropic Micro- and Nanostructures

Natural surfaces such as lotus leaves and compound eyes can be uniformly distributed with micro- and nanostructures over the whole surface; these structures can, thereby, be called isotropic micro- and nanostructures and exhibit self-cleaning effects. As shown in Fig. 1.37a–d, the surface of the lotus leaf is composed of micropapillae and nanohair waxy crystals (Fig. 1.37a–b). Trapped air exists in the micro- and nanostructures, which raise the solid–liquid–air TCLs (Fig. 1.37c). Such properties can be observed with general models of micropillars, microcylinders, and any nanostructures (Fig. 1.37d). The compound eye of mosquitoes can be considered a microsphere array, where each sphere is covered with nanopapillae arrays (Fig. 1.37e–g). Figure 1.37h illustrates the top view of the micro- and nanostructure arrays. Figure 1.37i indicates the model of the array in the compound eye, composed of the hemispheres with nanopapillae.

1.3.2 Anisotropic Structures

Surfaces such as butterfly wings, rich leaves, and shark skin incorporate anisotropic structures, including strings, grooves, ratchets, oriented lamellae, and so on. These features of micro- and nanostructures can promote the control of directional wettability due to different solid–liquid–air TCLs in different directions, i.e., anisotropic structures, for directional wettability.

As shown in Fig. 1.38a, based on the surface feature of the butterfly wing, the structure models of anisotropy can be demonstrated in two directions (Fig. 1.38c–e). If the oriented structure is formed along one direction (e.g., RO direction, Fig. 1.38c), the anisotropic wettability induces unidirectional droplet roll-off (i.e., along the RO direction). If a groove or stripe structure is formed, the anisotropic wettability induces different roll-off properties; i.e., along the X-direction, there is a continuous solid–liquid–air TCL; along the Y-direction, there is a discontinuous TCL (see Section 1.2.7, Fig. 1.34).

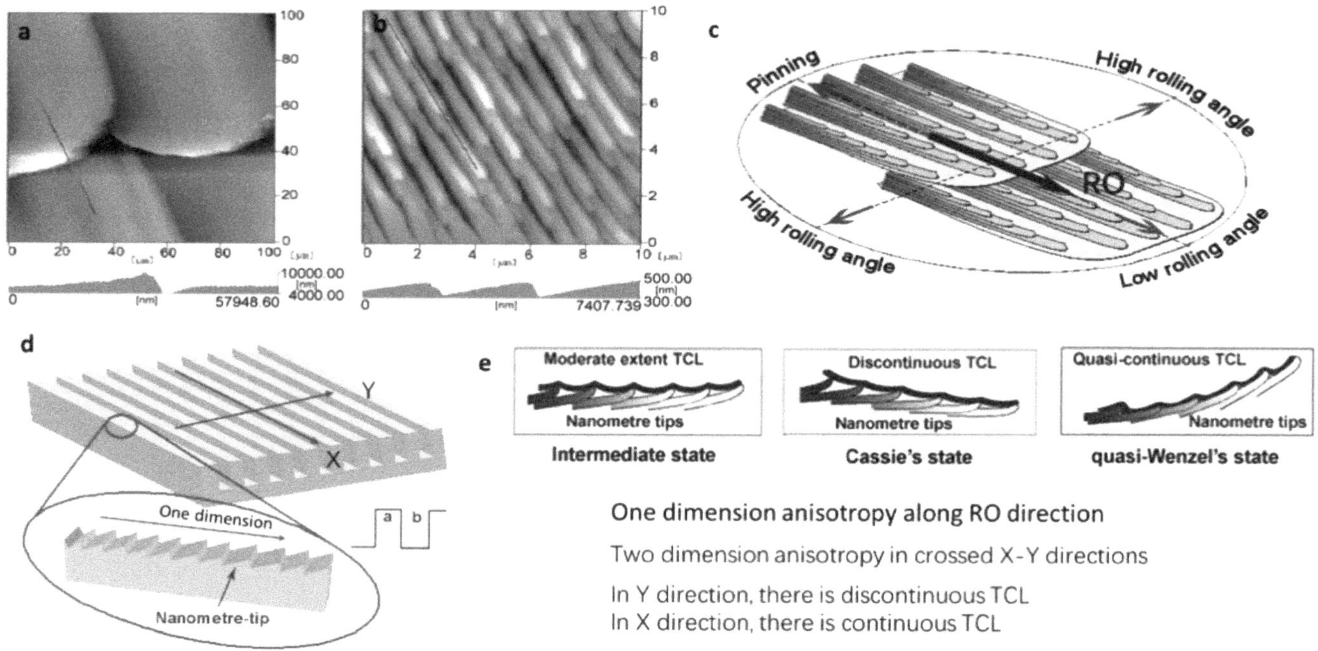

Figure 1.38 (a–b) Atomic force microscopy (AFM) images of butterfly wings, along with oriented ratchets at the microlevel and nanolevel. (c) Illustration of directional adhesion along different orientations, leading to different rolling properties. The RO direction induced a low rolling angle for the water droplet. (d) Illustration of two-directional (X, Y) anisotropy in water repellency. (e) One-directional anisotropy along/against the RO direction. It includes three kinds of wetting states: intermediate state (left), Cassie state (middle), and quasi-Wenzel state (right).

1.3.3 Heterogeneous Wettable Patterns

The beetle back has alternating hydrophobic and hydrophilic regions for droplet transport, where the hydrophilic region captures the fog droplets and hydrophobic regions repel the fog droplets, which overall can be defined as a heterogeneous wettable pattern. As shown in Fig. 1.39a, the hydrophilic and hydrophobic regions form a heterogeneous wettable pattern. Such surfaces can be modeled as circular (Fig. 1.39b) or star-shaped (Fig. 1.39c) hydrophilic regular regions on hydrophobic surfaces for most bioinspired surface designs.

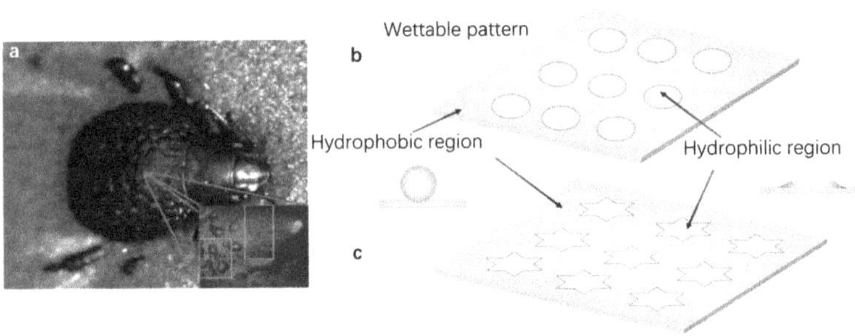

Figure 1.39 (a) The different wetting states on the beetle back including hydrophilic and hydrophobic states (see inset). (b–c) Illustration of general designed patterns; the wettable patterns with circular (b) or star-shaped (c) regions on hydrophobic surfaces.

1.3.4 Gradient Roughness and Conical Geometry

The spines of cactus and grass leaves can have a conical geometry and rough surface along the spine from the tips to ends, forming gradients in terms of roughness and geometry, defined as gradient roughness, for fog harvesting. As shown in Fig. 1.40a, the spine of the cactus has a long and conical geometry. The models of spines can be separated into those with a rough curvature or surface-tension gradients with a conical geometry, where droplets can be transported to locations with large-diameter regions (Fig. 1.40b); arrays of spines can be designed on surfaces to increase the amount of fog harvesting (Fig. 1.40c).

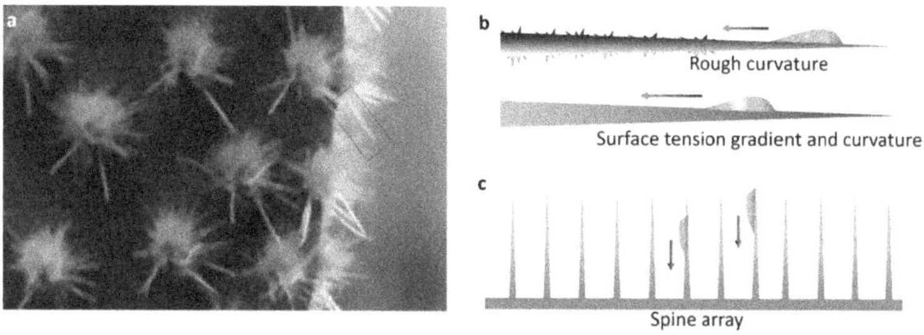

Figure 1.40 (a) Optical image of a cactus with spines. (b–c) Model of spines with a rough curvature or surface-tension gradient (b) and curvature or arrays (c) of spines on a surface.

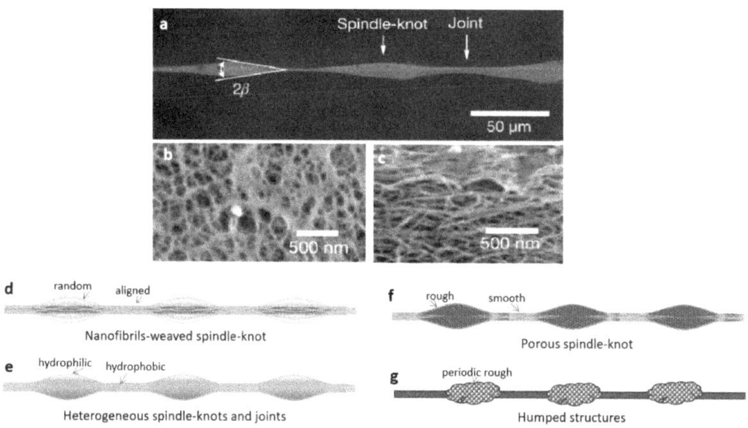

Figure 1.41 (a–c) SEM images of wetted spider silk composed of spindle-knots (cone angle of 2β) and joints (a) and random nanofibrils on a spindle-knot (b) and aligned nanofibrils (c). (d–g) The design of bioinspired spider silk via cooperative rough and curvature gradients on fibers: nanofibril-weaved spindle-knot (d), heterogeneous spindle-knots and joints (e), porous spindle-knot (f), and humped structures (g).

1.3.5 Cooperative Rough and Curvature Gradients

Wetted spider silk displays spindle-knots and joints or periodic humped structures and can be defined as having rough curvature gradients for water-condensed droplet transport with high efficiency. As shown in Fig. 1.41a, wetted spider silk is composed of spindle-knots and joints that result from random nanofibrils (Fig. 1.41b) and aligned nanofibrils (Fig. 1.41c), respectively. The cooperative rough and curvature gradient feature can form on wetted spider silk with spindle-knots and joints (Fig. 1.41 d–g), generating the cooperative force of both gradients to drive droplet motion. The spindle-knots and joints can be developed into porous structures with surface-tension gradients and nanoparticle humps, as well as heterogeneous features on their surfaces.

1.3.6 Liquid-Infused Interfaces

Nepenthes alata has a surface with liquid-infused interfaces (Fig. 1.42a–b), which induces low adhesion and lubrication effects for efficient liquid transport. This surface can be modeled as a surface with porous and liquid-infused features (Fig. 1.42c), forming a liquid–solid–liquid interface and achieving a slippery surface. These surfaces can serve as inspiration for the design of artificial surfaces and functional structures, which can be extended to some applications, including microfluidics, microdevices, liquid transport in control, anti-icing surfaces, and so on.

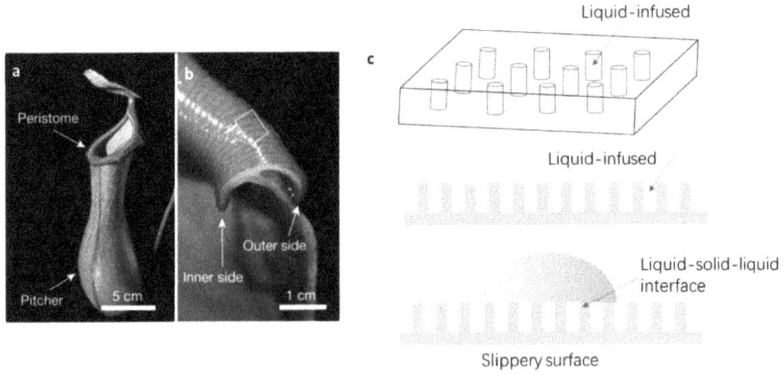

Figure 1.42 (a–b) Optical images of *Nepenthes alata* (a) and the unique structure of the peristome with liquid-infused slipperiness (b). (c) Illustration of a slippery surface with liquid–solid–liquid interfaces due to the liquid-infused layer in structures, lowering the adhesion of the surface.

1.4 Bioinspired Methods and Techniques to Fabricate Surfaces

1.4.1 Electrochemistry

Electrochemistry techniques can be developed for the fabrication of nanostructures with gradients. For instance, the cathodic electrodeposition method can be applied to realize a wettability gradient, as shown in Fig. 1.43a. In this design, a copper plate serves as an anode and provides a current (I) with a gradient ($I_3 > I_2 > I_1$). A drain is used to adjust the deposition time (t) with a gradient ($t_3 > t_2 > t_1$). By combining both gradients via the control of the voltage (V) and the flow velocity of electrolyte (f), the surface generates gradients of geometrical structure (i.e., density of structures) and chemical composition (including different types of Cu, Cu^{+}, and Cu^{2+}), which induce wettability gradient differences.

Figure 1.43b shows micro- and nanostructured surfaces with increasing superhydrophobicity in terms of wettability (from a water contact angle of 128° to 163° along the T to B direction) in type I (the conditions of V = 3 V, f = 3.5 L/h), where the micropapillae grow larger (Fig. 1.43b, frames 2–5). Figure 1.43c shows the surface in type II, which can be achieved by applying a voltage of 5 V. The micro- and nanostructures cause the contact angle of water to decrease from 162° to 59° (from T to B) due to the combined effect of the gradient structure and chemical composition. Apparently, wettability gradients on surfaces can be effectively realized electrochemically by modulating the currents and treatment times, which are adjusted by changing the distance between the anode and cathode or the flow velocity of the electrolyte.

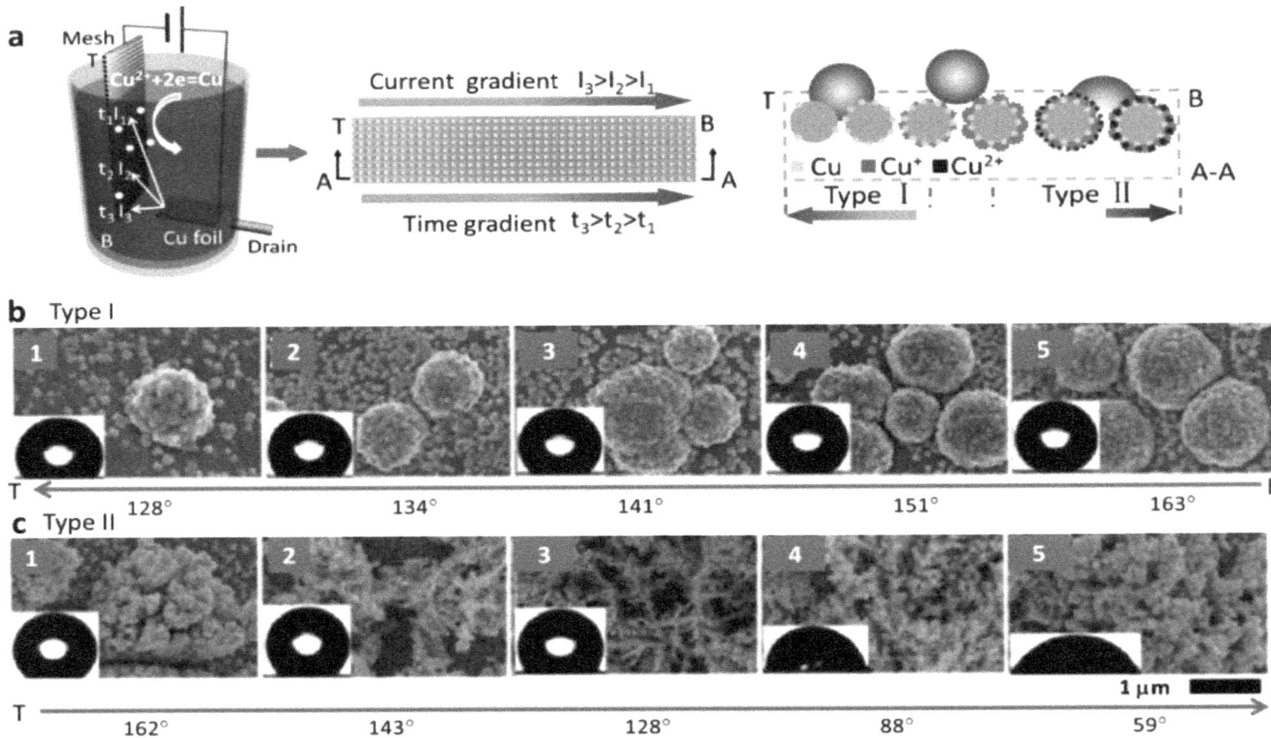

Figure 1.43 (a) Illustration and formation of various wettable gradient surfaces via electrochemistry under controlled current and time gradients (Hou et al., 2016). (b–c) The gradients of wettability on the surfaces: B→T, contact angles change 163°→128° (b), and T→B, contact angles change 162°→59°.

1.4.2 Soft Lithography

Biological structures are usually complex in configuration, representing a challenge to biomimetic research. Recently, soft lithography has been used to fabricate biological structures by using various polymer materials, e.g., polystyrene (PS), poly(vinyl alcohol) (PVA), polydimethylsiloxane (PDMS), poly(etherimide) (PEI), and polyvinylidene fluoride (PVDF). In studies on grass leaves with ratchet structures, the soft lithography technique can be used to fabricate biomimetic surfaces (Guo et al., 2012). Figure 1.44a shows a photograph of a fresh ryegrass leaf with a dew droplet. The surface of ryegrass leaves has a taper-ratchet stripe-style array with a periodicity of ~100 μm. The taper-ratchets have oriented tips (apex angle of ~5°, tilted angle of ~25°) in a periodicity of 30–50 μm along the array (Fig. 1.44b), with a length of ~10 μm and maximum diameter of ~5 μm.

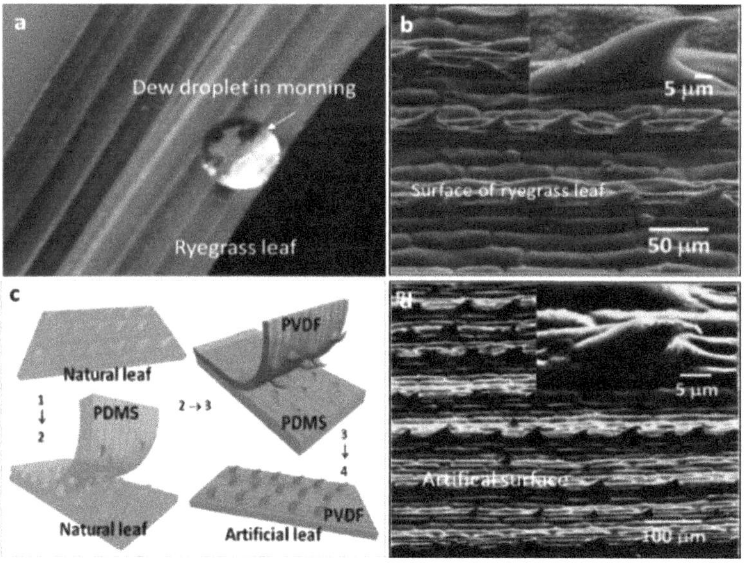

Figure 1.44 (a) Fresh ryegrass leaf with a dew droplet. (b) SEM images of a fresh ryegrass leaf surface. (c) Fabrication process. (d) SEM image of the artificial surface (Guo et al., 2012).

Such a complex structure can be fabricated by using PVDF via modified soft lithography, as shown in Fig. 1.44c. First, the PDMS prepolymer is used to copy negative ratchet arrays of fresh ryegrass leaves after treatment at a temperature $<60°C$ for 3 h. Then, PVDF polymer is used to emboss the positive ratchet arrays from negative ratchet arrays of the PDMS surface after a pressing process at a temperature of $\sim 270°C$, with a heat-pressing force of $\sim 2 \times 10^6$ Pa. As shown in Fig. 1.44d, the artificial PVDF surface is quite similar to that of fresh ryegrass leaves (Guo et al., 2012).

1.4.3 Dip Coating

The dip-coating method can be further developed into special coatings by controlling the thickness. For instance, inspired by the feature of wetted spider silk for water collection, some bioinspired fibers with a hump geometry have been fabricated by utilizing dip coating and the Rayleigh instability method (Bai et al., 2011a,b; Tian et al., 2011a). Figure 1.45a illustrates the fabrication process. A nylon fiber (diameter of 18 μm) is first soaked in a solution composed of poly(methyl methacrylate) (PMMA)/N,N-dimethylformamide (DMF) and then drawn out (velocity of 0.1–3 mm/s). The solution can be coated on fibers to form a cylindrical polymer liquid film. Subsequently, this polymer liquid film instantly breaks up into periodic polymer droplets along the fiber under the action of Rayleigh instability. Finally, the polymer droplets change into the desired humps on the fiber after evaporation of the DMF solvent. In the process of Rayleigh instability, the periodicity (λ) of the broken-up polymer droplets is related to the thickness of the polymer film, which can be determined by the drawing velocity and concentration of the polymer solution (i.e., viscosity). The film thickness of the coating is affected by the capillary number (Ca), i.e., $Ca = \eta V/\gamma$, where η is the solution viscosity, V is the fiber-drawing velocity, and γ is the solution surface tension. Thus, the size of the geometric humps can be further optimized. The bottom inset of Fig. 1.45a presents an optical microscopic image of a typical bioinspired fiber, which was fabricated in a 10% PMMA/DMF solution at a velocity of ~ 3 mm/s. The humps have a height (H) of ~ 47 μm and periodicity (λ) of ~ 400 μm on average.

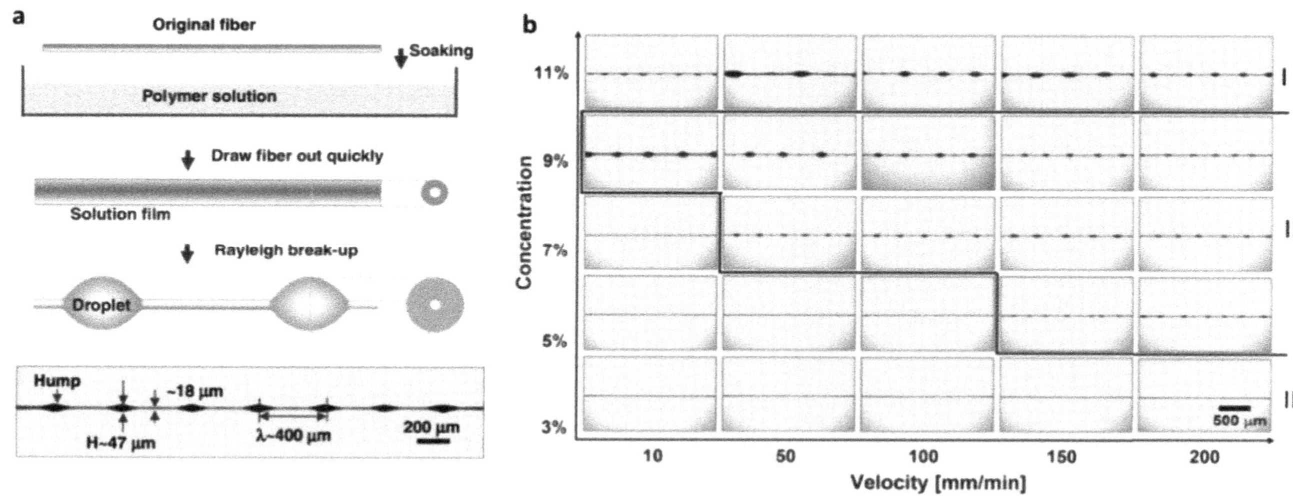

Figure 1.45 (a) Illustration of fabricating a bioinspired humped fiber (Tian et al., 2011a). (b) The humped features in regions I, II, and III are due to different concentrations and velocities of fluid coating (Bai et al., 2011a).

Figure 1.45b shows a series of bioinspired artificial fibers fabricated under drawing-out velocities (10–200 mm/min) by using PMMA polymer solution (weight percentage concentration of 3–11%), where periodic humped structures can be formed on fibers via different concentrations and drawing velocities. The humped features can be classified into three regions in terms of concentration (I, II, and III). The high-concentration region is defined as region I, and the humps have a large size. The low-concentration region is defined as region III, and small humps are achieved. The proper concentration region is defined as region II, with a range of 5–8%, and the hump features are distinct along with uniform geometry.

1.4.4 Microfluidics

Microfluidic systems were initially used to develop microdevices for medical applications (Whitesides and Stroock, 2001; Whitesides, 2006). Recently, based on fluid features controlled in microfluidic channels, microfluidic methods have been developed to fabricate novel materials (Kang et al., 2011). The example of an as-designed microfluidic system is presented, which was developed by studying the spinning apparatus of the spider and its ability to mix fluids from glands in its abdomen via an open–close valve to restart spinning after internal rupture (Fig. 1.46a); the microfluidic system was designed for the fabrication of bioinspired silks. Figure 1.46b illustrates the as-designed microfluidic chip; the integration of small pneumatic valves achieved the formation of triple-helix fibers. Figure 1.46c shows a photograph of the microfluidic device.

In this device, two alternating fluid streams are mixed via an open–close valve to generate continuous fibers with periodic nanoporous spindle-knots, where an alginate solution containing salt (0.5% w/w) is flowed into one of channels at a high flow rate, and the length of the spindle-knot can be controlled by altering the open-valve time. Salt dissolution would lead to the formation of nanoporous structures on the spindle-knots, leading to a higher surface energy to attract water more than the joints can. The driving force of the Laplace pressure difference of the self-collected water droplets can occur at conical spindle-knots on the fiber. These microfluidic systems can be used to develop novel materials for functional bioinspired silks via design of the chemical components.

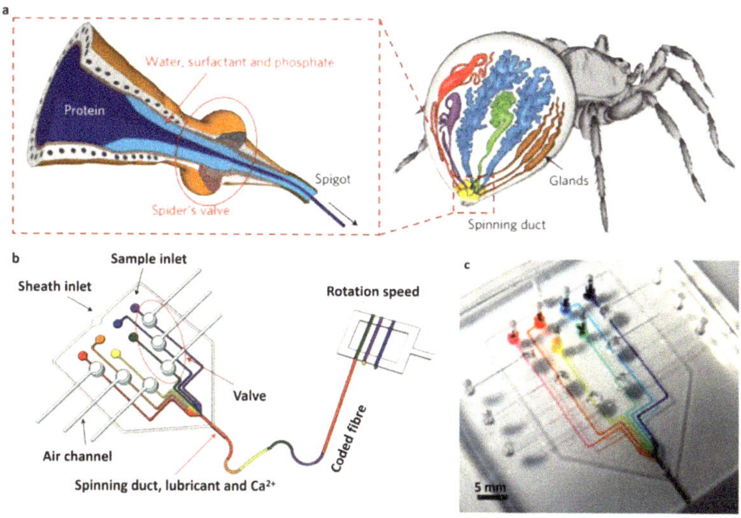

Figure 1.46 (a) Illustration of the spinning apparatus of the spider and mixing fluids from glands in its abdomen. (b) Bioinspired microfluidic system. (c) Photograph of the microfluidic device (Kang et al., 2011).

1.4.5 Electrospinning

Electrospinning can be used to fabricate unique structures, such as bead-on-string (Tian et al., 2011b), similar to the capture silk of ecribellate spiders with a uniform coating of glycoprotein glue. Figure 1.47a shows an experimental setup that was designed by two blunt metallic needles in the same axis. The conductive inner needle was connected to the cathode of a high-voltage generator, and an aluminum foil collecting substrate was connected to the anode. This strategy uses a sprayable outer fluid to imprint a series of heterogeneous beads on a fiber after combining with the spinnable inner fluid. The spinnable inner fluid was highly viscous and was composed of a 35% (w/v) polystyrene (PS, Mw = 350 000) solution in DMF. The sprayable outer fluid had a low viscosity and was composed of a 20% (w/v) poly(ethylene glycol) (PEG, Mw = 20 000) solution in a mixed solvent of N,N-dimethylformamide (DMF) and methylene chloride (MC) (v/v = 1:1). The distance of the needles and collecting substrate was ~22 cm, and the applied voltage was

19 Kv. The flow rate was tuned from 3 to 6 Ml/h. Figure 1.47b shows the bead-on-string profiles observed by ESEM. The achievement of bead-on-string structures is considered to be a result of the surface energy difference between the core and shell components, i.e., poly(ethylene glycol) with high surface energy and polystyrene with low surface energy, which were components in the outer and inner fluids, respectively. The higher surface energy of the shell polymer favors the spontaneous Rayleigh breakup of the outer fluid; thus, the beads imprint on the fiber. This technique can be used to develop novel fiber materials for engineering applications.

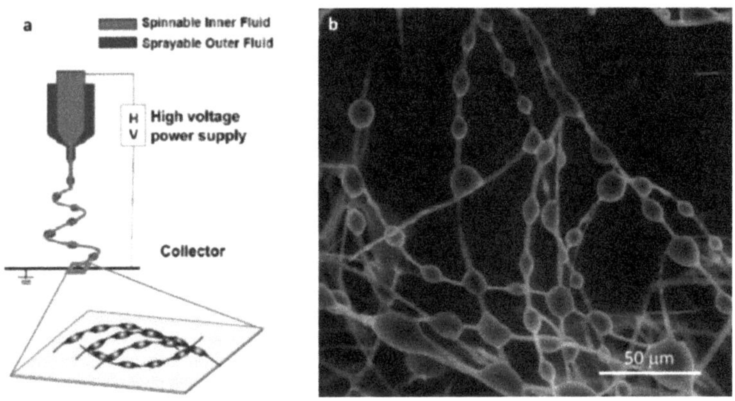

Figure 1.47 (a) Illustration of the combined electrospinning/electrospraying approach. (b) ESEM image of bead-on-string fibers with condensed water droplets on beads (Tian et al., 2011b).

1.4.6 Imprinting

Nanoimprint lithography (NIL) is an alternative approach to fabricate nanostructures from sub 100 nm (Chou et al., 1995) to 6 nm half-pitch lines (Perret et al., 2004; Auatin et al., 2005), presenting advantages of high throughput, high resolution, and low cost (Guo, 2004). The challenges of such techniques include imprinting on irregular or nonflat surfaces (Kuo et al., 2003; Hulme et al., 2006). Biological nanostructures exist on complex surfaces (Aizenberg et al., 2001; Parker et al., 2003), such as porous nanostructures on microscales of blue color *Morpho sulkowskyi* butterfly wings

(Kumazawa and Tabata, 2001; Kinoshita and Yoshioka, 2005), waxy nanohairs on microscale papillae of plant leaves with a self-cleaning effect (Barthlott and Neinhuis, 1997; Blossey, 2003), and arrays of tapered nanopillars on cicada wings and moth eyes exhibiting anti-reflectivity (Watson, 2004; Kaless et al., 2005). These biological microstructures are important for mimicking the special functions and have inspired the fabrication of artificial surfaces, such as artificial structures of cicada wings (Zhang et al., 2006).

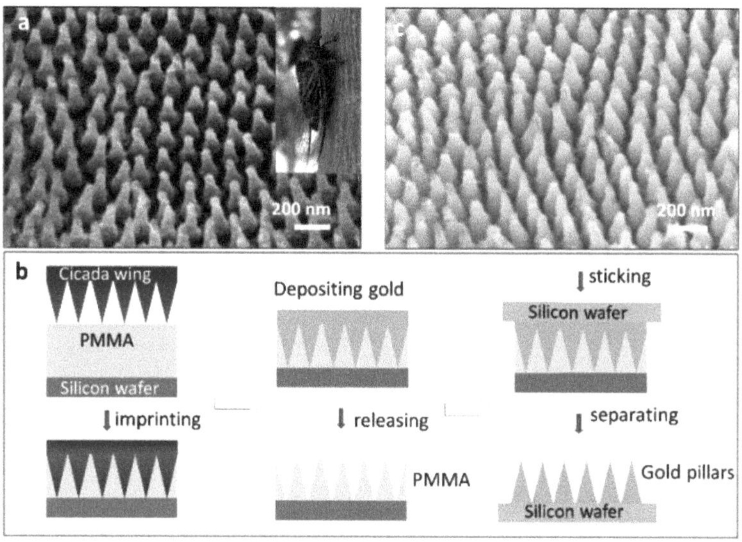

Figure 1.48 (a) The wing of a cicada (*Cryptympana atrata* Fabricius) compared with ordered hexagonal close-packed arrays of pillars. The inset is a photograph of a cicada. (b) Illustration of the process of nanoimprint lithography with cicada-wing stamps for an artificial structure based on cicada wings. (c) The SEM images show that the features of the artificial structure are consistent with those of the original stamps (Zhang et al., 2006).

Figure 1.48a shows the microscopic structures on the wing of a cicada (*Cryptympana atrata* Fabricius), where the ordered HCP arrays of pillars have a spacing of ~190 nm, height of ~400 nm, and diameters of 80–150 nm. Figure 1.48b illustrates the process of nanoimprint lithography with cicada-wing stamps, fabrication of gold pillar arrays, and transferring the pattern to a silicon substrate by using PMMA (weight-average molecular weight Mw = 95 000). The imprinting pressure applied was ~40 bar for ~180 s.

The temperature was ~190°C > T_g (glass-transition temperature, at 120°C), and PMMA had a low-enough viscosity. Figure 1.48c shows the as-fabricated gold pillar arrays, the features of which are consistent with those of the original stamps (Zhang et al., 2006).

1.5 General Materials for Bioinspired Fabrication

1.5.1 Spindle-Knotted Structured Fibers

The dip-coating method can be controlled to achieve a gradient geometry on structures, such as gradient spindle-knots on a fiber. Figure 1.49a illustrates the tilt-angle dip-coating system. Figure 1.49b shows optical images of the fabrication process of bioinspired fibers with gradient spindle-knots via the tilt-angle dip-coating route (Chen et al., 2013).

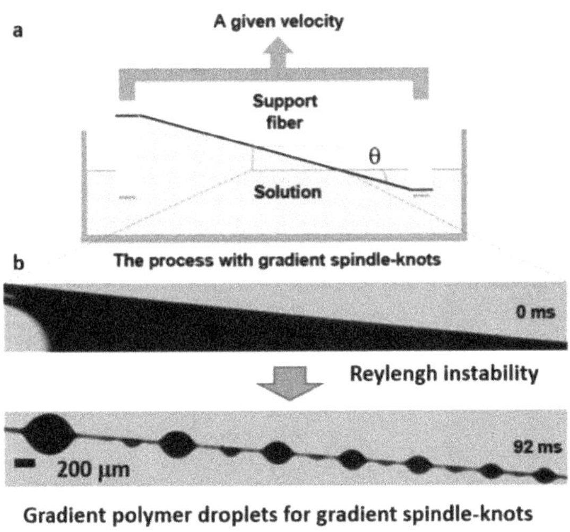

Figure 1.49 (a) Fabrication using the tilt-angle dip-coating method. (b) Optical images of gradient polymer droplets for gradient spindle-knots via broken up by Rayleigh instability in 92 ms (Chen et al., 2013).

A uniform nylon fiber (~18 mm) at a tilt angle θ was immersed into a polymer solution consisting of polyvinylidene fluoride (PVDF) dissolved in N,N-dimethylformamide (DMF) and then was drawn

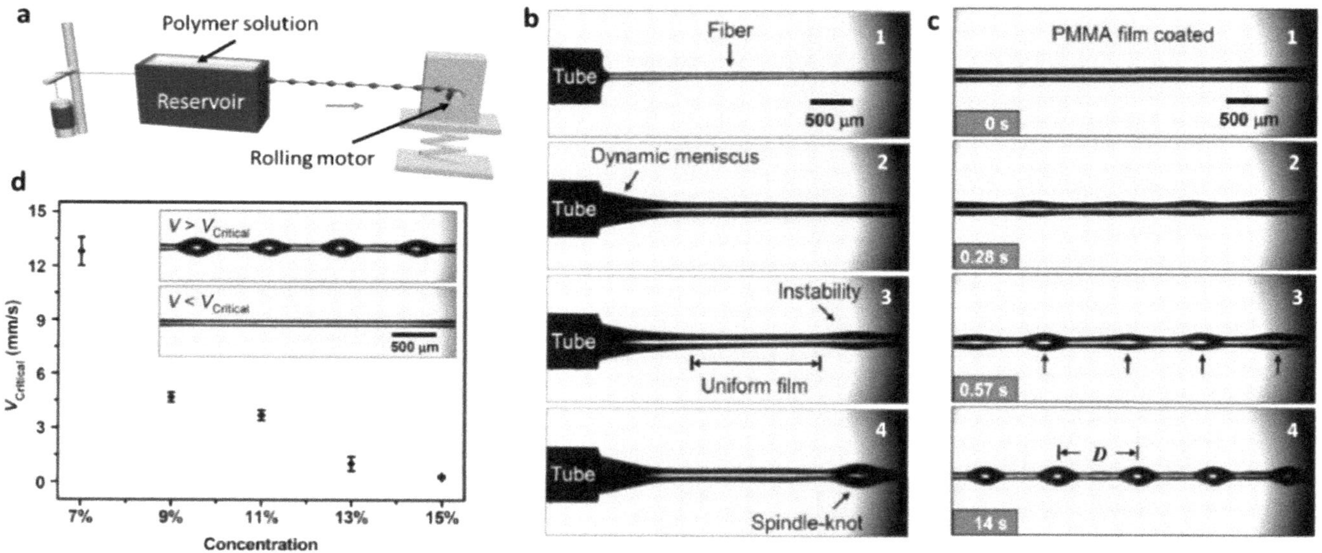

Figure 1.50 (a) Illustration of the fluid-coating system. (b) The dynamic process of fluid coating and initial formation of Rayleigh instability. (c) Result of Rayleigh instability and formation (in 14 s) of spindle-knots on a fiber (Bai et al., 2011b).

out at a tilted angle and a given velocity. Thus, a PVDF film with a gradient thickness covered the fiber surface (at 0 ms). After Rayleigh instability, the different gradient thickness film breaks into different polymer droplets along the direction of the gradient thickness. As a result, gradient polymer droplets are formed (at ~92 ms). It is estimated that the heights of polymer droplets range from ~186 to ~51 mm at ~92 ms, and their widths range from ~289 to ~160 mm. After the solvent evaporates, PVDF gradient spindle-knots are formed along the fiber.

The fluid-coating technique can be developed to fabricate novel structures on fibers (Bai et al., 2011b). As illustrated in Fig. 1.50a, a fiber is withdrawn from the reservoir polymer solution out of the tube exit by a motor with controllable fluid-coating velocity. Figure 1.50b shows the process of fluid coating on a fiber (75 μm in diameter) by a polymer solution composed of PMMA dissolved in N,N-dimethylformamide (DMF) with a weight percentage of 11% and a draw-out velocity of ~3.21 mm/s. At a location of uniform solution film thickness (e.g., ~53.8 μm), the Rayleigh instability starts to cause the film to be broken up into polymer droplets (Fig. 1.50a, Frame 4). Subsequently, the droplets are continuously formed (Fig. 1.50c) after ~0.57 s. The complete polymer droplet pattern forms within 14 s, where spindle-knots have a periodicity (D) of ~921.7 μm and joint radii (R) ≈ 91.3 μm, i.e., D ≈ 10 R for the array. The critical velocity (V_{Critical}) of the fluid-coating process is defined as the velocity at which the initial formation of spindle-knots occurs at a given solution concentration (Fig. 1.50c). According to the relationship Ca = $\eta V/\gamma$, a smaller velocity can coat a thick film with a more viscous solution. As a result, V_{Critical} decreases from ~12.81 to ~0.29 mm/s as the solution concentration increases from 7% to 15% for the fabrication of excellent spindle-knot fibers (Bai et al., 2011a).

A velocity-change coating method can be introduced into fluid coating to fabricate unique gradient-step spindle-knots (GSs) on fibers (Xue et al., 2014), as shown in Fig. 1.51a. Figure 1.51b shows optical images of in situ GS formation on a nylon fiber (110 μm in diameter) during coating with a solution of PMMA in N,N-dimethylformamide (DMF) (10 wt.%) by changing the drawing velocity from the initial velocity (V_0 ~ 0 mm/s) to the terminal velocity (V_t ~ 21.4 mm/s) with constant accelerated speed a = 8.3 mm/s^2. After a short time (1.2 s), the cone solution film appears to be convex–concave, subsequently producing periodic polymer droplets with a gradient-step height from ~208.8 μm to ~595 μm at 2.5 s. Figure 1.51c–f shows SEM images of the local gradient morphology of the spindle-knots, with a highly porous structure at the center of the spindle-knot, a gradient porous structure at the middle, and a relatively smooth structure at the side. Such unique gradient spindle-knots can facilitate directional droplet transport.

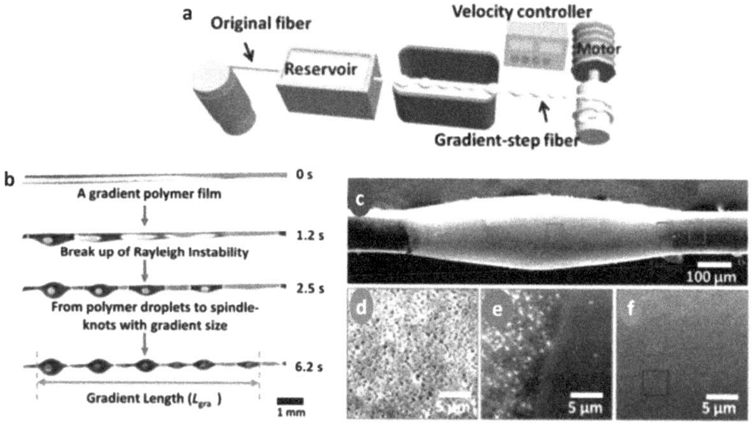

Figure 1.51 (a) Fluid-coating system with changeable drawing velocity. (b) Optical images of in situ observations of the dynamic fabrication process. Gradient spindle-knots form along the fiber in 6.2 s due to droplet breakup via Rayleigh instability. (c–f) SEM images of the gradient microstructure on the fiber (Xue et al., 2014).

Self-assembly processes have been developed to fabricate novel structures by controlling the wetting time and water condensation, defined as wet assembly (Zhao et al., 2014). Based on the method of wet assembly, a kind of humped fiber can be fabricated by using composite materials composed of PMMA and titanium tetrachloride ($TiCl_4$)-hydrolyzed nanoparticles by integrating PMMA electrospinning in nanoparticle-containing fog with a humidity of 50–90% and a wetting time from 2 s to 16 s. Figure 1.52a–c shows the process of wet assembly on humped (or bead) structures. Nanoparticles (NPCTi), 0.282–0.072 μm in size, formed via hydrolysis of 0.8 mL of $TiCl_4$ are captured on PMMA fibers in a humid atmosphere. With water condensation, small water droplets coalesce to assemble nanoparticles into beads along the fiber, which introduces a driving force resulting from cooperation of both the surface-energy gradient and Laplace pressure for coalescence and directional transport of nanoparticle-included droplets. Fig. 1.52d–f shows the SEM images of bead structures in the wet-assembly process, where the bead is composed of nanoparticles. Along with rough features, such a bead can be defined as a hump. Such humps can be controlled in size by different cycles of humidity and time.

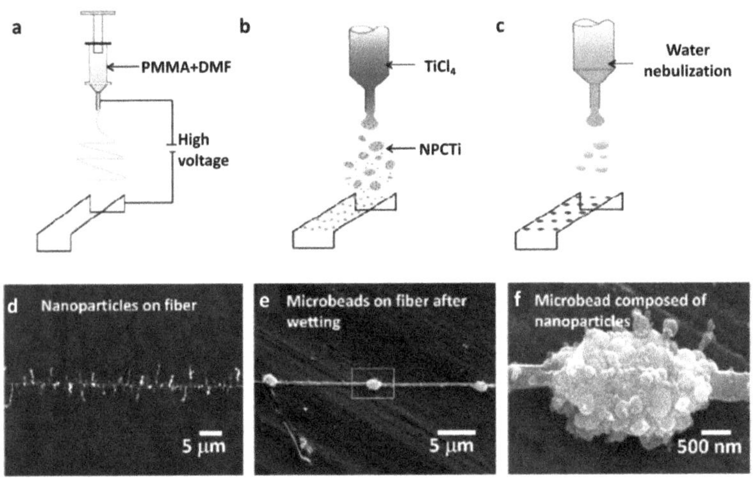

Figure 1.52 (a–c) Fabrication of beads in a wet-assembly system. (d–f) SEM images of nanoparticles (d), microbeads (e), and microbeads composed of nanoparticles (f) from in situ observations of the dynamic process of wet assembly (Zhao et al., 2014).

1.5.2 Micro- and Nanostructure of Lotus Leaves

Plant leaves repel water due to their hierarchical micro-/nanostructured topography, which provides an air layer between the solid surface and the liquid droplets. Taking the plant leaf as a template, a biomimetic surface has been achieved by integrating soft lithography and hydrothermal methods, resulting in a flexible superhydrophobic surface with a hierarchical structure composed of composite materials, namely, poly(dimethylsiloxane) (PDMS) and zinc oxide (ZnO). PDMS was used to duplicate the microstructure of the fresh lotus leaf via soft lithography, and zinc oxide (ZnO) nanohairs were fabricated on the PDMS microstructure via an enhanced hydrothermal technique, mimicking the nanohairs of the lotus leaf (Wang et al., 2016b).

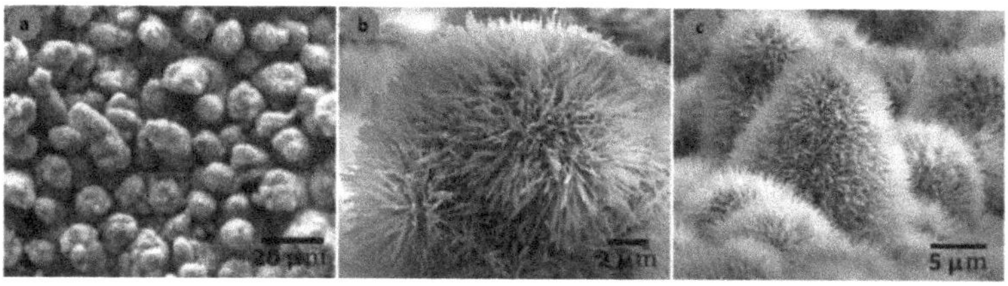

Figure 1.53 SEM images of the biomimetic lotus leaf structures (Wang et al., 2016b). (a) The surface on a large scale. (b) Micropapillae covered with ZnO nanohairs. (c) Side view of micropapillae.

Figure 1.53a–c shows SEM images of the biomimetic surface. The large-scale surface is covered with micropapillae. The diameters and heights of the papillae lied in the range of 8–12 μm and 15–20 μm, respectively (Fig. 1.53b,c). The ZnO nanohairs are covered with micropapillae, similar to natural nanohairs of lotus leaves. After treatment with heptadecafluorodecyl tri-propoxy silane (FAS-17), a low-surface-energy surface, similar to that of the lotus leaf, is effectively achieved. Figure 1.54a–f presents optical images of the water droplet removal of dust. The biomimetic surface can be made dirty due to the excessive presence of dust, and droplets can easily remove the dust to clean the surface, as observed when the biomimetic surface has a temperature of −20°C with a relative humidity of 90% in the chamber.

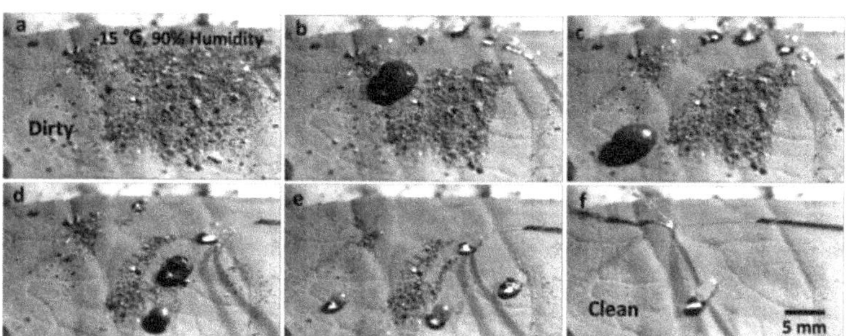

Figure 1.54 Optical images of the dynamic process of the self-cleaning biomimetic surface under low temperature (−20°C) and high humidity (90%). (a) The dirty surface at −15°C with 90% humidity. (b–e) The water droplets absorb the dirt by rolling. (f) The surface is cleaned after the droplets roll off.

Some hybrid structures can be obtained by various methods, such as electrochemical deposition (Zhang et al., 2004), electrohydrodynamics (Jiang et al., 2004), self-assembly (Zhu et al., 2007), and chemical vapor deposition (Gao et al., 2007). By using electrohydrodynamics, a bioinspired lotus-leaf-like self-cleaning superhydrophobic (polystyrene, PS) film consisting of porous microspheres and nanofibers was fabricated by electrospinning, where the porous microspheres contributed to the superhydrophobicity by increasing the surface roughness while nanofibers were interweaved to form a 3D network, binding the porous microspheres, which reinforced the composite film (Zhang et al., 2017). In addition, self-cleaning superhydrophobic polyaniline/polystyrene (PANI/PS) composite films show similar lotus-leaf-like micro-/nanostructures fabricated by a simple electrospinning method (using 3.72 wt.% PS: azoben-zenesulfonic acid (ABSA)/dimethyl-formamide (DMF) solution). Figure 1.55a,b shows an SEM image of the surface, and Fig. 1.55c,d shows an AFM image. The contact angle of water on this surface reaches 166.5°, and the sliding angle is less than 5°, indicating that the surface is self-cleaning (Zhu et al., 2006).

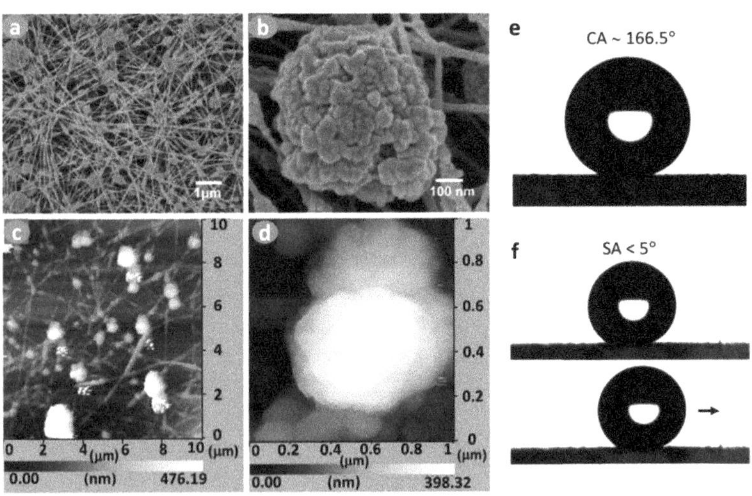

Figure 1.55 (a–b) SEM images and (c–d) AFM images of bioinspired lotus-like superhydrophobic surfaces prepared by electrohydrodynamics. (Zhu et al., 2006) (e–f) Optical microscopy images of a water droplet with a high water contact angle (CA) of 166.5° (e) and a low sliding angle (SA) of less than 5° (f).

Self-assembly methods are widely performed by using conducting and superhydrophobic rambutan-like hollow spheres of PANI by means of a self-assembly method in the presence of perfluorooctane sulfonic acid, which is employed as a dopant and soft template (Zhu et al., 2007). The water sphere formed on a surface composed of rambutan-like hollow spheres has a CA as high as 162.5°, where the water floats on the hollow spheres. Another artificial lotus leaf structure can be fabricated on cotton substrates via the controlled assembly of carbon nanotubes, which have been endowed with superhydrophobic properties—water contact angles greater than 150° (Liu et al., 2007). The layer-by-layer (LBL) assembly method has been used to create a novel poly(vinylpyrrolodone)/poly(urushiol)-CuS multilayer micro- and nanoscale hierarchical superhydrophobic film with a water contact angle of 152° (Zheng et al., 2011).

1.5.3 Compound-Eye Structured Surface

As is known, with a decrease in temperature and increase in humidity, the state of droplets on the surface will easily change from the Cassie state to the Wenzel state, which causes antifogging to fail. A critical factor in antifogging is avoiding the entry of the microdroplets into the micro- and nanostructure during water condensation. In addition, the duration of antifogging performance and repeatability are key performance indicators to measure the material antifogging property. Inspired by micro- and nanostructures of mosquito compound eyes (Gao et al., 2007; Zhang et al., 2017), an antifogging film (i.e., ZnO-on-SGH) was fabricated by the process illustrated in Fig. 1.56a. First, an airless spray method was adopted to obtain soda lime borosilicate glass hollow (SGH) sphere microstructure. A homogeneous sol composed of both SGH microspheres and zinc oxide (ZnO) seed crystals was sprayed by airless spray equipment onto the substrate surface on a large scale. Subsequently, ZnO nanostructures (e.g., nanorods) were fabricated on the microstructure SGH sphere surface by the crystal growth method. The surface structures were observed by SEM, as shown in Fig. 1.56b,c. It is noted that the SGH microspheres form uniform ~40 μm layers spread out on the entire film, and every SGH microsphere was densely covered with ZnO nanoarrays with a diameter ranging from 50 to 100 nm and a height of ~2 μm on average. The density of ZnO nanoarrays on the microspheres is similar to that of mosquito compound eyes. After being modified by using FAS-17, the surface reaches a water contact angle of 163° and a rolling angle of 6°, which results in a lower adhesion property.

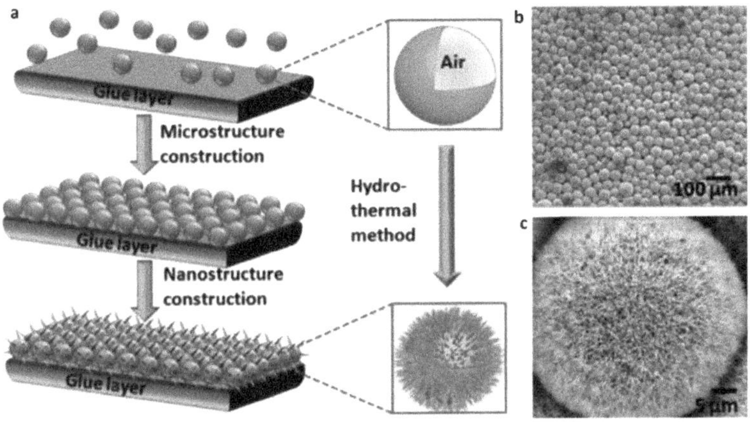

Figure 1.56 (a) Scheme of the fabrication process of artificial mosquito compound eyes. (b–c) The structure and morphology of the micro- and nanostructure (Zhang et al., 2017).

Thinking Questions

1. Can you give some examples of the interesting things about biological surfaces?
2. What are bioinspired materials and surfaces? Can you give some examples that you understand?

References

Adamec L. Mineral nutrition of carnivorous plants: A review. *Bot. Rev.* **63**, 273 (1997).

Aizenberg J., A. Tkachenko, S. Weiner, L. Addadi, and G. Hendler. Calcitic microlenses as part of the photoreceptor system in brittlestars. *Nature* **412**, 819 (2001).

Andrew R. P. and C. R. Lawrence. Water capture by a desert beetle. *Nature* **414**, 33 (2001).

Aussillous P. and D. Quéré. Liquid marbles. *Nature* **411**, 924 (2001).

Austin M. D., W. Zhang, H. Ge, D. Wasserman, S. A. Lyon, and S. Y. Chou. 6 nm half-pitch lines and 0.04 μm^2 static random access memory patterns by nanoimprint lithography. *Nanotechnology*, **16**, 1058 (2005).

Autumn K. and A. Peattie. Mechanisms of adhesion in geckos. *Integrative Comp. Biol.* **42**, 1081 (2002).

Autumn K., M. Sitti, A. Peattie, W. Hansen, S. Sponberg, Y. A. Liang, T. Kenny, R. Fearing, J. Israelachvili, and R. J. Full. Evidence for van der Waals adhesion in gecko setae. *Proc. Natl. Acad. Sci. USA* **99**, 12252 (2002).

Autumn K., Y. A. Liang, S. T. Hsieh, W. Zesch, W.-P. Chan, W. T. Kenny, R. Fearing, and R. J. Full. Adhesive force of a single gecko foot-hair. *Nature* **405**, 681 (2000).

Backer N., et al. Molecular nanosprings in spider capture-silk threads. *Nat. Mater.* **2**, 278 (2003).

Bai H., J. Ju, R. Sun, Y. Chen, Y. Zheng, and L. Jiang. Controlled fabrication and water collection ability of bioinspired artificial spider silks. *Adv. Mater.* **23**, 3708–3711 (2011a).

Bai H., R. Sun, J. Ju, X. Yao, Y. Zheng, and L. Jiang. Large-scale fabrication of bioinspired fibers for directional water collection. *Small* **7**, 3429–3433 (2011b).

Barthlott W. and C. Neinhuis. Purity of the sacred lotus, or escape from contamination in biological surfaces. *Planta* **202**, 1 (1997).

Bauer U., H. F. Bohn, and W. Federle. Harmless nectar source or deadly trap: *Nepenthes pitchers* are activated by rain, condensation and nectar. *Proc. R. Soc. B* **275**, 259 (2008).

Bauer U. and W. Federle. The insect-trapping rim of *Nepenthes pitchers*: Surface structure and function. *Plant Signal Behav.* **4**, 1019 (2009).

Bhushan B. *Biomimetics: Bioinspired Hierarchical-Structured Surfaces for Green Science and Technology*, Springer, Heidelberg, Germany, 2012.

Bhushan B. Biomimetics: Lessons from nature—an overview. *Philos. Trans. R. Soc. London, Ser. A*, **367**, 1445 (2009).

Bico J., C. Tordeux, and D. Quéré. Rough wetting. *Europhysics Lett.* **55**, 214 (2001).

Bico J., U. Thiele, and D. Quéré. Wetting of textured surfaces. *Colloids Surf. A* **206**, 41 (2002).

Bixler G. D. and B. Bhushan. Fluid drag reduction with shark-skin riblet inspired microstructured surfaces. *Adv. Funct. Mater.* **23**, 4507 (2013).

Blossey R. Self-cleaning surfaces—virtual realities. *Nat. Mater.* **5**, 301–306 (2003).

Bohn H. F. and W. Federle. Insect aquaplaning: *Nepenthes pitcher* plants capture prey with the peristome, a fully wettable water-lubricated anisotropic surface. *Proc. Natl. Acad. Sci. USA* **101**, 14138 (2004).

Bonser R. H. C. and C. Dawson. The mechanical properties of down feathers from Gentoo penguins (*Pygoscelis papua*). *J. Zool.* **251**, 545 (2000).

Boreyko J. B. and C.-H. Chen. Restoring superhydrophobicity of lotus leaves with vibration-induced dewetting. *Phys. Rev. Lett.* **103**, 174502 (2009).

Brandon S., A. Wachs, and A. Marmur. Simulated contact angle hysteresis of a three-dimensional drop on a chemically heterogeneous surface: A numerical example. *J. Colloid Interface Sci.* **191,** 110 (1997).

Caponigro M. A. and C. H. Erilsen. Surface film locomotion by the water strider, Gerris remiges Say. *Am. Midland Nat.* **95**, 268 (1976).

Carroll B. Equilibrium conformations of liquid drops on thin roll-up process cylinders under forces of capillarity. *Langmuir* **2**, 248 (1986).

Cassie A. B. D. and S. Baxter. Wettability of porous surfaces. *Trans. Farad. Soc.* **40**, 546 (1944).

Chen H., P. Zhang, l. Zhang, H. liu, Y. Jiang, D. Zhang, Z. Han, and L. Jiang. Continuous directional water transport on the peristome surface of *Nepenthes alata*. *Nature* **532**, 85 (2016).

Chen Y., L. Wang, Y. Xue, L. Jiang, and Y. Zheng. Bioinspired tilt-angle fabricated structure gradient fibers: Micro-drops fast transport in a long-distance. *Sci. Rep.* **3**, 2927 (2013).

Chou S. Y., P. R. Krauss, and P. J. Renstrom. Imprint of sub-25 nm vias and trenches in polymers. *Appl. Phys. Lett.* **67**, 3114 (1995).

Chu K.-H., R. Xiao, and E. N. Wang. Uni-directional liquid spreading on asymmetric nanostructured surfaces. *Nat. Mater.* **9**, 413 (2010).

Crawford C. S. *Biology of Desert Invertebrates*, Springer, Berlin, 1981.

Dawson C., J. F. V. Vincent, G. Jeronimidis, G. Rice, and P. Forshaw. Heat transfer through penguin feathers. *J. Theor. Biol.* **199**, 291 (1999).

De Gennes P.-G. Wetting: Statics and dynamics. *Rev. Mod. Phys.* **57**, 827 (1985).

Dickinson M. Animal locomotion: How to walk on water. *Nature* **424**, 621 (2003).

Duprat C., S. Protière, A. Y. Beebe, and H. A. Stone. Wetting of flexible fibre arrays. *Nature*, **482**, 510 (2012).

Edmonds D. T. and F. Vollrath. The contribution of atmospheric water vapour to the formation and efficiency of a spider's capture web. *Proc. R. Soc. Lond. B* **248**, 145 (1992).

Ellison A. M. Nutrient limitation and stoichiometry of carnivorous plants. *Plant Biol.* **8**, 740 (2006).

ElSherbini A. I. and A. M. Jacobi. Liquid drops on vertical and inclined surfaces I. An experimental study of drop geometry. *J. Colloid Interface Sci.* **273**, 55 (2004a).

ElSherbini A. I. and A. M. Jacobi. Critical contact angles for liquid drops on inclined surfaces. *J. Colloid Interface Sci.* **273**, 556 (2004b).

Emile O., A. L. Floch, and F. Vollrath. Biopolymers: Shape memory in spider draglines. *Nature* **440**, 621 (2006).

Eral H. and J. Oh. Contact angle hysteresis: A review of fundamentals and applications. *Colloid Polym. Sci.* **291,** 247 (2013).

Extrand C. W. Retention forces of a liquid slug in a rough capillary tube with symmetric or asymmetric features. *Langmuir* **23**, 1867 (2007).

Federle W., M. Riehle, A. S. G. Curtis, and R. J. Full. An integrative study of insect adhesion: mechanics and wet adhesion of pretarsal pads in ants. *Integrative Comp. Biol.* **42**, 1100 (2002).

Feng L., S. Li, Y. Li, H. Li, L. Zhang, J. Zhai, Y. Song, B. Liu, L. Jiang, and D. Zhu. Super-hydrophobic surfaces: From natural to artificial. *Adv. Mater.* **14**, 1857 (2002).

Gao X. and L. Jiang. Water-repellent legs of water striders. *Nature* **432**, 36 (2004).

Gao X., X. Yan, X. Yao, L. Xu, K. Zhang, J. Zhang, B. Yang, and L. Jiang. The dry-style antifogging properties of mosquito compound eyes and artificial analogues prepared by soft lithography. *Adv. Mater.* **19**, 2213 (2007).

Guadarrama-Cetina J., A. Mongruel, M.-G. Medici, E. Baquero, A. R. Parker, I. Milimouk-Melnytchuk, W. González-Viñas, and D. Beysens. Dew condensation on desert beetle skin. *Euro. Phys. J. E.* **37**, 109 (2014).

Guo L. J. Recent progress in nanoimprint technology and its applications. *J. Phys. D*, **37**, R123–R141 (2004).

Guo P., Y. Zheng, C. Liu, J. Ju, and L. Jiang. Directional shedding-off of water on natural/bio-mimetic taper-ratchet array surfaces. *Soft Matter* **8**, 1770 (2012).

Hamilton W. J. and M. K. Seely. Fog basking by the Namib Desert beetle, *Onymacris unguicularis*. *Nature* **262**, 284 (1976).

Hancock M. J., K. Sekeroglu, and M. C. Demirel. Bioinspired directional surfaces for adhesion, wetting, and transport. *Adv. Funct. Mater.* **22**, 2223 (2012).

Hanna G. and W. J. P. Barnes. Adhesion and detachment of the toe pads of tree frogs. *J. Exp. Biol.* **155**, 103 (1991).

Hansen W. R. and K. Autumn. Evidence for self-cleaning in gecko setae. *PNAS*, **102**, 385 (2005).

Hartung R. Energy metabolism in oil-covered ducks. *J. Wildl. Mgmt* **31**, 798 (1967).

Hauksbee F. An experiment touching the ascent of water between two glass plates in a hyperbolick figure. *Phil. Trans. R. Soc. Lond.* **27**, 539 (1712).

Holdgate M. W. The wetting of insect cuticles by water. *J. Exp. Biol.* **32**, 591 (1955).

Hou Y., S. Feng, L. Dai, and Y. Zheng. Droplet manipulation on wettable gradient surfaces with micro-/nano-hierarchical structure. *Chem. Mater.* **28**, 3625 (2016).

Hu D. L., B. Chan, and J. W. M. Bush. The hydrodynamics of water strider locomotion. *Nature* **424**, 663 (2003).

Huang Z., Y. Chen, Y. Zheng, and L. Jiang. Capillary adhesion of wetted cribellate spider capture silks for larger pearly hanging-drops. *Soft Matter*, **7**, 9468 (2011).

Hulme J. P., J. Gwak, and Y. Miyahara. Biomolecular embossing. *J. Am. Chem. Soc.* **128**, 390 (2006).

Jiang L., Y. Zhao, and J. Zhai. A lotus-leaf-like superhydrophobic surface: A porous microsphere/nanofiber composite film prepared by electrohydrodynamics. *Angew. Chem. Int. Ed.* **43**, 4338 (2004).

Jiao Y., S. Gorb, and M. Scherge. Adhesion measured on the attachment pads of *Tettigonia viridissima* (Orthoptera, insecta). *J. Exp. Biol.* **203**, 1887 (2000).

Ju J., H. Bai, Y. Zheng, T. Zhao, R. Fang, and L. Jiang. A multi-structural and multi-functional integrated fog collection system in cactus. *Nat. Commun.* **3**, 1247 (2012).

Juniper B. E. and J. K. Burras. How pitcher plants trap insects. *New Sci.* **13**, 75 (1962).

Juniper B. E., R. J. Robins, and D. M. Joel. *The Carnivorous Plants*. Academic Press, 1989.

Kaless A., U. Schulz, P. Munzert, and N. Kaiser. NANO-motheye antireflection pattern by plasma treatment of polymers. *Surf. Coat. Technol.* **200**, 58–61 (2005).

Kang E., G. Jeong, Y. Choi, K. H. Lee, A. Khademhosseini, and S. Lee. Digitally tunable physicochemical coding of material composition and topography in continuous microfibres. *Nat. Mater.* **10**, 877 (2011).

Kinoshita S. and S. Yoshioka. Structural colors in nature: The role of regularity and irregularity in the structure. *ChemPhysChem* **6**, 1442 (2005).

Koch C. Comprehensive notes on the tenebrionid fauna of the Namib Desert. *Ann. Trans. Mus.* **24**, 61 (1962).

Koch K., B. Bhushan, and W. Barthlott. Diversity of structure, morphology and wetting of plant surfaces. *Soft Matter*, **4**, 1943 (2008).

Koch K., B. Bhushan, Y. C. Jung, and W. Barthlott. Fabrication of artificial lotus leaves and significance of hierarchical structure for superhydrophobicity and low adhesion. *Soft Matter*, **5**, 1386 (2009).

Koch K. and W. Barthlott. Superhydrophobic and superhydrophilic plant surfaces: An inspiration for biomimetic materials. *Phil. Trans. R. Soc. A* **367**, 1487 (2009).

Kumazawa K. and H. Tabata. A three-dimensional fluorescence analysis of the wings of male Morpho sulkowskyi and Papilio xuthus butterflies. *Zool. Sci.* **18**, 1073–1079 (2001).

Kuo C. W., J. Y. Shiu, Y. H. Cho, and P. Chen. *Adv. Mater.* **15**, 1065–1068 (2003).

Lafuma A. and D. Quéré. Slippery pre-suffused surfaces. *EPL* **96**, 56001 (2011).

Liu C., J. Ju, Y. Zheng, and L. Jiang. Asymmetric ratchet effect for directional transport of fog drops on static and dynamic butterfly wings. *ACS Nano* **8**, 1321 (2014).

Liu C., Y. Xue, Y. Chen, and Y. Zheng. Effective directional self-gathering of drops on spine of cactus with splayed capillary arrays. *Sci. Rep.*, **5**, 17757 (2015).

Liu Y., J. Tang, R. Wang, H. Lu, L. Li, Y. Kong, et al. Artificial lotus leaf structures from assembling carbon nanotubes and their applications in hydrophobic textiles. *J. Mater. Chem.* **17**, 1071–1078 (2007).

Liu Y., Z. Shao, and F. Vollrath. Relationships between supercontraction and mechanical properties of spider silk. *Nat. Mater.* **4**, 901 (2005).

Loik M. E. The effect of cactus spines on light interception and photosystem II for three sympatric species of Opuntia from the Mojave Desert. *Physiol. Plant* **134**, 87 (2008).

Lorenceau É., C. Clanet, and D. Quéré. Capturing drops with a thin fiber. *J. Colloid Interface Sci.* **279**, 192 (2004).

Louw G. N. The role of advective fog in the water economy of certain Namib Desert animals. *Symp. Zool. Soc. Lond.* **31**, 297 (1972).

Machin L. The study of evaporation from small surfaces by the direct measurement of water vapour pressure gradients. *J. Exp. Biol.* **53**, 753 (1970).

Malvadkar N. A., M. J. Hancock, K. Sekeroglu, W. J. Dressick, and M. C. Demirel. An engineered anisotropic nanofilm with unidirectional wetting properties. *Nat. Mater* **9**, 1023 (2010).

Malvadkar N. A., M. J. Hancock, K. Sekeroglu, W. J. Dressick, and M. C. Demirel. An engineered anisotropic nanofilm with unidirectional wetting properties. *Nat. Mater.* **9**, 1023 (2010).

Mauseth J. D. Structure–function relationships in highly modified shoots of cactaceae. *Ann. Bot.* **98**, 901 (2006).

McGill D. J. and W. W. King. *Statics and an Introduction to Dynamics*, 2nd edn. PWS-Kent, Boston, 1989.

McHale G., M. I. Newton, and N. J. Shirtcliffe. Dynamic wetting and spreading and the role of topography. *J. Phys.: Condens. Matter* **21**, 464122 (2009).

Mei H., D. Luo, P. Guo, C. Song, C. Liu, Y. Zheng, and L. Jiang. Multi-level micro-/nanostructures of butterfly wings adapt at low temperature to water repellency. *Soft Matter* **7**, 10569 (2011).

Miguel S., A. Hehn, and F. Bourgau. Nepenthes: State of the art of an inspiring plant for biotechnologists. *J. Biotechnol.* **265**, 109 (2018).

Mosco A. Micro-morphology and anatomy of Turbinicarpus (Cactaceae) spines. *Rev. Mex. Biodivers.* **80**, 119 (2009).

Nilsson D. E. *Facets of Vision* (D. G. Stavenga and R. C. Hardie, eds.). Springer-Verlag, Berlin Heidelberg, 1989, pp. 30–73.

Nishimotoab S. and B. Bhushan. Bioinspired self-cleaning surfaces with superhydrophobicity, superoleophobicity, and superhydrophilicity. *RSC Adv.* **3**, 671 (2013).

Nosonovsky M. and B. Bhushan. *Multiscale Dissipative Mechanisms and Hierarchical Surfaces: Friction, Superhydrophobicity, and Biomimetics.* Springer, Heidelberg, Germany, 2008.

O'Hara P. D. and L. A. Morandin. Effects of sheens associated with offshore oil and gas development on the feather microstructure of pelagic seabirds. *Mar. Pollut. Bull.* **60**, 672 (2010).

Ogburn R. M. and E. J. Edwards. Anatomical variation in Cactaceae and relatives: Trait lability and evolutionary innovation. *Am. J. Bot.* **96**, 391 (2009).

Parker A. R., V. L. Welch, D. Driver, and N. Martini. Structural colour: opal analogue discovered in a weevil. *Nature* **426**, 786–787 (2003).

Perret C., C. Gourgon, F. Lazzarino, J. Tallal, S. Landis, and R. Pelzer. Characterization of 8-in. wafers printed by nanoimprint lithography. *Microelectron. Eng.* **73**, 172–177 (2004).

Peter H. M. The spinning apparatus of Uloboridae in relation to the structure and construxtion of capture threads (Arachnida, Araneida). *Zoomorphology* **104**, 96 (1984).

Porter D. and F. Vollrath. Nanoscale toughness of spider silk. *Nano Today* **2**, 6 (2007).

Prakash M., D. Quéré, and J. W. M. Bush. Surface tension transport of prey by feeding shorebirds: The capillary ratchet. *Science* **320**, 931 (2008).

Prakash M. and J. W. Bush. Interfacial propulsion by directional adhesion. *Int. J. Nonlinear Mech.* **46**, 607 (2011).

Reif W.-E. Morphogenesis and function of the squamation in sharks. *Neues Jahrb. Geol. Palaontol. Abh.* **164**, 172 (1982).

Renvoisé P., J. W. M. Bush, M. Prakash, and D. Quéré. Drop propulsion in tapered tubes. *EPL (Europhysics Letters)* **86**, 64003 (2009).

Rijke A. M. and W. A Jesser. The feather structure of dippers: Water repellency and resistance to water penetration. *Wilson J. Ornithol.* **122**, 563 (2010).

Rijke B. Y. A. M. The water repellency and feather structure of cormorants, Phalacrocoracidae. *J. Exp. Biol.* **48**, 185 (1968).

Roberts C. S., M. K. Seely, D. Ward, D. Mitchell, and J. D. Campbell. Body temperatures of Namib Desert tenebrionid beetles: Their relationship in laboratory and field. *Physiol. Entomol.* **16**, 463 (1991).

Rueda L. C., L. G. Ortega, N. A. Segura, V. M. Acero, and F. Bello. *Lucilia sericata* strain from Colombia: Experimental colonization, life tables and evaluation of two artificial diets of the blowfly *Lucilia sericata* (Meigen) (Diptera: Calliphoridae), Bogotá, Colombia strain. *Biolog. Res.* **43**, 197 (2010).

Ruibal R. and V. Ernst. The structure of the digital setae of lizards. *J. Morphol.* **117**, 271 (1965).

Russell A. P. A contribution to the functional analysis of the foot of the Tokay, Gekko gecko (Reptilia: Gekkonidae). *J. Zool.* (London) **176**, 437 (1975).

Schleich H. H. and W. Kästle. Concentration of setae on foot. *Amphibia-Reptilia* **7**, 141 (1986).

Song C. and Y. Zheng. Wetting-controlled strategies: From theories to bio-inspiration. *J. Colloid Interface Sci.* **427**, 2 (2014).

Srygley R. B. and A. L. R. Thomas. Unconventional lift generating mechanisms in free-flying butterflies. *Nature* **420**, 660 (2002).

Stavenga D. G., S. Foletti, G. Palasantzas, and K. Arikawa. Light on the moth-eye corneal nipple array of butterflies. *Proc. R. Soc. B* **273**, 661 (2006).

Stork N. E. A comparison of the adhesive setae on the feet of lizards and arthropods. *J. Nat. History* **17**, 829 (1983).

Su B., Y. Tian, and L. Jiang. Bioinspired interfaces with superwettability: From materials to chemistry. *J. Am. Chem. Soc.* **138**, 1727 (2016).

Sun Z., T. Liao, K. Liu, L. Jiang, J. H. Kim, and S. X. Dou. Fly-eye inspired superhydrophobic anti-fogging inorganic nanostructures. *Small* **10**, 3001 (2014).

Tanaka H., K. Hoshino, K. Matsumoto, and I. Shimoyama. Flight dynamics of a butterfly-type Ornithopter. *IEEE/RSJ International Conference on Intelligent Robots and Systems*; Edmonton, Alberta, Canada, 2005; IEEE: Piscataway, NJ, 2005; 27062711.

Tanaka M., K. Matsumoto, and I. Shimoyama. Design and performance of micromolded plastic butterfly wings on butterfly Ornithopter. *IEEE/RSJ International Conference on Intelligent Robots and Systems*; Nice, France, 2008; IEEE: Piscataway, NJ, 2008; 30953100.

Taylor B. Concerning the ascent of water between two glass plates. *Phil. Trans. R. Soc. Lond.* **27**, 538 (1712).

Taylor J. R. E. Thermal insulation of the down and feathers of pygoscelid penguin chicks and the unique properties of penguin feathers. *Auk* **103**, 160 (1986).

Tian X., H. Bai, Y. Zheng, and L. Jiang. Bio-inspired heterostructured bead-on-string fibers that respond to environmental wetting. *Adv. Funct. Mater.* **21**, 1398 (2011a).

Tian X., Y. Chen, Y. Zheng, H. Bai, and L. Jiang. Controlling water capture of bioinspired fibers with hump structures. *Adv. Mater.* **23**, 5486–5491 (2011b).

Vollrath F. and D. Porter. Spider silk as archetypal protein elastomer. *Soft Matter* **2**, 377 (2006).

Vollrath F. and D. T. Edmonds. Modulation of the mechanical properties of spider silk by coating with water. *Nature* **340**, 305 (1989).

Vollrath F., W. Fairbrother, R. J. P. Williams, E. K. Tillinghast, D. T. Bernstein, K. Gallagher, and M. Townley. Compounds in the droplets of the orb spider's viscid spiral. *Nature* **345**, 526 (1990).

Vollrath F. Strength and structure of spiders' silks. *Rev. Mol. Biotechnol.* **74**, 67 (2000).

Wagner T., C. Neinhuis, and W. Barthlott. Wettability and contaminability of insect wings as a function of their surface sculptures. *Acta Zool.* **77**, 213 (1996).

Wanasekara N. D. and V. B. Chalivendra. Role of surface roughness on wettability and coefficient of restitution in butterfly wings. *Soft Matter* **7**, 373 (2011).

Wang L., Q. Gong, S. Zhan, L. Jiang, and Y. Zheng. Robust anti-icing performance of a flexible superhydrophobic surface. *Adv. Mater.* **28**, 7729 (2016a).

Wang Q., X. Yao, H. Liu, D. Quéré, and L. Jiang. Self-removal of condensed water on the legs of water striders. *PNAS* **112**, 9247 (2015).

Wang S., Z. Yang, G. Gong, J. Wang, J. Wu, S. Yang, and L. Jiang. Icephobicity of penguins *Spheniscus humboldti* and an artificial replica of penguin feather with air-infused hierarchical rough structures. *J. Phys. Chem. C* **120**, 15923 (2016b).

Watanabe K. Fluid frictions of shark skin and lotus leaf-how dose the drag reduction occur? *J. Jap. Soc. Tribol.* **45**, 354 (2000).

Watson G. S. and J. A. Watson. Natural nano-structures on insects: Possible functions of ordered arrays characterized by atomic force microscopy *Appl. Surf. Sci.* **235**, 139 (2004).

Wen L., J. C. Weaver, and G. V. Lauder. Biomimetic shark skin: Design, fabrication and hydrodynamic function. *J. Exp. Biol.* **217**, 1656 (2014).

Wenzel R. N. Resistance of solid surfaces to wetting by water. *Ind. Eng. Chem.* **28**, 988 (1936).

Whitesides G. M. The origins and the future of microfluidics. *Nature* **442**, 368 (2006).

Whitesides G. M. and A. D. Stroock. Flexible methods for microfluidics. *Phys. Today* **54**, 6 (2001).

Williams E. E. and J. A. Peterson. Convergent and alternative designs in the digital adhesive pads of scincid lizards. *Science* **215**, 1509 (1982).

Wisdom K. M., K. M. Wisdom, J. A. Watson, X. Qu, and C.-H. Chen. Self-cleaning of superhydrophobic surfaces by self-propelled jumping condensate. *Proc. Natl. Acad. Sci. USA* **110**, 7992 (2013).

Wong T. S., T. S. Wong, S. H. Kang, S. K. Y. Tang, E. J. Smythe, B. D. Hatton, A. Grinthal, and J. Aizenberg. Bioinspired self-repairing slippery surfaces with pressure-stable omniphobicity. *Nature* **477**, 443 (2011).

Xue Y., Y. Chen, T. Wang, L. Jiang, and Y. Zheng. Directional size-triggered microdroplet target transport on gradient-step fibers. *J. Mater. Chem. A*, **2**, 7156 (2014).

Zhang G., J. Zhang, G. Xie, Z. Liu, and H. Shao. Cicada wings: A stamp from nature for nanoimprint lithography. *Small* **2**, 1440 (2006).

Zhang M., L. Wang, S. Feng, and Y. Zheng. A strategy of antifogging: Air-trapped hollow microsphere nanocomposites. *Chem. Mater.* **29**, 2899 (2017).

Zhang X., F. Shi, X. Yu, H. Liu, Y. Fu, Z. Wang, L. Jiang, and X. Li. Polyelectrolyte multilayer as matrix for electrochemical deposition of gold clusters: Toward super-hydrophobic surface. *J. Am. Chem. Soc.* **126**, 3064 (2004).

Zhao L., C. Song, M. Zhang, and Y. Zheng. Bioinspired heterostructured bead-on-string fibers via controlling the wet-assembly of nanoparticles. *Chem. Commun.* **50**, 10651 (2014).

Zheng X.-L., J.-B. Weng, B.-H. Hu, X.-Z. Lv, D.-L. Meng, and A. S. C. Chan. Fabrication of a stable superhydrophobic film constructed by poly(vinylpyrrolidone)/poly(urushiol)-CuS through layer-by-layer assembly. *Mater. Chem. Phys.* **130**, 1054 (2011).

Zheng Y., D. Han, J. Zhai, and L. Jiang. In situ investigation on dynamic suspending of microdroplet on lotus leaf and gradient of wettable micro- and nanostructure from water condensation. *Appl. Phys. Lett.* **92**, 084106 (2008).

Zheng Y., H. Bai, Z. Huang, X. Tian, F. Nie, Y. Zhao, J. Zhai, and L. Jiang. Directional water collection on wetted spider silk. *Nature* **463**, 640 (2010).

Zheng Y., X. F. Gao, and L. Jiang. Directional adhesion of superhydrophobic butterfly wings. *Soft Matter* **3**, 178 (2007).

Zhu Y., D. Hu, M. X. Wan, L. Jiang, and Y. Wei. Conducting and superhydrophobic rambutan-like hollow spheres of polyaniline. *Adv. Mater.* **19**, 2092 (2007).

Zhu Y., J. Zhang, Y. Zheng, Z. Huang, L. Feng, and L. Jiang. Stable, superhydrophobic, and conductive polyaniline/polystyrene films for corrosive environments. *Adv. Funct. Mater.* **16**, 568 (2006).

Chapter 2

Water-Repellent Surfaces

In this chapter, biological water-repellent surfaces are introduced, especially regarding special wettability on biological surfaces, such as dynamic water condensation on lotus leaves, two superhydrophobic states on butterfly wings, anisotropic wettability on water strider legs, directional structures on grass leaves, and droplet rebounding on biological surfaces. Based on the water repellency, the necessity of low-surface-energy, waxy and nonwaxy coverings for solid–liquid–air interfaces, tunable water-repellent coatings, hot-water repellency, nanostructure-induced dynamic water repellency, and micro- and nanostructures to decrease the surface energy are introduced and discussed. The design of roughness is introduced, including phototunable rough structures, micro- and nanostructured surfaces, nanostructure design, and multilevel roughness for water repellency. The roles of micro- and nanostructures are presented, such as rebounding on micro- and nanostructured surfaces, tower-like structures with nanoroughness, textured surfaces with pancake rebound, transparent superhydrophobic films with hierarchical design, microsphere-fiber interpenetrated structures, and droplet rotational bounding on an adhesive spiral surface. Examples of coatings for self-cleaning applications are discussed, for example

Bioinspired Materials Surfaces: Nature-Inspired Micro-/Nanostructures
Yongmei Zheng

ISBN 978-981-5129-34-2 (Hardcover), 978-1-003-51388-9 (eBook)
www.jennystanford.com

dentistry, self-cleaning transparent heat mirrors, bactericidal surfaces, spray coating of stable emulsions, highly transparent and hazy cellulose nanopaper, cotton fabric modified by a simple immersion technique, controlling liquid splash, fluorine-free superwetting systems, high-stress-resistant super-repellent materials, coatings from identical precursors, and anti-fingerprint coating.

2.1 Biological Water-Repellent Surfaces

Typical biological surfaces with water repellency include lotus leaves, rice leaves, grass leaves, butterfly wings, water strider legs, and compound eyes of insects (Fig. 2.1). Two factors can induce water repellency: (1) roughness resulting from micro- and nanostructures and (2) a low-surface-energy surface resulting from a waxy covering. In particular, the arrangements and styles of micro- and nanostructures control directional and anisotropic wettability or water repellency. Water repellency is characterized by a low sliding angle (<10°), low rolling angle (<10°), large contact angle (>150°), short contact time of the droplet impacting the surface, and so on. Based on such properties and functions of microstructures, bioinspired surfaces have been designed.

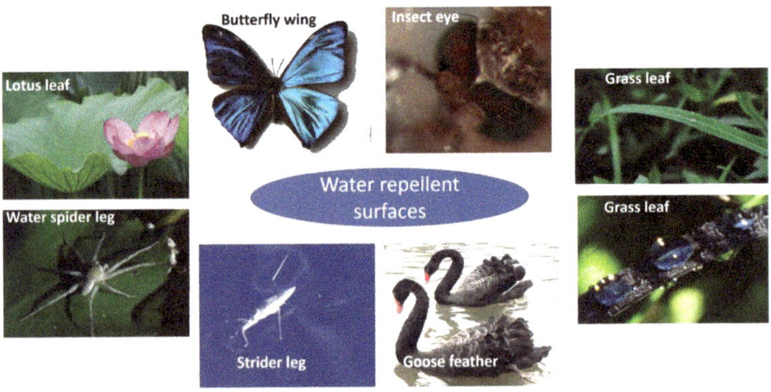

Figure 2.1 Water-repellent surfaces (e.g., lotus leaf, butterfly wing, insect eye, grass leaf, goose feather, water strider leg, water spider leg) in nature.

2.1.1 Dynamic Water Condensation on Lotus Leaves

The lotus is famous for blooming "in the mud, yet neatly and cleanly out of it." The self-cleaning property of lotus leaves is so significant that the word "lotus effect" was created to describe all surfaces with self-cleaning properties. In the late 1970s, German biologists Barthlott and Neihuis studied over 300 plant leaves (Patankar, 2004) and found a rough structure and hydrophobic wax cover on the surfaces exhibiting self-cleaning properties. When the wax of the surface was removed, the surface lost its self-cleaning property. However, researchers consider the structures of surfaces to be important and have been inspired to design promising functional engineering surfaces. Recently, many studies have also focused on revealing micro- and nanostructures as well as low-surface-energy surfaces (Feng et al., 2002; Zheng et al., 2008).

Figure 2.2 (a) Photograph of a lotus leaf and water droplet, (b) the gas cushion between a lotus leaf and water, and (c) the morning dew droplets observed on a lotus leaf (see light points). (d) Condensed water droplets on the lotus leaf observed by ESEM (gray spheres of different sizes, whereas light white spheres are papillae of lotus leaf) (Zheng et al., 2008).

Figure 2.2a–c shows photographs of a lotus leaf and water droplets on the leaf's surface, where a gas cushion is observed between the lotus leaf and water (Fig. 2.2b). The morning dew droplet reflects sunlight (Fig. 2.2c), and environmental scanning electron microscopy (ESEM) reveals condensed water droplets with spherical shapes similar to those on lotus leaves (Fig. 2.2d).

The dynamic process of the as-condensed droplets suspended on the top of papillae was observed by in situ ESEM (Fig. 2.3a–c), where the drop standing in the basin among the papillae is shown in Fig. 2.3b (see also the bottom). The water drop is suspended by the papillae with nanohairs (Fig. 2.3c, bottom). It is also found that the top of the papillae contains a thin liquid-like layer (Fig. 2.3d), which implies the formation of a high-surface-energy region. The wettability gradient along papillae also plays a role in droplet suspension onto the surface of the lotus leaf, in addition to the cooperation between microscale papillae and nanohairs with low-surface-energy components. A microdroplet first contacts the nanohairs in the valley during water condensation (Fig. 2.3a, and the bottom). When droplets grow to contact the locations with high surface energy on papillae (the top of papillae), the droplet is driven to move to the top of the papillae and becomes suspended above the bottom of the papillae. For the grown droplet, the contact angle (θ_{top}) at the top of the papillae is less than the contact angle (θ_{bottom}) at the bottom of the papillae (Fig. 2.3d, bottom). The driving force (F_{D}) can be expressed as (Chaudhury and Whitesides, 1992):

$$F_{\text{D}} = \int_0^L k\gamma\left(\cos\theta_{\text{top}} - \cos\theta_{\text{bottom}}\right)dl \tag{2.1}$$

where γ is the liquid tension, dl is the change in the contact length of the drop along the papillae surface, and k is the number of papillae suspending the microdroplet (k = 3, 4, 5, etc., corresponding to a configuration that is triangular, quadrangular, pentagonal, etc., respectively). F_{D} is obtained by integrating over the whole length (L) of the microdroplet moving along the papillae. Such an in situ investigation of microdroplets suspended on lotus leaves can reveal the self-cleaning mechanism of lotus leaves at the micro- and nanolevels. During water condensation, the micro- and nanostructures can introduce a wettability gradient of the unbalanced surface tension as a kind of driving force; as a result, the as-condensed growing microdroplet would move to the top of the surface.

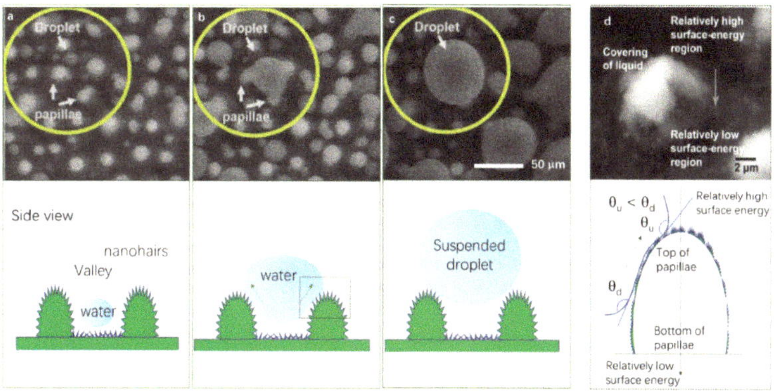

Figure 2.3 Investigation of droplets suspended on lotus leaves and wettability gradients by ESEM. (a–c) In situ observations of the growth of condensed water droplets on the surface with papillae; the bottom images indicate the dynamic growth of droplets on the surface in the side view. (d) An ESEM image of the formed surface energy gradient along papillae; the bottom image illustrates the difference in surface energy along papillae (Zheng et al., 2008).

2.1.2 Two Superhydrophobic States on Butterfly Wings

Butterfly wings have water-repellent properties, in addition to structural color. The water contact angles on the wings are above 150°, resulting in superhydrophobicity. The self-cleaning property also appears on butterfly wings, and contaminating particles can be easily washed away with a small tilt angle (~3°). These properties result from the unique structures. Take the example of the blue wings of the butterfly *Morpho rhetenor*. The wing is composed of microscales with two layers (Fig. 2.4a,b), where the upper layer scale is called the glass scale and the lower layer scale is called the ground scale. On the glass scale, there are larger ridges with a periodicity of 3 μm on average (Fig. 2.4c). The ridge is composed of tilted seven- to eight-layer lamellae (Fig 2.4f). On the ground scale, there is a close ridge with a periodicity of 1 μm (Fig. 2.4d), and the ridge is composed of tilted seven- to eight-layer lamellae

(Fig. 2.4e). These multilevel structures represent the basis of the optical properties and superhydrophobicity of butterfly wings. The butterfly wings of *Morpho nestira* and *Morpho menelaus terrestris* have structures similar to those of *M. rhetenor*, with two-layer scales, but the wings of *Morpho sulkowskyi* (Fig. 2.4g–i) and *Morpho aega* (Fig. 2.4j–l) butterflies have ground scales without glass scales. The scale is composed of one layer, along with close ridges composed of tilted 8–12-layer lamella. The surfaces of these overlapping scales with oriented tilted lamella are defined as directional anisotropic structured surfaces.

Figure 2.4 Morpho butterflies with multilevel structures. (a–f) SEM images of microstructures on the wing of the *M. rhetenor* butterfly. (g–i) SEM images of microstructures on the wing of the *M. sulkowskyi* butterfly. (j–l) SEM images of microstructures on the wing of the *M. aega* butterfly.

As is well known, tropical butterflies from Central and South America have very fine scales on the surface of their wings; for example, the blue Morpho butterflies keep their wings clean and dry, the inspiration of which can be applied to self-cleaning fabric coatings. Imagine playing rugby in the rain and finishing the match to find that your kit is both dry and sparklingly clean; it may sound farfetched but could be possible in the future. The robust microlevel superhydrophobicity on the surface of butterfly wings can be further observed by ESEM. When the wing is placed in a moist environment, tiny spherical water drops begin to condense on the overlapping scales, with different sizes ranging from 1 μm to 100 μm. Figure 2.5 shows ESEM images of water condensation (gray balls) on the wing surfaces of two different butterflies, *M. rhetenor* (left, Fig. 2.5a) and *M. aega* (right, Fig. 2.5b) (Zheng, 2015).

Figure 2.5 (a–b) ESEM images of condensed water droplets on the wings of the *M. rhetenor* butterfly (a) and *M. aega* butterfly (b).

As a macrolevel superhydrophobicity example, the *M. aega* butterfly wing reacts differently to water depending on the directional structure of the wing (Zheng et al., 2007) and even at low temperature (Mei et al., 2011). As the butterfly tips its wing downward in flight, droplets of water roll away from the body, carrying dust and dirt with them. As the butterfly wing is tipped upward, the direction of the scales prevents the droplet from moving toward the butterfly's body. Such directional water repellency is attributed to the directional structure of the wing surface. Directional adhesion can be achieved on butterfly wings with overlapping microscales including nanoscale lamellae. Figure 2.6a shows that the oriented micro- and nanostructures of the butterfly wing cause the droplet to easily roll off along the RO direction, but against the RO direction, the droplet is pinned, even though the surface is vertically oriented. The retention (F_r) can be induced by the contact angle hysteresis, i.e., $F_r \approx w\gamma(\cos\theta_A - \cos\theta_R)$, where w is the droplet diameter, γ is the surface tension of liquid, and θ_A and θ_R are the intrinsic advancing and receding contact angles, respectively. A balance exists between the downward component of weight (F_d = mg sinβ) and β, the tilt angle of the surface (roll-off angle). F_r is always related to surface adhesion along the opposite direction of droplet movement. A surface with low adhesion would have a small F_d value for a droplet (2 μL) rolling off a surface with a slightly tilted angle, as shown in Fig. 2.6b. Along the RO direction, F_r would be $<5 \times 10^{-3}$ mN, corresponding to small tilted angles <30°. In contrast, against the RO direction, F_r would be greater than 30×10^{-3} mN, corresponding to a large titled angle greater than 70°. Apparently, the surface can generate two superhydrophobic states (Fig. 2.6b, insets). When the surface is tilted toward the RO direction, the composite contact mode with an enlarged discontinuous three-phase contact line (TCL) is enlarged more than that in the normal state; thus, the droplet is in the Cassie state (the bottom of inset). When the surface is tilted against the RO direction, the wet contact mode with a close continuous TCL results in adhesion; thus, the droplet is in the quasi-Wenzel state (the top of inset). Multilevel micro-/nanostructures of *M. nestira* butterfly wings also adapt to water repellency at low temperature.

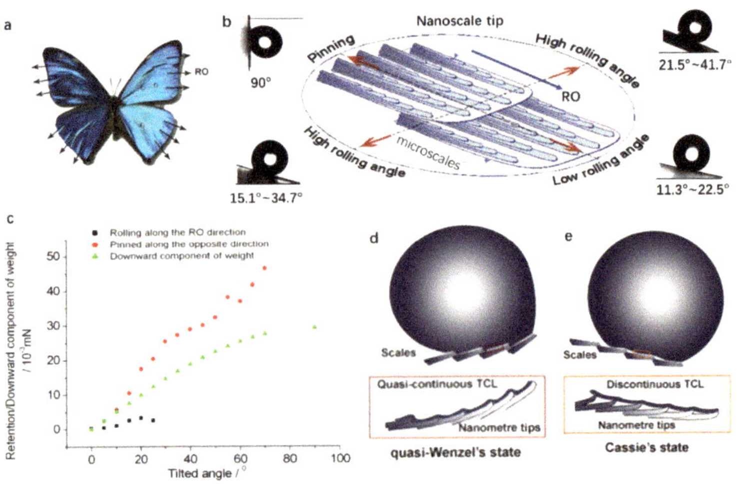

Figure 2.6 Directional adhesion on the same surface with two superhydrophobic states. (a) Photograph of the *M. aega* butterfly; RO indicates the radial outside direction along the wings. (b) Optical image of droplets with sliding angles on the wing along the different directions: low sliding angle of 11.3~22.5° along the RO direction and high sliding angle of pinning against the RO direction. (c) Curve of droplets sliding off the wing surface along different directions. (d–e) Two superhydrophobic states for droplets in the quasi-Wenzel state against the RO direction (d) and in the Cassie state along the RO direction (Zheng et al., 2007).

2.1.3 Anisotropic Wettability on Water Strider Legs

Water striders (*Gerris remigis*) walk fast on water, which is attributed to the combination of their nonwetting legs with secreted wax (Caponigro and Erilsen, 1976; Dickinson, 2003; Hu et al., 2003) and the hierarchical structure of the legs covered by large numbers of oriented tiny hairs (microsetae) with fine nanogrooves (Tadanaga et al., 1997). Figure 2.7a shows a water strider on water. Figure 2.7b shows the leg covered with chitinous setae at an oriented angle of ~20° to the base of the leg. The surface of the setae has nanogrooves (Fig. 2.7c), revealing that the surfaces of the legs are hierarchical structures (Gao and Jiang, 2004). Figure 2.7d,e shows three representative images of a water droplet rolling off the leg surface with a gradual tilting process. The water droplet rolls off at a tilting angle of 11.1° in the AS direction (i.e., along the oriented tiny hairs—microsetae), while the water droplet is pinned at a tilting angle of 18.6° in the OS direction (i.e., against the oriented tiny hairs—microsetae). The droplet exhibits a distinct difference in rolling angles between the AS and OS directions, corresponding to different adhesion properties depending on the direction, where the AS direction generates low adhesion.

Figure 2.7f shows schematic and force curves of the interaction between the water strider leg and water droplet in immersion experiments. The force curves between the water strider leg and water droplet in different directions (OS and AS) reveal the interactions that occur. It is evident that when gradually immersed in water, the leg experiences a larger wetting force in the OS direction than in the AS direction. The oriented structures exhibit an anisotropic wet adhesion effect; this effect is attributed to the flexibly oriented setae on the water strider leg, which induces asymmetric adhesion to enable the water-walking propulsion of the water strider (Xu et al., 2012). Water exists on the tips of the oriented setae without the wetting side surface of the setae, where θ is the original inclination angle (Fig. 2.8a). When a pressure p is applied (Fig. 2.8b), water may enter the gaps between the setae, which partially wets the setae. θ decreases to θ_1 when relative motion occurs between the leg and water in the AS direction (Fig. 2.8c). θ increases to θ_2 when relative motion happens between the leg and water in the OS direction (Fig. 2.8d). The arrows of AS (Fig. 2.8c) and OS (Fig. 2.8d) denote the direction of relative motion between the water and leg, respectively.

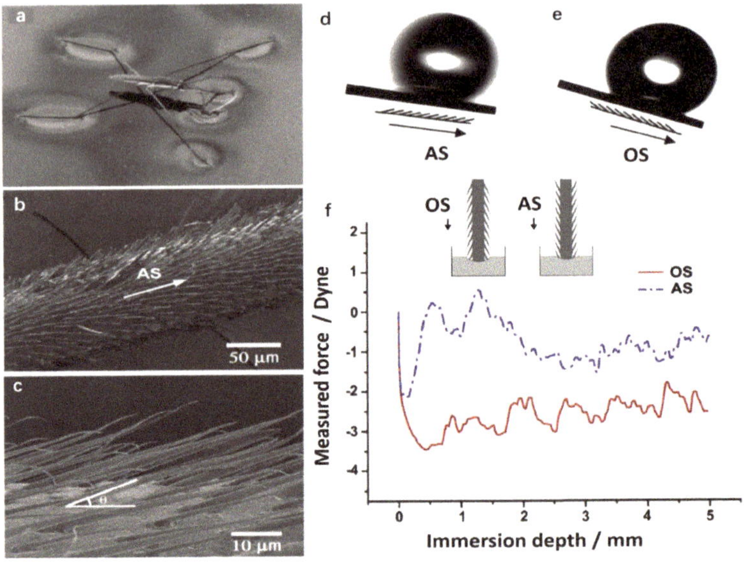

Figure 2.7 Anisotropic wet adhesion effect on the water strider leg (Xu et al., 2012). (a) Photograph of a water strider standing on the surface of water. (b–c) The leg covered with chitinous setae (b) along the AS direction at an oriented angle θ of ~20° to the base of the leg (c). (d) Easy roll-off of the droplet in the AS direction. (e) Pinning of droplets in the OS direction. (f) Force curves of the interaction between the water strider leg and water droplet in immersion experiments along the OS and AS directions.

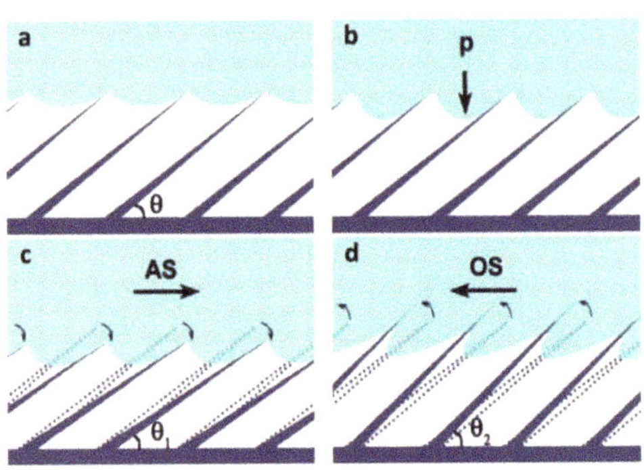

Figure 2.8 Illustration of the anisotropic wet adhesion effect on the water strider leg along the AS and OS directions. (a) Initial state between the tips of oriented setae at an angle θ. (b) A pressure p induces setae to be partially wetted. (c) Change in the oriented tilt angle $\theta \rightarrow \theta_1$ along the AS direction. (d) Change in the oriented tilt angle $\theta \rightarrow \theta_2$ along the OS direction (Xu et al., 2012).

2.1.4 Directional Structures on Grass Leaves

Due to their self-cleaning ability, plant leaves are easily cleaned by pure water to remove any contaminants (Guo et al., 2012). Recently, directional self-cleaning has been observed on some leaves, including fresh ryegrass leaves (widths of 1–2 cm and lengths of 30–50 cm), which were obtained at the Lincoln garden in Beijing, China.

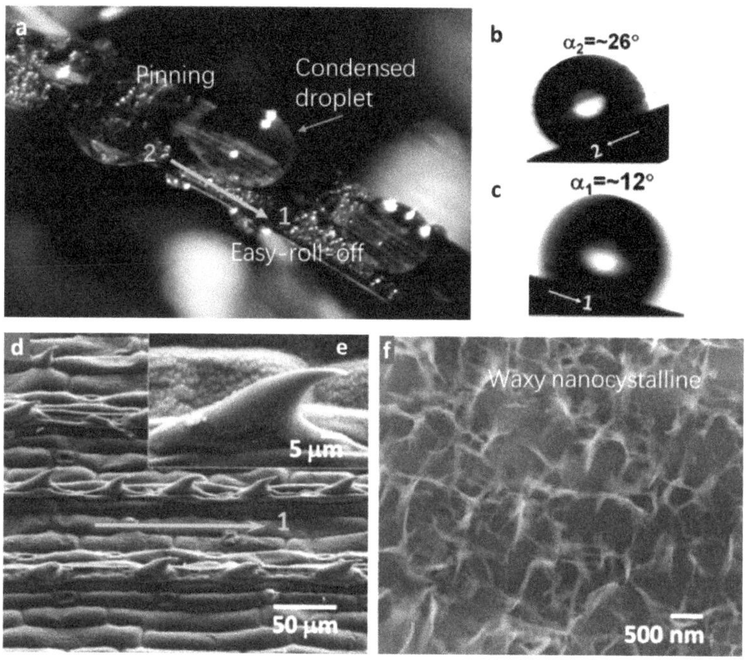

Figure 2.9 (a) Photograph of a ryegrass leaf with condensed droplets in an ellipsoid shape. The easy roll-off of the droplet is generated in direction 1, and pinning is generated in direction 2. (b–c) Optical images of droplets rolling off the tilted surface, with roll-off angles of $\alpha_2 \approx 26°$ in direction 2 and $\alpha_1 \approx 12°$ in direction 1. (d–f) Anisotropic structures composed of a ratchet array (d), ratchet oriented toward direction 1 (e) and waxy nanocrystalline covering (f) on the surface of ryegrass leaves (Guo et al., 2014).

An ellipsoid condensed drop forms on ryegrass leaves (Fig. 2.9a) in morning humid air, which implies that anisotropic wettability depends on leaf orientation (Fig. 2.9b). The typical directions are defined as follows: direction 1 is the direction parallel to the leaf edge toward the end of the leaf, and direction 2 is the opposite direction

(Fig. 2.9a). It is found that the drop rolls off easily along direction 1, with the roll-off angle $\alpha_1 \approx 12°$ (Fig. 2.9b), while it hardly rolls off along direction 2, with the roll-off angle $\alpha_2 \approx 26°$ (Fig. 2.9c). The drop generates distinct anisotropy in one dimension, which is attributed to the unique one-dimensional asymmetric microstructure of the ryegrass leaf.

Figure 2.9d–f shows scanning electron microscopy (SEM) images of the microstructures on the ryegrass leaf. The surface of the ryegrass leaf is composed of taper ratchets in a stripe-style array, with a periodicity of ~100 μm between each stripe (Fig. 2.9d). The tops of taper ratchets are tips oriented toward direction 1, with a periodicity of 30–50 μm between the neighboring tips. The tips are all conical, with an apex angle of ~5°, tilted up to an angle of ~25° to the base surface (Fig. 2.9e, and the inset), as well as a length of ~10 μm and a maximum diameter of ~5 μm (Fig. 2.9f). In addition, waxy crystals are spread over the bottoms of the taper ratchets, except for the tips (Fig. 2.9f). Therefore, the surface of the ryegrass leaf (with anisotropic wettability formed in one direction) shows differences from that of the rice leaf, which is composed of crossed-arranged periodic "papillae" without oriented tips (with anisotropic wettability forming in cross-directions). Apparently, the surface of the ryegrass leaf provides a unique micro-/nanostructure that is biologically fabricated with taper ratchets to promote anisotropic wettability.

2.1.5 Droplet Rebounding on Biological Surfaces

Water repellency can be regulated by microstructures; for example, the wings of the Morpho butterfly (*Morpho didius*) (Fig. 2.10a,b) and the leaves of the nasturtium plant (*Tropaeolum majus* L.) (Fig. 2.10c,d) have multiple superhydrophobic ridges or veins. When droplets impact the ridges of these surfaces, the droplets can be split into two parts (Fig. 2.10b and d). For lotus leaves, droplet impact induces axisymmetric recoil (Fig. 2.10e,f).

In general, a superhydrophobic surface causes the impacting droplet to rebound and have a short contact time (i.e., at the solid–liquid interface) due to water repellency. A stronger water repellency would result in a shorter solid–liquid contact time, which is related to the structures of the surfaces. The parameter of droplet contact time, t_c, is independent of the dimensionless Weber number ($W_e = \rho U^2 R/\gamma$), where R is droplet radii, U is velocity of a droplet, γ is the liquid tension, and ρ is the liquid density. The inertial-capillary timescale $\tau = \sqrt{\rho R^3 / \gamma}$. The contact time t_c can be related to the parameters of W_e, τ, and microstructures of the surface. A strong water repellency induces the droplet to have a short contact time as the droplet impacts surfaces (Bird et al., 2013), which can be used to guide the designs of superwettable surfaces for some applications, including anti-icing, ice-phobic, antifogging, etc.

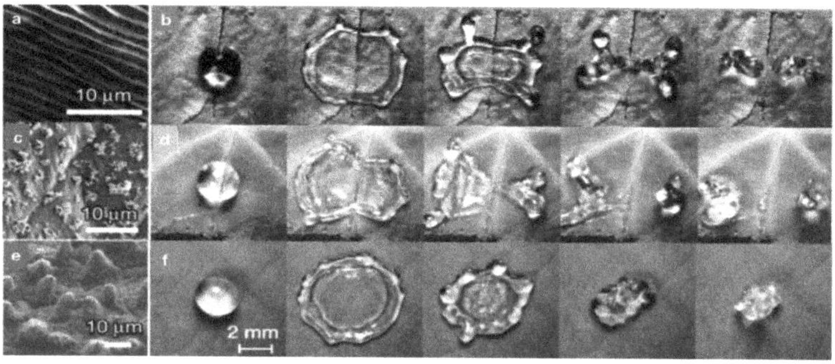

Figure 2.10 The impacting droplet is rebounded on different biological surfaces (Bird et al., 2013). (a–b) The wings of the Morpho butterfly (*Morpho didius*) and (c–d) the leaves of the nasturtium plant (*Tropaeolum majus* L.), where one ridge cuts the impacting droplet into two parts. (e–f) Surface of a lotus leaf, where the droplet impact induces axisymmetric recoil.

2.2 Necessity of Low Surface Energy

2.2.1 Waxy and Nonwaxy Coverings for Solid–Liquid–Air Interfaces

A waxy covering can be considered a low-surface-energy surface. For example, the surface of a lotus leaf would be considered a waxy covering due to the nanohairs; i.e., the micropapillae along with the nanohairs covering them collectively achieve superhydrophobicity. This feature can be directly demonstrated by observations by ESEM and measurements by atomic force microscopy (AFM). The as-condensed droplets on fresh lotus leaves can be observed in situ by ESEM. In particular, the solid–liquid interface can be revealed in detail. Figure 2.11a,b shows a millimeter-sized water droplet in the Cassie state on the surface of a lotus leaf as observed by ESEM, where the solid–liquid–air TCL exhibits much more air trapped between the papillae and the water droplet (Fig. 2.11b). Another method is in situ direct detection of solid–liquid–air interfaces by AFM, where the probe can be placed on the solid–liquid interface to directly reflect the contact extent, which is called the "droplet-inside-view" method (Chen et al., 2009), as shown in Fig. 2.11c–h; the solid–liquid–air interfaces on the lotus leaf can be measured by AFM in MAC-mode.

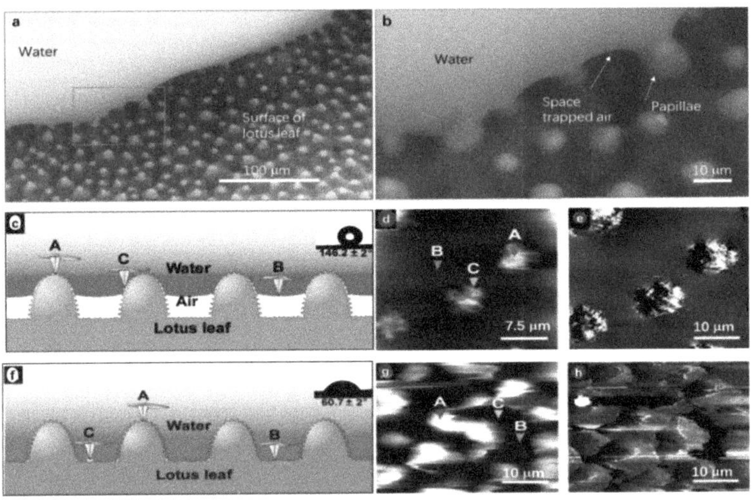

Figure 2.11 (a–b) ESEM images of a water droplet on the lotus leaf surface (with waxy nanohairs). The water droplet is supported by micropapillae (a), where the discontinuous solid–liquid–gas contact line implies trapped air in spaces of micropapillae (b) (Chen et al., 2009). (c) Illustration and AFM images of droplets on lotus leaves with waxy nanohairs; the air is inside the micropapillae and nanohairs. Sites A, B, and C can be measured, indicating that the droplet is supported by micropapillae and that the droplet is in the Cassie state. (d) Illustration and AFM images of droplets on lotus leaves with nonwaxy nanohairs; the air is lost, sites A, B, and C show wet contact, and the droplet is thus in the quasi-Wenzel state.

Figure 2.11c illustrates the case of lotus leaves with papillae and nanohairs (the inset is an optical image of the water droplet with a contact angle of ~146.2°). The as-marked sites A, B, and C indicate the contact sites of AFM tips (site A is at the top of papillae, site B is at the water surface, and site C is at the side of papillae). Figure 2.11d,e shows the topographic and deflection images corresponding to sites A, B, and C. This observation indicates that the droplet is in the Cassie state. In contrast, Fig. 2.11f illustrates the treated hydrophilic lotus leaf underneath the water droplet observed by contact-mode AFM (inset is the optical image of the droplet on the hydrophilic lotus leaf, with a contact angle of ~60.7°) (Cheng and Rodak, 2005). Sites A, B, and C are located at the top and valleys of papillae. Figure 2.11g,h shows that the droplet completely wets the papillae without trapped air, indicating that the droplet is in the quasi-Wenzel state. The sites marked A, B, and C are different colors in Fig. 2.11h, indicating the solid–liquid wet contact case (Adamson and Gast, 1997; Cheng and Rodak, 2006).

2.2.2 Tunable Water-Repellent Coatings

Learning from the structured features on the surface of lotus leaves, some studies have focused on investigating coatings to achieve water repellency for different applications. The design of materials is key to achieving robust water-repellent surfaces, including chemical components to select for low-surface-energy properties and the formation or self-assembly of micro-/nanostructures, as well as, in particular, the cooperation of both interactions together on a surface. Here, as an example (Fig. 2.12), polypropylene (PP), which includes atactic PP (A-PP)(Fig. 2.12a) and isotactic PP (I-PP) (Fig. 2.12b), each with distinct tacticity, is one of the most widely used general plastics. Recently, researchers reported that ultra-water-repellent PP coatings with tunable adhesion to water could be prepared via a simple casting method. The pure A-PP coating shows a hierarchical morphology with micro-/nanobinary structures, exhibiting a water contact angle (CA) greater than 150° and a sliding angle less than 5° (for a 5 μL water droplet) (Fig. 2.12c).

In contrast, the pure I-PP coating has a less rough morphology, with a water contact angle of approximately 130°, and the water droplets stick on the coating at any tilted angle (Fig. 2.12e). For the composite I-PP/A-PP coatings, ultra-water repellency with CA > 150° as well as tailorable water adhesion from slippery to sticky can be realized (Fig. 2.12d), depending on the amounts of A-PP and I-PP. The different wetting behaviors are due to the various microstructures of the composite coatings, which result from the distinct crystallization ability of A-PP and I-PP. Furthermore, the presence of A-PP in the composite coatings enhances the mechanical properties compared to those with only the I-PP coating. The proposed method is feasible for modifying various substrates, and potential applications in no-loss liquid transportation, slippery surfaces, and patterned superhydrophobic surfaces have been demonstrated (Zhu et al., 2017).

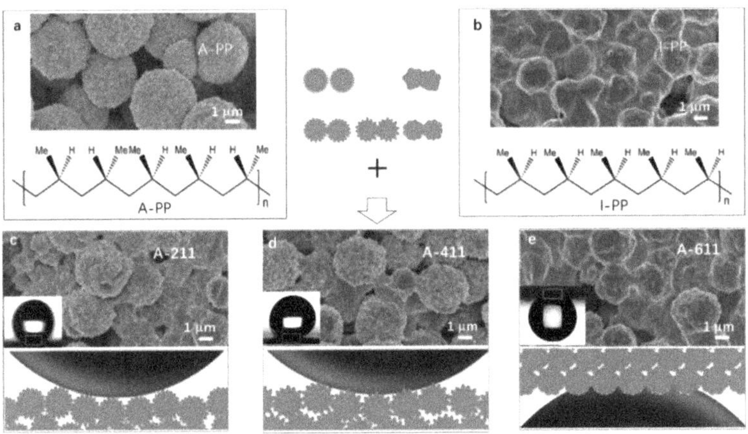

Figure 2.12 (a) A-PP structures. (b) I-PP structures. (c–e) Tunable water repellency and adhesion by A-PP and I-PP (Zhu et al., 2017).

2.2.3 Hot-Water Repellency

Surfaces with hot-liquid repellency have attracted much attention due to the demand for their application in our daily lives and in industry, such as in food containers (Liu et al., 2009), electrosurgical instruments (Li and Zhang, 2016), engines (Urata et al., 2013), and hot-water pipelines (Li and Zhang, 2016). In particular, there can be significant waste of hot liquid food such as retort curry or drinks as residue in food containers (Wang et al., 2016). This waste results from adhesion of liquid food to the inside of food packages and prevents recycling of the containers.

In the example shown in Fig. 2.13, a fluorine-free coating is achieved, which presents hot-water repellency and improved stability against boiling. In that study, oleic acid is chosen as a lubricant because its –COOH groups interact with the π-electrons in the base layer through OH–π interactions (Fig. 2.13a). Furthermore, to improve the boiling stability, the density of the phenyl groups is controlled by avoiding the use of tetramethoxysilane (TMOS) (Fig. 2.13b) and enhancing the interactions between the lubricant and the base layer (Fig. 2.13c). As a result, the fluorine-free surface is successfully fabricated, which allows hot water droplets to slide off at very low sliding angles (<2°) and exhibits improved boiling stability (Togasawa et al., 2018). As shown in Fig. 2.13d, as the water temperature is increased, the contact angle and sliding angle are stable (Fig. 2.13d). The comparison experiment shows that hot-liquid repellency is distinct for the TMOS base layer (Fig. 2.13e, the bottoms), in contrast to the TMOS-free base layer (Fig. 2.14e, the tops).

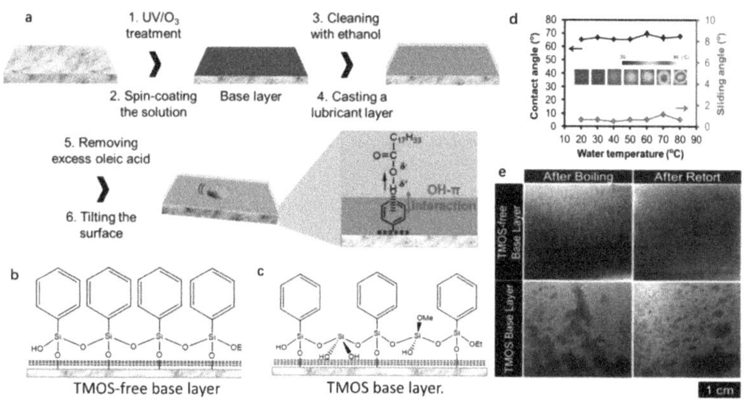

Figure 2.13 (a) Illustration of the fabrication process to achieve a hot-liquid-repellent surface. (b–c) Illustration of the chemical components and structures of the TMOS-free base layer (b) and TMOS base layer (c). (d) The stability of the contact angle and sliding angle versus water temperature. (e) Comparison of liquid repellency on both coatings of the TMOS-free base layer (top) and TMOS base layer (bottom) after boiling (left) and retorting (right) (Togasawa et al., 2018).

2.2.4 Nanostructure-Induced Dynamic Water Repellency

Recently, the dynamic processes of impacting droplets on solid surfaces have become more practically meaningful and have received increasing attention (Mertaniemi et al., 2012; Davies et al., 2014; Ellinas et al., 2015; Li et al., 2016). This dynamic process can be different, depending on the types of microstructures. For instance, different dynamic behaviors were observed on three kinds of superhydrophobic nanostructure surfaces, namely, nanotube (NT), nanomesh (NM), and nanowire (NW) surfaces (Shen et al., 2018), which were fabricated by using TiO_2 nanomaterials. All of these resultant surfaces were endowed with robust superhydrophobicity (Fig. 2.14a–c) and low liquid adhesion, with sliding angles from 7.5° to 3°. Interestingly, the bouncing dynamics of droplets impacting on these surfaces showed remarkable distinctions, with different capacities to rebound and different contact times (Fig. 2.14d–f). For the NT surface, the rebounding of the impacting droplet involves a long contact time of more than 19.6 s (Fig. 2.14d). For the NM surface, the rebounding of the impacting droplet involves a slightly shorter contact time of ~19.0 s (Fig. 2.14e). Interestingly, for the NW surface, the rebounding of the impacting droplet involves a shorter contact time of ~12.4 s (Fig. 2.14f). The resulting observations and comparisons imply that the nanostructure may control the dynamic water-repellent behaviors.

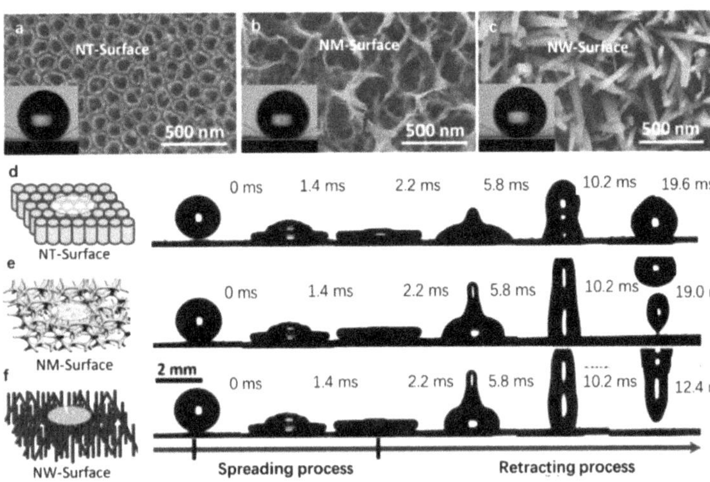

Figure 2.14 (a–c) SEM images of three surfaces, namely, NT (a), NM (b), and NW (c) surfaces, along with high contact angles from optical images in the insets. (d) Illustration of the NT surface and rebounding of an impacting droplet, with a long contact time of more than 19.6 s. (e) Illustration of the NM surface and rebounding of an impacting droplet, with a slightly shorter contact time of ~19.0 s. (f) Illustration of the NM surface and rebounding of the impacting droplet with a shorter contact time of ~12.4 s (Shen et al., 2018).

It can be explained that the higher capillary-induced adhesion force interaction occurs on the NT or NM surface during the dynamic solid–liquid contact processes for the impacting droplets than for the static droplets (Fig. 2.15a,b); accordingly, the NT or NM surface ultimately causes the impacting droplet to not bounce off. In contrast, the impacting droplet can successfully bounce off the nanowire (NW) structure surface, which is mainly because there is almost no capillary adhesion force interaction due to the open structure system on the superhydrophobic NW surface. This investigation offers insight into the design of novel structures (e.g., open structures) on material surfaces for some engineering applications, e.g., smart liquid-repellent surfaces.

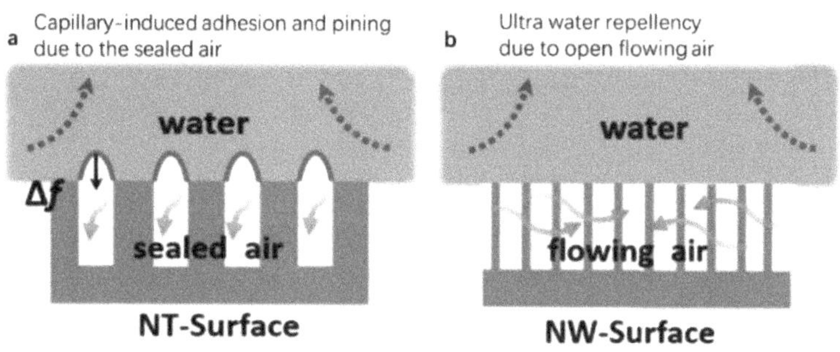

Figure 2.15 The different modes of air in different nanostructures for different water repellency effects (Shen et al., 2018). (a) On the NT surface, capillary-induced adhesion and pinning due to the sealed air. (b) On the NW surface, ultra-water repellency due to open flowing air.

2.3 Design of Roughness

2.3.1 Phototunable Rough Structures

Applications of phototunable rough structures would be significant in some fields, including microfluidics, liquid transport, etc. Here, an example of a photoresponsive rough superhydrophobic surface is provided, which can be prepared by using diarylethene via irradiation with UV light and storage in the dark at a given temperature (Uchida et al., 2010). As illustrated in Fig. 2.16a, the cycloreversion reaction from 1c (left top of Fig. 2.16a) to 1o (right top of Fig. 2.16a) proceeded under visible light irradiation to the surface, where the surface is formed by cube-shaped crystals of 1o. Furthermore, the multistructures could be fabricated by irradiation with UV light and storage at 70°C for 3 h in the dark; small rod-shaped crystals of 1c grew from the newly generated 1c on the surface (see Fig. 2.16a, steps (1)–(9)) (Nishimura et al., 2016).

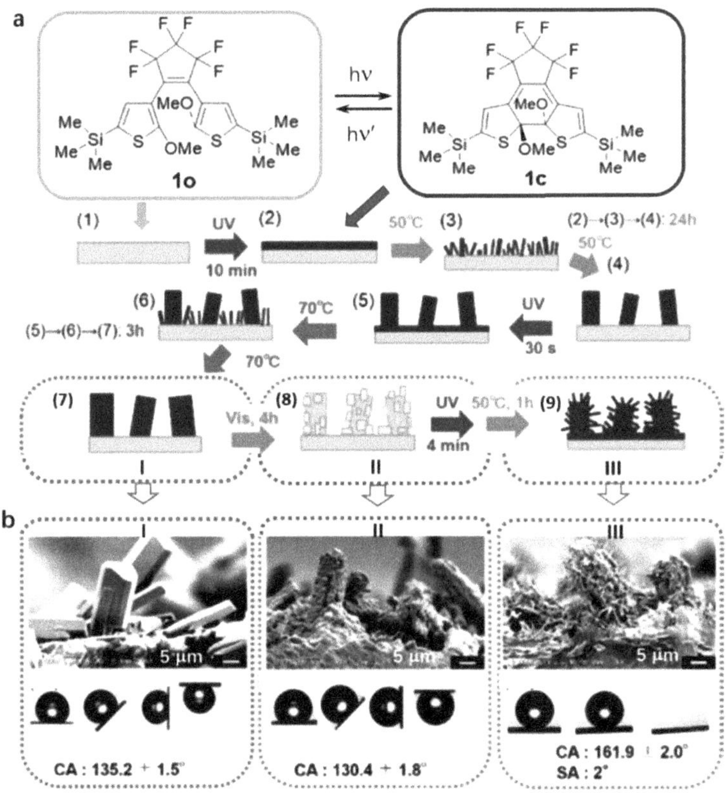

Figure 2.16 (a) Illustration of the fabrication process for material surfaces. (b) The phototunable rough structured surfaces for distinct wettability (contact angle (CA) changes: Frame I, CA ~135.2°; Frame II, CA ~130.4°; Frame III, CA ~161.9°), along with high adhesion and low slide-off angles of droplets due to changeable rough structures by controlling light irradiation (Nishimura et al., 2016).

The as-fabricated structures can be observed by SEM, as shown in Fig. 2.16b, steps I–III, corresponding to Fig. 2.17a, steps (7)–

(9). The structural differences on these surfaces induce different droplet states. The flat and small rough structured surfaces (see Fig. 2.16b, Frames I and II) are moderately superhydrophobic, with contact angles (CA) of ~135° and ~130°, respectively, and exhibit high surface adhesiveness. The multistructured surface (Frame III, Fig. 2.16b) composed of microcubic and nanorod shapes generates a high contact angle of ~161°, along with a slide-off angle of 2°, and exhibits low surface adhesiveness (Nishimura et al., 2016). This finding offers insight into the design of novel surfaces with in situ control of roughness; accordingly, the wettability can be controlled effectively, which would be promising for certain applications.

2.3.2 Micro- and Nanostructured Surfaces for Water Repellency

A micro- and nanostructured surface with low surface energy can display strong water repellency. The micro- and nanostructures are important to maintain stable water repellency. Here, as an example, the water-repellent properties of different structures are compared. Figure 2.17 shows a water droplet (7.6 μL, at h = 1.8 mm) impacting three surfaces, namely, natural lotus leaf (Fig. 2.17a), diarylethene microcrystalline surfaces with a single structure (Fig. 2.17b), and a micro- and nanostructured surface (Fig. 2.17c), where the droplets display different water-repellent properties. The micro- and nanostructured surface makes the droplet rebound, similar to the case with the lotus leaf. However, the single-structured surface pins the droplet, and the droplet does not rebound. This distinct repellency property is attributed to the balance between the Laplace pressure (P_L) and dynamic pressure (P_d). P_L is determined by the height H of the surface structures and the effective pitch p_{eff} between perpendicularly standing needles (red lines). $P_L = 16\gamma_L H / \left(\sqrt{2}p - D\right)^2$, where γ is the surface tension; H and D are the height and diameter of the structures, respectively; and p is the periodicity of the structure. P_d = $(1/2)\rho V^2$, where ρ is the density and V is the velocity (Jung and Bhushan, 2008). If $P_L > P_d$, droplet bouncing occurs on a surface. A smaller P_{eff} generates higher superhydrophobicity, along with a higher water contact angle and stronger rebounding of the impacting droplet.

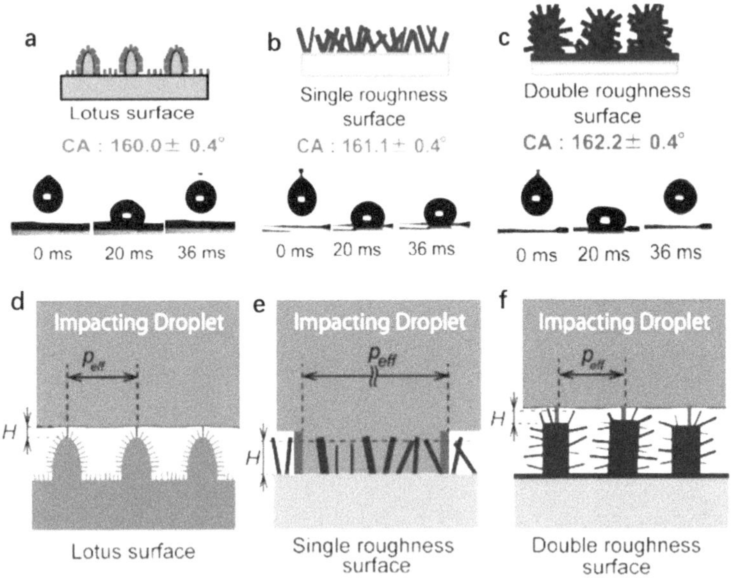

Figure 2.17 (a–c) The micro- and nanostructures of a lotus leaf (a), single roughness surface (b), and double roughness surface (c) and water droplets rebounding. (d–f) Illustration of the mechanism on the lotus surface (d), single roughness surface (e), and double roughness surface (f), along with the micro- and nanostructures, where P_{eff} is generated differently by different structures. A smaller P_{eff} generates higher superhydrophobicity, along with a higher water contact angle and stronger rebounding of the impacting droplet (Nishimura et al., 2016).

2.3.3 Nanostructure Design

Nanostructures vary and can be designed in different types; accordingly, the wettability is different in different cases. For instance, nanostructured surfaces can be fabricated via hydrothermal treatment using zinc acetate ($Zn(CH_3COO)_2{\cdot}6H_2O$), ammonium hydroxide ($NH_3{\cdot}H_2O$), zinc nitrate ($Zn(NO_3)_2{\cdot}6H_2O$), hexamethylenetetramine ($C_6H_{12}N_4$), etc. (Shen et al., 2019). As a result, nanostructured surfaces, including both sealed (Fig. 2.18a,b) and open microscopic (Fig. 2.18c,d) structures, can be achieved. After modification via fluorination, the same superhydrophobicity is obtained, as shown in the insets of Fig. 2.18a (contact angle of ~158.82°) and Fig. 2.18d (contact angle of ~160.25°), respectively.

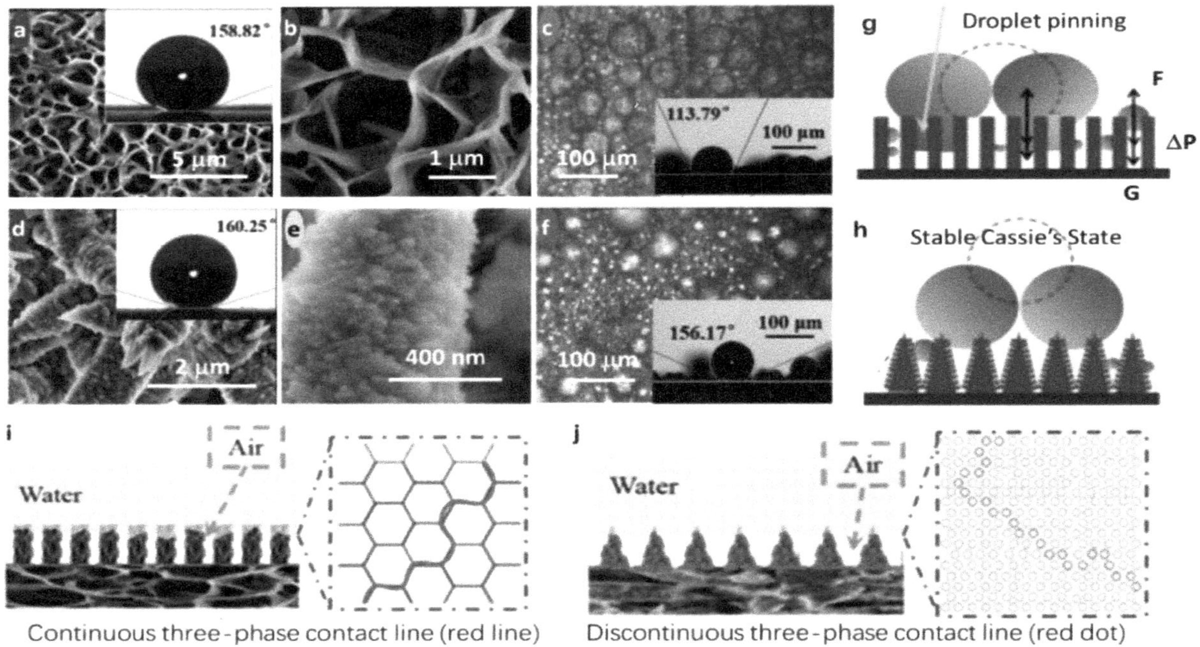

Figure 2.18 (a–c) SEM image of a sealed structured surface with a water contact angle of 158.82° (inset) and optical image of a condensed water droplet pinned with a low contact angle of 113.79° (c). (d–f) SEM image of an open structured surface with a water contact angle of 160.25° (inset) and optical image of a condensed water droplet suspended at a contact angle of 156.17°. (g–h) Illustration of different rough contact modes of sealed (g) and open (h) structured surfaces. (i–j) Illustration of the mechanism on a sealed structure with a continuous three-phase contact line (i) and on an open structure with a discontinuous three-phase contact line (j) (Shen et al., 2019).

For the sealed structure surface (Fig. 2.18a,b), the condensed water droplet is pinned to the surface (Fig. 2.18c), and the coalesced droplet presents a contact angle of 113.79° (the inset of Fig. 2.18c). For the open structured surface (Fig. 2.18d,e), the condensed water droplet appears spherical (Fig. 2.18f), along with a contact angle of ~156.17° (the inset of Fig. 2.18f). It can be understood that the structural differences induce differences in the contact state.

The sealed structures form a negative pressure, thereby leading to the pinning of the droplet (Fig. 2.18g). The open structures suspend the droplet after droplet coalescence, and the droplet is in a stable Cassie state (Fig. 2.18h). There are different modes in terms of the solid–liquid–air TCLs: the continuous TCL by the sealed structure (red line, Fig. 2.18i) and discontinuous TCL by the open structure (red dots, Fig. 2.18j). This study indicates that the design of structures for rough surfaces is necessary to induce the desired wettability.

2.3.4 Multilevel Roughness for Water Repellency

The cone structure along multilevel nanorough surfaces generates strong water repellency during water condensation. When the droplets coalesce together, the droplet jumps up (Fig. 2.19a, b, d and Frames I–III). The force relationship can be expressed as follows (Shen et al., 2019):

$$F_{\mathrm{P}} = F_{\mathrm{V}} + \left[mg + m\left(\frac{d^2h}{dt^2}\right)\right]\left(-\frac{dh}{dr}\right) \tag{2.2}$$

where $F_P = 2\pi\sigma(\cos\theta* - \cos\theta')$, which is defined as the propelling force of the TCL (Lv et al., 2015); $F_V = 2\pi\sigma f\,(1 + \cos\theta)$, which is calculated from the potential energy mg·*dh*; *h* is the droplet height; σ is the surface tension; θ, $\theta*$, and θ' are the intrinsic, apparent, and balanced contact angles, respectively; *f* is the roughness factor; *r* is the contact radius (where for the jumping droplet, $r = 0$); and the bounding height *H* is given by $H = \frac{1}{2g}\left(\frac{dh}{dt}\right)^2$. The bounding velocity (V_e) of the coalesced droplet is $V_e = \frac{dh}{dt}$ (Fig. 2.19c,d and Frame III). If the release of surface energy is not enough to compel the droplet to rise and depart from the surface, i.e., $r > 0$, the coalesced droplet sweeps the other droplets along the surface (Fig. 2.19d, Frame IV).

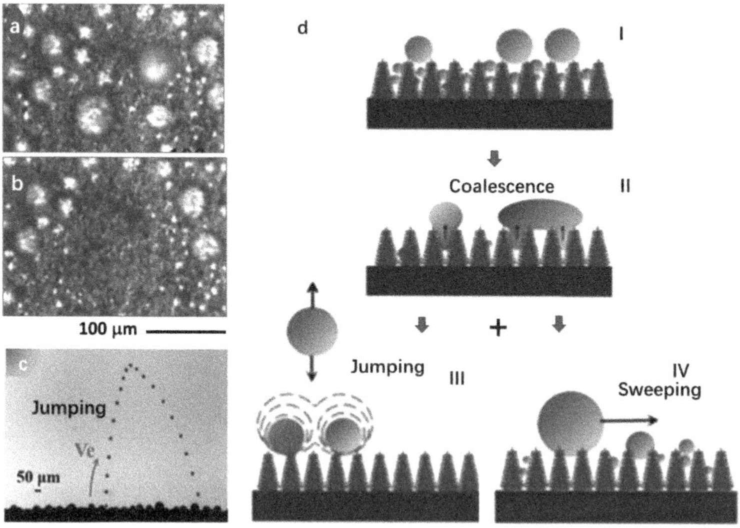

Figure 2.19 (a–b) Optical images of the as-condensed droplets coalescing and jumping or sweeping away from the surface. (c) Observation of the jumping of droplets in the side view. (d) Illustration of water repellency on a multilevel rough surface via the as-condensed droplet coalescence and jumping and sweeping after droplet coalescence (Shen et al., 2019).

2.4 Role of Micro- and Nanostructures

2.4.1 Rebounding on Micro- and Nanostructured Surfaces

Controlling droplet dynamics on a surface has been widely studied for potential applications in pesticide sprays (Bergeron et al., 2000), self-cleaning (Tuteja et al., 2007; Deng et al., 2012), and anti-icing (Mishchenko et al., 2010). Water droplets impacting a superhydrophobic surface often bounce off (Deng et al., 2009; Vasileiou et al., 2016). It is desirable to have surfaces on which impacting droplets exhibit short contact times and then completely rebound, the development of which would find potential industrial applications in anti-icing for aircraft wings, protection of high-voltage transmission lines from freezing rain, etc.

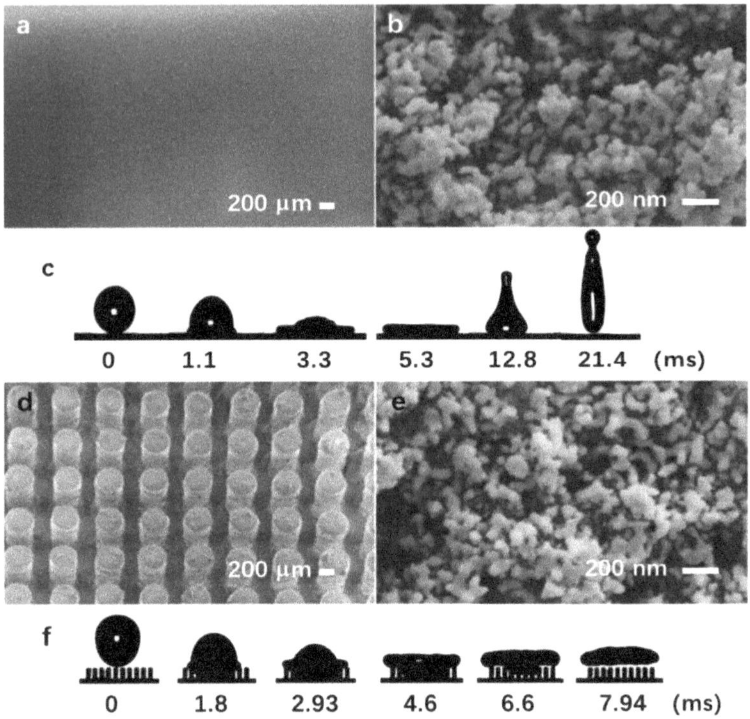

Figure 2.20 (a–c) A superhydrophobic surface (a) with a nanostructure alone (b) achieves a long contact time of 21.4 ms (c) during droplet impact. (d–f) A superhydrophobic surface (d) with micropillars covered with a nanostructure (e) achieves a short contact time of 7.94 ms (f) during droplet impact (Song et al., 2017a).

Generally, there are different bouncing phenomena of water droplets on different surfaces, such as superhydrophobic flat surfaces or superhydrophobic pillar arrays (Song et al., 2017a). Figure 2.20a shows a low-resolution SEM image of a superhydrophobic surface, while the high-resolution image shows the nanostructure (Fig. 2.20b) with a diameter of 10–20 nm. When a water droplet (17.9 μL, r_0 = 1.62 mm) impacts the superhydrophobic surface at W_e = 13.3, the droplet rebounds (at ~21.4 ms), with the contact time ranging from 1.11 to 12.8 ms (Fig. 2.20c). In contrast, when focusing on another superhydrophobic surface with distinct microstructures, such as micropillars, the behavior of droplets impacting such a structured surface would understandably be greatly different.

Figure 2.20d shows an SEM image of a superhydrophobic pillar array with the same nanostructure (Fig. 2.20e); the pillars have a diameter (D) of 0.3 mm, spacing (S) of 0.2 mm, and height (H) of 1 mm. Interestingly, it can be observed that a water droplet (17.9 μL, r_0 = 1.62 mm) impacting the superhydrophobic pillar arrays at W_e = 13.3 rebounds at ~7.9 ms, with a short contact time of 1.8–6.6 ms. The control experiment indicates that the optimal structured (e.g., D, S, H) surface can produce smart water repellency. This finding indicates that in contrast to a superhydrophobic surface with a nanostructure alone, the cooperative role of micro- and nanostructures would achieve stronger water repellency with an ultrashort contact time.

2.4.2 Tower-Like Structures with Nanoroughness for Condensed Droplet Rebounding

Superhydrophobic tower-like arrayed surfaces were fabricated onto Al surfaces via a simple method, i.e., immersing Al into hot water with ammonia (He et al., 2012, 2013), for example hot water (75°C) with 2 wt.% ammonia for 2 h. Then, the tower-like arrayed surfaces were immersed into a solution of polydimethylsiloxane (PDMS) and curing agent with n-hexane as the solvent for 1 h, followed by curing at 100°C. After the surface was modified by 1 wt.% PDMS/n-hexane, the tower-like array morphology changed little (defined St, Fig. 2.21a). A thick PDMS layer formed on the microtower surface when the concentration of PDMS/n-hexane exceeded 2 wt.% (defined as Sc, Fig. 2.21b). St and Sc achieve water contact angles of ~171° and ~167°, respectively, and contact angle hysteresis of 4.7° and 5.4°. During water condensation on the two surfaces, the tower-like surface (St) makes the droplet coalesce on the top of the structures and facilitates droplet rebounding (Fig. 2.21c), where the droplets are condensed and grown (Frame I), and then coalesced (Frame II). Finally, the droplets depart away from the surface (Frame III of Fig. 2.21c). As for the cone-like surface (Sc) (Fig. 2.21d), the droplets are condensed, grown (Frame I), and coalesced (Frame II). Finally, the droplets stick on the surface (Frame III of Fig. 2.21d). This observation indicates that the nanoscale roughness (e.g., rough wall side of the tower-like structure) controls the droplet behavior in terms of suspending and rebounding (He et al., 2016). This result demonstrates that the nanoroughness plays a role in inducing the condensed droplet to rebound, even though both surfaces are superhydrophobic surfaces.

Figure 2.21 (a) SEM image of the superhydrophobic tower-like (St) surface, with a water contact angle of 171.1°. (b) SEM image of the superhydrophobic cone-like (Sc) surface, with a water contact angle of 167.3°. (c) Optical images of condensed droplets exhibiting rebounding (i.e., departing droplets) on the St surface. (d) Optical images of condensed droplets exhibiting pinning (i.e., sticking of droplets) on the Sc surface (He et al., 2016).

2.4.3 Textured Surfaces with Pancake Rebound

The pancake rebound of a droplet is a unique case that happens when the droplet impacts a superhydrophobic structured surface. Here, an example of a cone pillar superhydrophobic surface is discussed. Cone pillar superhydrophobic surfaces can be obtained by combining 3D mold printing, soft lithography with flexible polymeric materials (Graeber et al., 2018), and spray coating. Figure 2.22a shows an SEM micrograph of the printed mold. Figure 2.22b shows a micrograph of PDMS replication. After spray coating with poly(tetrafluoroethylene) (PTFE; Sigma-Aldrich, powder, 1 μm particle size) and acetone, a superhydrophobic surface with a particle-coated texture can be achieved (Fig. 2.22c), where the coating surface can be seen in Fig. 2.22d. Water repellency can be observed on the surface, in contrast with the smooth surface (Fig. 2.22e, Frames I–IV). The Weber number can be described with $W_e = \rho u^2 r_0/\gamma$, where ρ and γ are the density and the surface tension of the fluid, respectively; r_0 is the droplet radius; and u is the impacting velocity. This equation indicates that a superhydrophobic textured surface would facilitate the special pancake rebound of a droplet (when $W_e = 11.4$, Fig. 2.22g), but other surfaces lead to the conventional rebound of droplets (Fig. 2.22e,f).

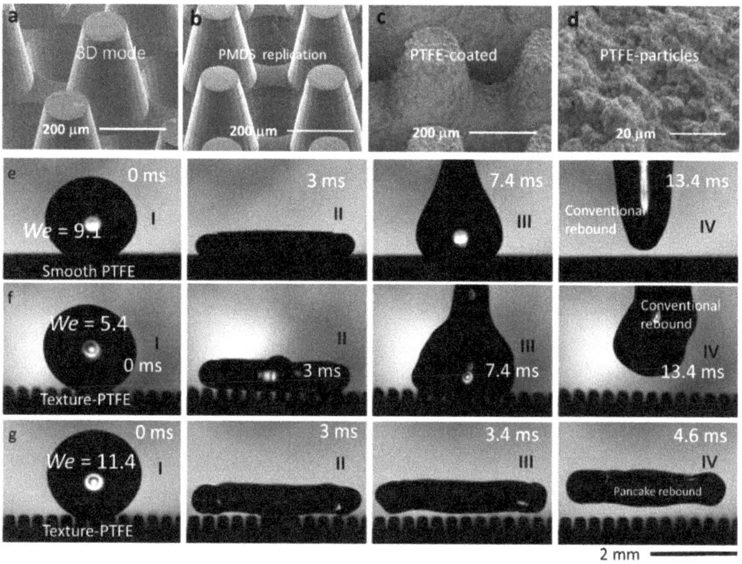

Figure 2.22 (a–d) SEM images of the 3D mold (a) and PDMS replication (b) and PTFE-coated (c) and PTFE particles (d). (e–g) Different styles of rebounding of impact droplets on three surfaces: a smooth PTFE surface at $W_e = 9.1$ (e), textured PTFE surface at $W_e = 5.4$ (f), and textured PTFE at $W_e = 11.4$ (g). The textured surface with high $W_e = 11.4$ can achieve excellent pancake rebound of the droplet (Graeber et al., 2018).

The solid–liquid contact time of the droplet is also an important concept to describe the pancake rebound of the droplet. It is demonstrated that there is conventional rebound when $W_e \leq 7$, and the droplet pancake shape appears for $7 < W_e < 20$, along with the reduced contact time $t_C/\tau \approx 1$, where t_c is the droplet–substrate contact time, τ is the inertial-capillary time, and $\tau = \sqrt{\rho r_0^3 / \gamma}$. The pancake-bouncing phenomenon was first described by Liu et al. (2014), who explained that when parts of the impacting droplet penetrate specially as-designed structured surfaces, capillary energy is stored and subsequently returned as droplet kinetic energy. It is important that the stored energy is sufficient to lift up the droplet.

In addition, the Ohnesorge number ($Oh = \eta / \sqrt{\rho r_0 \gamma}$, where r_0 is the initial droplet radius and η is the droplet dynamic viscosity) can be used to describe the cases when viscous droplets impact the texture-PTFE surface. Droplet pancake rebound occurs in a range of $0.003 \leq Oh \leq 0.013$. This investigation is helpful for understanding and designing structured surfaces with improved water-repellent properties.

2.4.4 Transparent Superhydrophobic Films with Hierarchical Design

A transparent superhydrophobic film is challenging to achieve due to the requirement of both thinness for transparency and roughness for superhydrophobicity. A recent report proposed that the hierarchical surface morphology is easily fabricated by the simple transfer of a porous alumina membrane (PAM) to the top surface of UV-imprinted polymeric micropillar arrays and subsequent chemical treatments (Fig. 2.23a). Through the optimization of the hierarchical surface design, the resultant superhydrophobic films show superior surface-wetting properties. The PAM is synthesized by standard anodic aluminum oxidation (AAO), along with a sacrificial dry film resist (DFR) layer on a supporting silicon substrate. A soft lithography technique can be used to obtain the desired micropillar array. After the photoresist (PR) mold is formed on the silicon substrate and the UV imprinting process is performed, a UV-curable polymer (UCP)-filled PDMS stamp can be obtained, which is then pressed together with a polyethylene terephthalate (PET) film. The periodic polymeric micropillar arrays are finally transferred to the PET film by carefully separating the PDMS stamp. The roughness can be fabricated on the top of pillars via UCP-attached PAM by UV curing and chemistry processes to remove the PAM. As shown in the circle inset of Fig. 2.24a, nanowires can be formed on the tops of pillars.

The dynamic process of droplet impact on different surfaces is observed, such as the flat tops of pillars (Fig. 2.23b) and the rough tops of pillars (Fig. 2.23c). The droplets are found to be rebounding on the rough tops of pillars, instead of on the flat tops of pillars. This observation demonstrates that micro- and nanostructures are important to achieve water repellency (Kim et al., 2015).

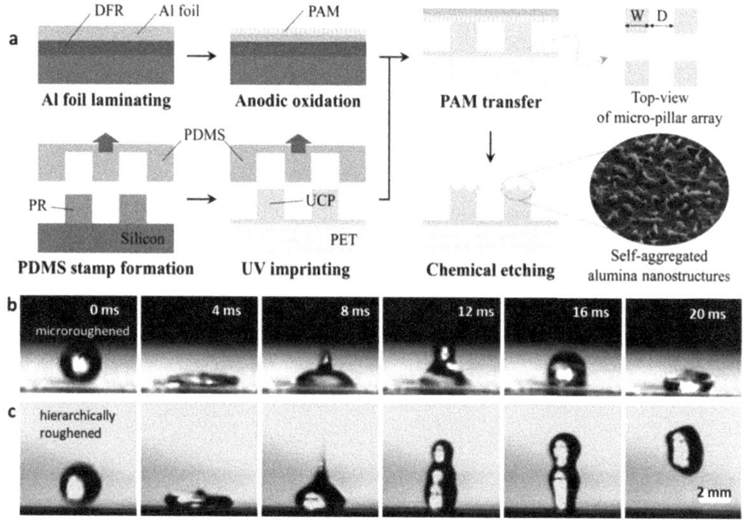

Figure 2.23 (a) The fabrication of micro- and nanostructures via PDMS stamp formation, UV imprinting, and chemical etching. The inset is the SEM image of self-aggregated alumina nanostructures. (b–c) Optical images comparing droplet rebound on a microroughened surface with flat pillar tops (b) and a hierarchically roughened surface with rough pillar tops (c). Observed at the same time (in 20 ms), droplet rebounding happens on the hierarchically roughened surface rather than on the microroughened surface (Kim et al., 2015).

2.4.5 Microsphere-Fiber Interpenetrated Structures

A superhydrophobic microporous membrane made up of pure poly(vinylidene fluoride) (PVDF) can be realized via electrospinning under high humidity; thus, special microsphere-fiber interpenetrated rough structures are obtained, which successfully exhibit waterproofness (Jiang et al., 2018). Figure 2.24a illustrates the fabrication process of the microsphere-fiber interpenetrated membrane. In detail, the PVDF solution is ejected from a high-voltage syringe needle and deforms into a conical shape of a Taylor cone under an electrostatic field formed between the syringe needle and the metallic rotating roller collector. The relative humidity is ~90% for this fabrication process. PVDF solutions are prepared by dissolving PVDF powder in N-methylpyrrolidone (NMP), N,N-dimethylformamide (DMF), and DMF/NMP. Figure 2.24b shows the rough structure of the membrane composed of interpenetrated microspheres and fibers by using a 0.15 g/mL PVDF/NMP solution (DMF/NMP is in v:v = 2:1). The magnified observation of the structure shows microspheres ~3.9 μm in diameter, along with nanoscale polymer fibers of microspheres interlinked with each other (Fig. 2.24c). Such structures are all on the membrane, as shown in Fig. 2.24d; the cross-sectional SEM image shows that the microsphere-fiber interpenetrated structure penetrates the membrane from top to bottom, with a thickness of ~55 μm. The wettability can be optimized by pore structure to control the breathable performance, with a hydrostatic pressure of 62 kPa and a water vapor transmission rate of 10.6 kg/m^2 d.

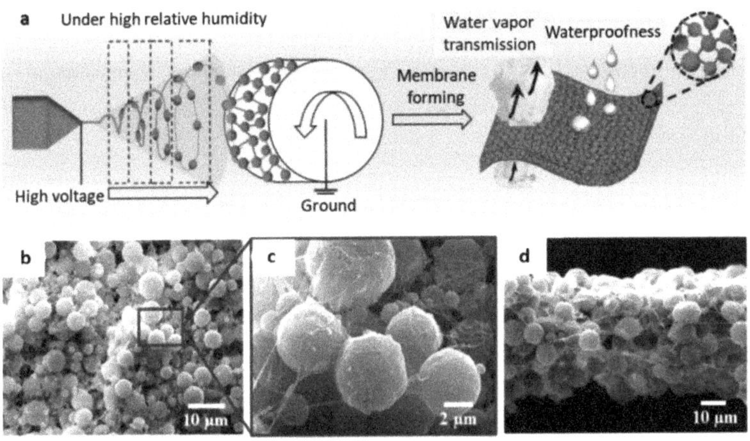

Figure 2.24 (a) Illustration of the fabrication process of a superhydrophobic PVDF membrane with waterproofness functions. (b–d) SEM images of a membrane composed of microsphere-fiber interpenetrated structures (b–c) and the cross-sectional view (d) (Jiang et al., 2018).

2.4.6 Droplet Rotational Bouncing on an Adhesive Spiral Surface

Droplets impacting and bouncing off solid surfaces play a vital role in various biological/physiological processes and engineering applications, such as self-cleaning (Blossey, 2003; Lu et al., 2015), anti-icing (Mishchenko et al., 2010; Kreder et al., 2016), and inkjet printing (Minemawari et al., 2011; Wu et al., 2015). Droplet behavior while impacting onto a surface can be controlled by the design of the surface wettability, such as droplet rotation rebounding. In general, when a water droplet impacts a hydrophobic and low-adhesive surface (Fig. 2.25a), it spreads into a liquid film at first and then uniformly recedes and rebounds from the substrate due to the water repellency, as shown in Fig. 2.25b. Interestingly, a hydrophobic solid surface with specially designed chemical heterogeneity makes a droplet bounce with gyration (Li et al., 2019). The chemically patterned surface consists of high-adhesive spirals surrounded by hydrophobic and low-adhesive regions (Fig. 2.25c).

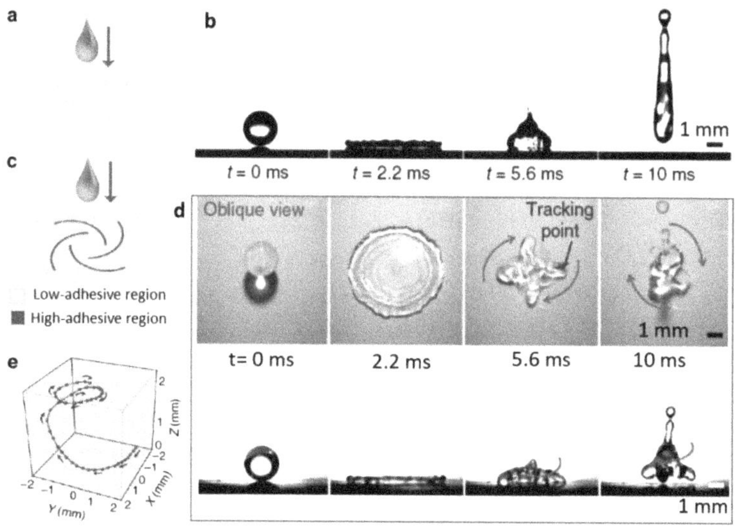

Figure 2.25 (a–b) Illustration (a) and optical images showing that an impacting droplet rebounds on a superhydrophobic surface (b). (c–d) Illustration (c) and optical images of the dynamic spiral behavior of an impacting droplet on a structured surface with spiral-patterned adhesion in the top view (d, the tops) and side view (d, the bottoms). (e) 3D analysis of droplet impact dynamics with trajectory on the adhesive spiral surface (Li et al., 2019).

As shown in Fig. 2.25d, the droplet initially spreads into a circular film similar to that on the homogeneous low-adhesive substrate; a distinctive behavior occurs at t = 5.6 ms such that the droplet recedes nonaxisymmetrically and forms a pinwheel-like morphology. After the retraction proceeds to the center, the droplet rebounds from the substrate with lobed gyrating behavior (t = 10 ms in Fig. 2.25d). A droplet (marked at t = 5.6 ms in Fig. 2.25d) trajectory can be illustrated in Fig. 2.25e, where the tracking point initially gyrates on the surface and then swirls upward. This gyration motion is in sharp contrast to the previous reports involving drop impact (Richard et al., 2002; Bird et al., 2013; Liu et al., 2014) or the collisions depicted in Newton's law of impact, where both the input and output movements are primarily translational.

2.5 Coating for Self-Cleaning Applications

2.5.1 Dentistry

A nanofilm of transparent clear overlay appliances (COAs) can be fabricated by using urushiol (a durable coating material derived from plants) via a layer-by-layer (LBL) assembly technique, i.e., a simple method involving sequential adsorption of interactive molecules in aqueous solutions (Decher, 1997). In particular, after thermal treatment, polymerized urushiol (PU) provides COAs with higher mechanical strength in large-scale assessments, lower cytotoxicity, and intrinsic hydrophobicity, which can be applied for antimicrobial use; furthermore, PU is expected to functionalize COAs with nanofilms for advanced biomedical use.

As illustrated in Fig. 2.26, urushiol-based durable nanofilms are used on COAs. Urushiol is the major component of ancient Korean lacquer, which is a durable coating material extracted from plants (*Toxicodendron vernicifluum*). The urushiol has a small molecular weight (MW), less than 500, and is composed of a catechol group substituted with long alkyl chains. The alkyl chains are irregularly distributed as saturated or unsaturated forms, with 1–3 double bonds (Watanabe et al., 2013). Though urushiol is very small, it can produce nanofilms with BPEI (branched poly(ethylene imine), MW ~ 25 000) by hydrogen bonding. Urushiol existing in nanoparticle form can produce durable intrinsic hydrophobic LbL multilayer films with BPEI. Through post-thermal treatment, urushiol can be converted into polyurushiol (PU) to hydrogenate unsaturated alkyl chains (Zheng et al., 2009; Jeong et al., 2016). PU exhibits higher mechanical properties but loses its mild toxicity, causing a skin allergy that is induced by unsaturated alkyl chains and catechol groups (McGovern and Barkley, 1998).

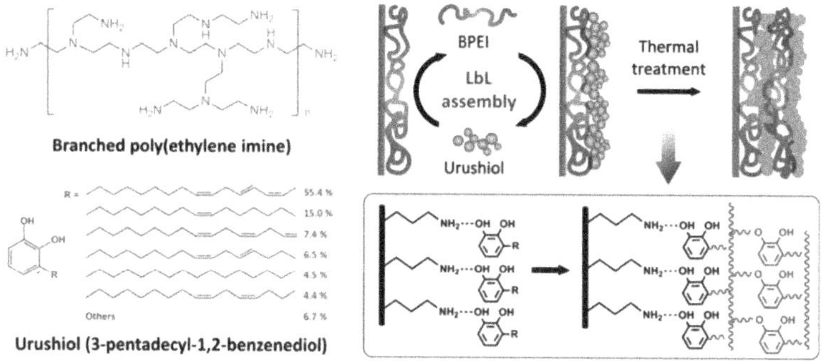

Figure 2.26 Illustration of the chemical structure and nanofilm fabrication via LBL assembly and thermal treatment processes (Jeong et al., 2016).

The multilayered films can successfully use urushiol as a building block. Controlling the wettability of the film can be simply realized by changing the outermost layer. Furthermore, the multilayered films are capable of exhibiting antibiofouling by controlling the wettability proportional to the number of layers and the effect of unsaturated alkyl groups (Jeong et al., 2015).

2.5.2 Self-Cleaning Transparent Heat Mirrors

As is well known, sunlight includes ultraviolet (0.2–0.4 μm), visible (0.4–0.7 μm), and infrared (IR) (0.7–2.0 μm) rays, which are in proportions of 5%, 40%, and 55%, respectively (Gueymard et al., 2002). It is significant that a transparent heat mirror (THM) transmits visible light in sunlight, reflects infrared (IR) rays, and conserves internal radiant heat. A novel self-cleaning transparent heat mirror (SC-THM) was recently reported, which was produced by depositing a plasma polymer fluorocarbon thin film on a silver-and-SiNx multilayer structure fabricated by a continuous roll-to-roll sputtering technique (Fig. 2.27a) (Kim et al., 2018). To achieve the desired effect, three-layer and five-layer SC-THMs with designs of optical simulation are integrated together. In the five-layer SC-THM, the visible light transmittance is 60.67% at a wavelength of 406 nm, and the IR transmittance is 6.86% at 1000 nm and 2.50% at 1500 nm (Fig. 2.27b). The wettability of the film can be obtained based on the contact angle ~111° of water after applying a low surface-energy fluorocarbon thin film to the top layer (Fig. 2.27c), which favors the self-cleaning property. The SC-THM film is applied to a window and achieves its effect in terms of visibility (Fig. 2.27d). The as-designed SC-THM film is suitable for use in various applications to improve the energy efficiency in order to effectively control sunlight and consistently maintain the self-cleaning nature of the surface (Kim et al., 2018).

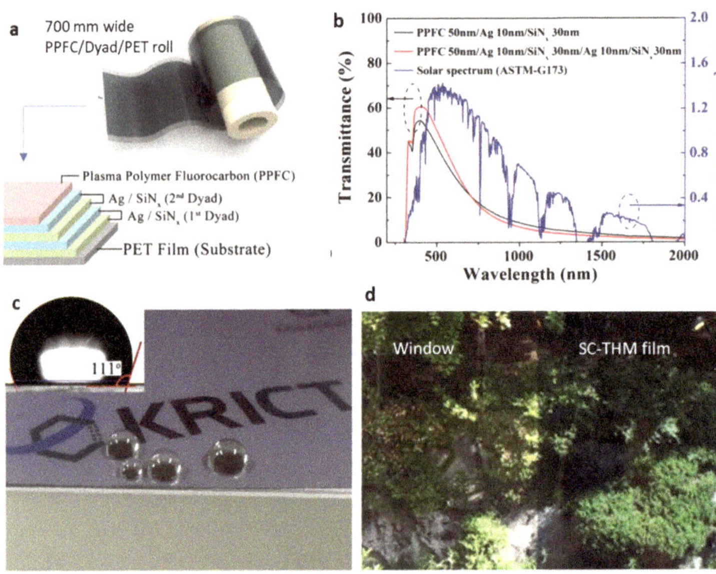

Figure 2.27 (a) Illustration of SC-THM fabrication via the design of multilayer films. (b) The transparency versus wavelength. (c) Hydrophobicity with a water contact angle of 111°. (d) Photograph of the SC-THM film on a window, along with the control of sunlight (Kim et al., 2018).

2.5.3 Bactericidal Surfaces

Bacterial adhesion and proliferation on surfaces are a challenge in medical and industrial fields. Hybrid materials with bactericidal performance and the ability to prevent bacterial attachment can be achieved. A simple one-step technique has been reported to fabricate self-cleaning and bactericidal surfaces (Hwang et al., 2017).

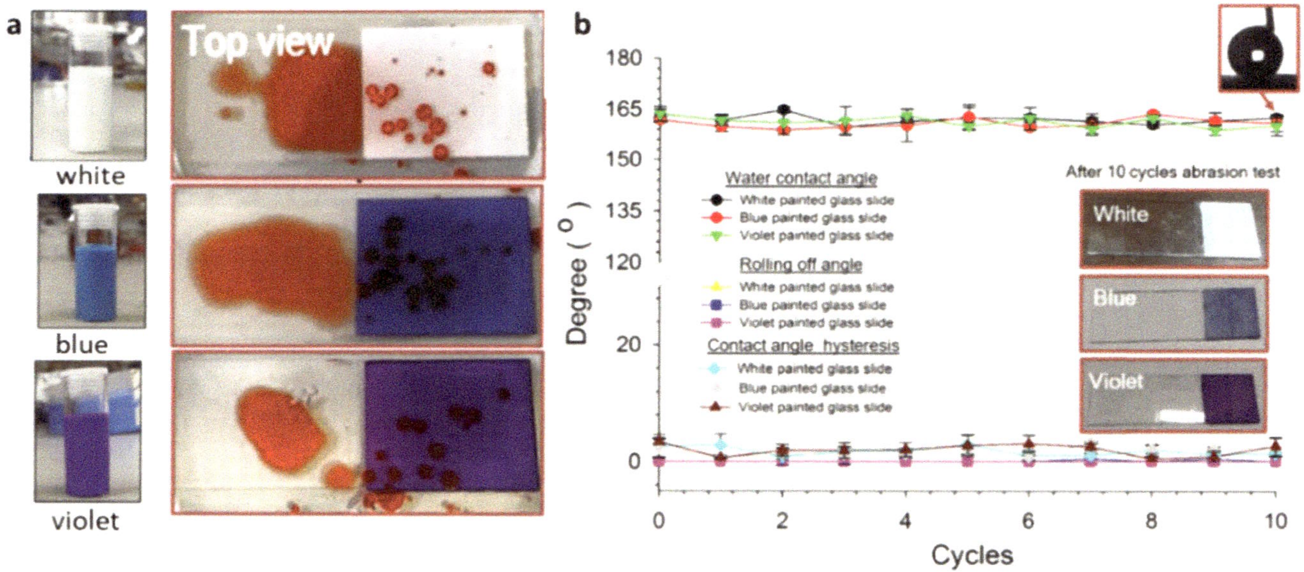

Figure 2.28 (a) Photographs of white (top), blue (middle), and violet (bottom) paints and superhydrophobicity (rights) on the paint surfaces. (b) The stability of superhydrophobic white/blue/violet paints with light-activated bactericidal properties (Hwang et al., 2017).

White, blue, and violet paints were produced by using titanium dioxide nanoparticles, 1H,1H,2H,2H-perfluorooctyltriethoxysilane, crystal violet, toluidine blue O (TBO), and ethanol solution (Fig. 2.28a). The painted surfaces are superhydrophobic in air, and even after hexadecane oil contamination, the water repellency and self-cleaning properties are effectively maintained. Significant reductions (>99.8%) in the number of adherent bacteria are achieved on the painted surfaces. After white light exposure, the paints show enhanced bactericidal activity compared to that of the same material in the dark. The blue and violet paints have the highest activities, and the activity of white paint is relatively weak. The potent photobactericidal activity of the blue and violet paints is due to TBO and crystal violet (CV), which are well-known white light-activated antimicrobial agents (WLAAAs) (Naik et al., 2011; Sehmi et al., 2015). For white paint, TiO_2 is only activated by UV light (Hashimoto et al., 2005). Figure 2.28b shows the stability of surfaces after abrasion cycles for certain applications, along with high contact angles (~165°), low roll-off angles, and a low contact angle hysteresis of far less than 10° in 10-cycle abrasion tests.

2.5.4 Spray Coating of Stable Emulsions

It is of tremendous interest to prepare cellulose-based materials that do not lose their original mechanical properties or durability after being imparted with self-cleaning properties, representing a disruptive waterproofing nanotechnology to eliminate moisture and microorganisms. For example, a stable colloidal emulsion can be formed with ethanol as the solvent and modified nano-SiO_2 as colloidal particles, which are synthesized from tetraethylorthosilicate (TEOS) and $NH_3 \cdot H_2O$ via a sol–gel method, followed by successive modification with PDMS and (heptadecafluoro-1,1,2,2-tetradecyl) trimethoxysilane (17F) (Wei et al., 2018).

The stable colloidal emulsion can be coated on paper and wood to achieve superhydrophobicity, along with a high water contact angle of more than 150° and a small slide-off angle of less than 5°. Figure 2.29a shows the droplets with high contact angles on the as-coated paper, with an illustration shown in Fig. 2.29b. The SEM images show the microstructure on the paper, where the rough fibers on the paper and the nanoparticles cover the entire surface of the fiber (Fig. 2.29c). The magnified SEM image shows that nanoparticles are spherical structures with diameters of ~110 nm on average, indicating SiO_2 distributions (Fig. 2.29d,e). Such a coating can be applied on various surfaces, including paper, wood, etc. Figure 2.29f presents a photograph of droplets on the as-coated paper, where the spherical droplets roll off the surface. Figure 2.29g shows the contrast result of droplet repellency on the as-coated region, whereas droplets spread on the uncoated region. The coatings achieve self-cleaning properties.

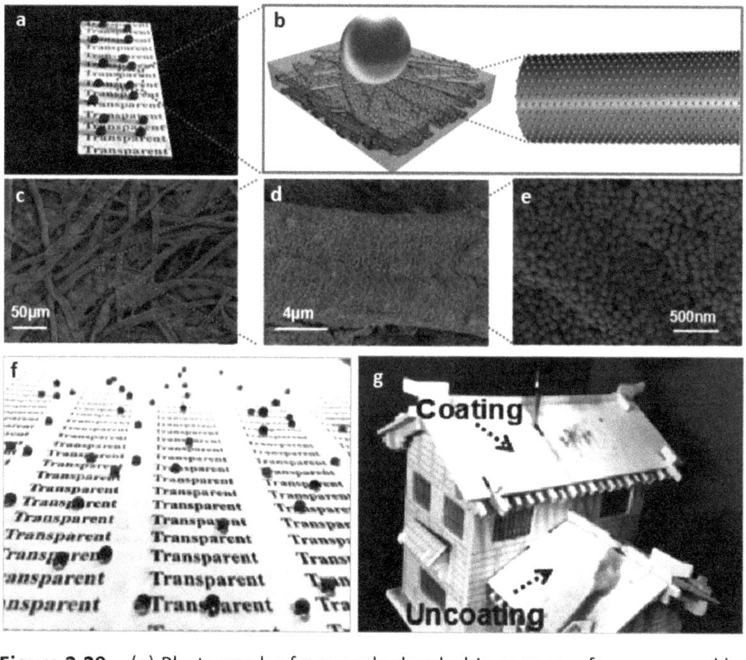

Figure 2.29 (a) Photograph of a superhydrophobic paper surface prepared by spray coating a stable emulsion. (b) Illustration of a superhydrophobic surface with rough structures. (c–e) SEM images of surfaces with microfibers and a covering of nanoparticles. (f–g) Photographs of superhydrophobicity on paper on a large scale (f) and wood for self-cleaning properties (g) (Wei et al., 2018).

2.5.5 Highly Transparent and Hazy Cellulose Nanopaper

A superhydrophobic, highly transparent, and hazy nanopaper made of TEMPO-oxidized cellulose nanofibrils (TOCNFs) and polysiloxanes achieved a static contact angle of 159.6°, sliding angle of 5.8°, transparency of 90.2%, and haze of 46.5% (Chen et al., 2018). Cellulose nanocrystals (CNCs) were obtained from wood pulp fibers by a well-established acid hydrolysis method with slight modifications (Fox et al., 2012). 2,2,6,6-Tetramethylpiperidinyl-1-oxyl (TEMPO)-oxidized cellulose was also synthesized (Saito et al., 2007). The native and oxidized cellulose fibers were then mechanically disintegrated by a high-pressure homogenizer (60 MPa) to produce suspensions of cellulose nanofibrils (CNFs) and TEMPO-oxidized cellulose nanofibrils (TOCNFs), respectively. Nanopapers were fabricated by vacuum filtering the cellulose/water dispersions (Chen et al., 2018).

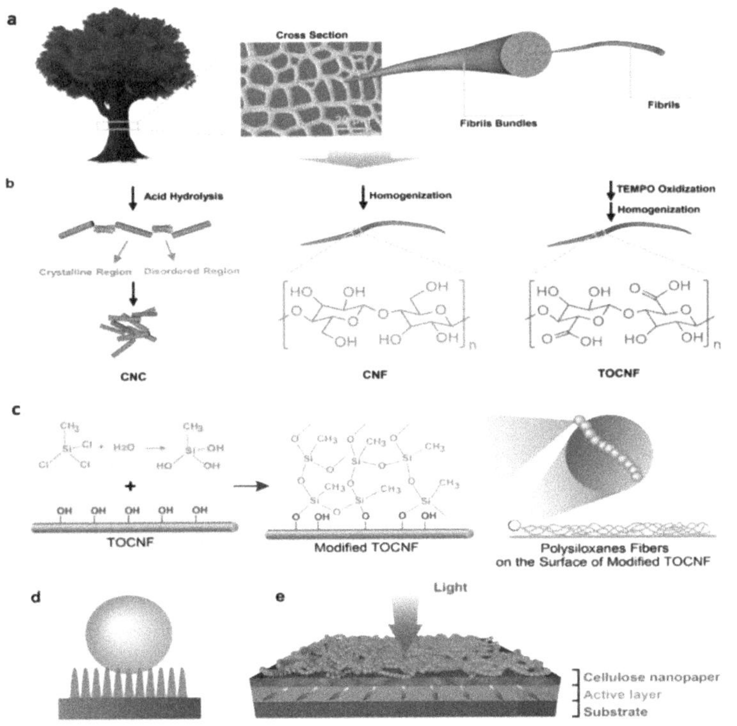

Figure 2.30 (a–b) Illustration of cellulose nanocrystals (CNCs), cellulose nanofibrils (CNFs), and TEMPO-oxidized cellulose nanofibrils (TOCNFs) from wood pulp fibers via acid hydrolysis (b, left), homogenization (b, middle), and TEMPO-oxidized homogenization (b, right). (c) The fabrication of nanopapers by using TOCNFs by modifying and combining polysiloxane fibers. (d) Illustration of droplet suspension on the microstructured surface. (e) Illustration of the light distribution through modified cellulose nanopaper incident on a solar cell for promising applications (Chen et al., 2018).

Figure 2.30a–b shows the fabrication process using CNCs, CNFs, and TOCNFs. Various papers can be achieved, including CNC-P, CNF-P, TOCNF-P, H40-TOCNF-P, etc. H40-TOCNF is the TOCNF-P sample that was further modified by immersion into a 0.5 M methyltrichlorosilane solution in anhydrous toluene at room temperature for 40 min. Thus, the surface of modified TOCNF-P features pearl-necklace-like polysiloxane fibers and a 3D network (Fig. 2.30c), along with superhydrophobicity (Fig. 2.30d). The light distribution through modified cellulose nanopaper incident on a solar cell indicates promising application prospects (Fig. 2.30e).

Figure 2.31a shows the optical properties of nanopapers (CNC-P, CNF-P, TOCNF-P, H40-TOCNF-P), including optical transmittance and haze, which are critical for substrates to have wide applications in optoelectronic devices. High transparency can be observed on H40-TOCNF-P. The light scattering effect can be achieved by the fuzzy school emblem (nanopapers approximately 5 cm away from the images). The self-cleaning property can be effectively achieved on H40-TOCNF-P, as shown in Fig. 2.31b, where a high contact angle of water can be achieved, reaching 159.6°.

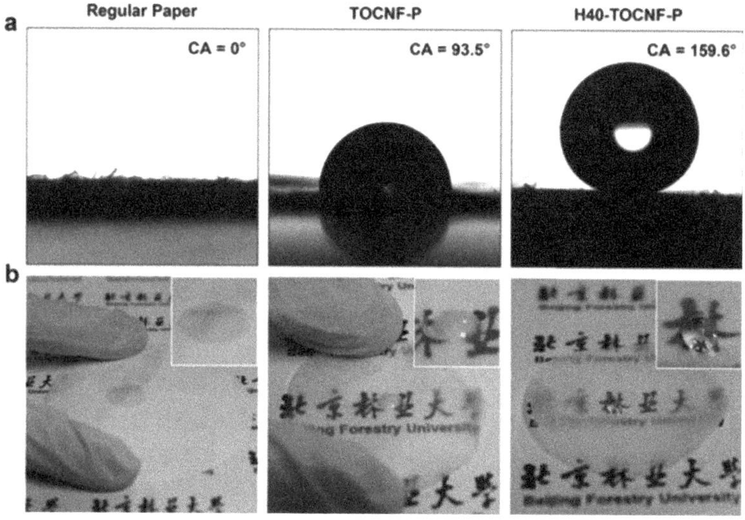

Figure 2.31 (a) Optical images of droplet behaviors on regular paper and TOCNF-P and H40-TOCNF-P, with contact angles (CAs) of 0°, 93.5°, and 159.6°, respectively. (b) Photographs of transparent properties for regular paper (nontransparent) and TOCNF-P (transparent) and H40-TOCNF-P (transparent) versus the hydrophobicity (Chen et al., 2018).

2.5.6 Cotton Fabric Modified by a Simple Immersion Technique

Cotton fabric materials have been extensively used in various household and industrial applications due to their unique properties such as good feel, biodegradability, renewability, and environmental friendliness. The presence of hydroxyl (–OH) groups on the cotton fabric surface makes it hydrophilic. Due to this hydrophilicity, cotton fabric can be quickly wetted by water and various liquids, including oil. However, this high affinity of water and various liquids to the cotton surface increases the probability of staining and bacterial growth, subsequently limiting the usage of cotton fabric in several applications. Imparting cotton fabric with self-cleaning properties is, thus, significant to widening the applications of these materials.

The following example demonstrates modification of a surface to make it superhydrophobic. As reported, superhydrophobic cotton fabrics can be fabricated by processing, for example immersing cotton fabrics in a 5% (v/v) hexadecyltrimethoxysilane (HDTMS) solution in ethanol for 5 h and then drying at 120°C for 6 h (Chauhan et al., 2019). Figure 2.32a shows various spherical droplets (e.g., ethylene glycol, honey, color water, tea, lemon, turmeric, milk, and glycerol) with high contact angles of more than 150°, indicating superhydrophobicity. The contrasting properties can be observed on the surfaces of both normal (untreated original) (Fig. 2.32b–d, Frame I) and superhydrophobic (treated) cotton fabrics (Fig. 2.32b–d, Frame II). The normal surface pins the droplets and would be considered a non-self-cleaning surface, in addition to being easily polluted (Frame I of Fig. 2.32b–d). Interestingly, the superhydrophobic surface exhibits water repellency and a self-cleaning ability when contaminated by muddy water (Frame II of Fig. 2.32b–d), which demonstrates the excellent self-cleaning property, with high stain resistance.

The superhydrophobic cotton fabric can separate oil from oil–water (kerosene–water, n-hexane–water) and oil–liquid (benzene ethylene glycol) mixtures, with a high separation efficiency of 99%. The cotton fabric also shows high antibacterial activity against *Escherichia coli* bacteria (Fig. 2.33), in which no bacterial growth is observed in 24 h. The superhydrophobic cotton fabric can be used in a variety of applications.

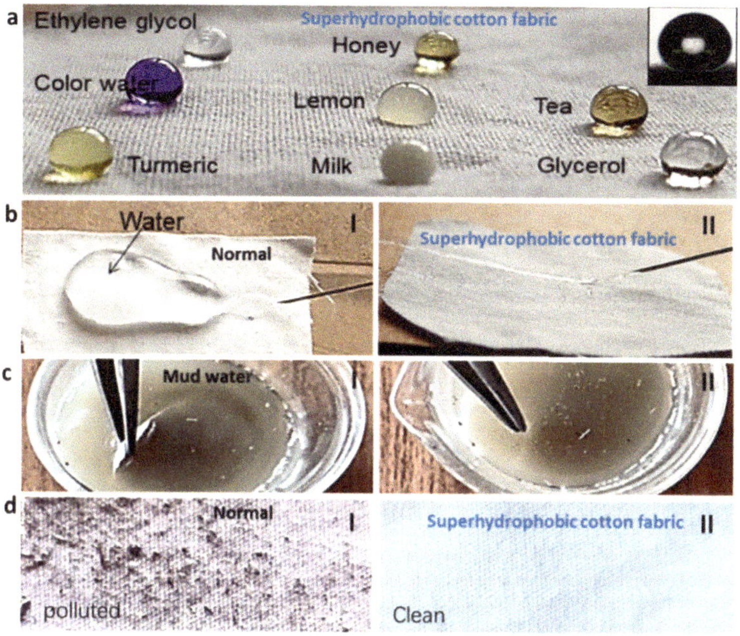

Figure 2.32 (a) Photograph of superhydrophobicity for various liquid droplets (ethylene glycol, color water, honey, lemon, turmeric, milk, tea, glycerol) on cotton fabric. (b–d) Comparison with normal cotton fabric. The superhydrophobic cotton fabrics (Frame II) display excellent self-cleaning properties, in contrast to the normal cotton fabrics (Frame I) (Chauhan et al., 2019).

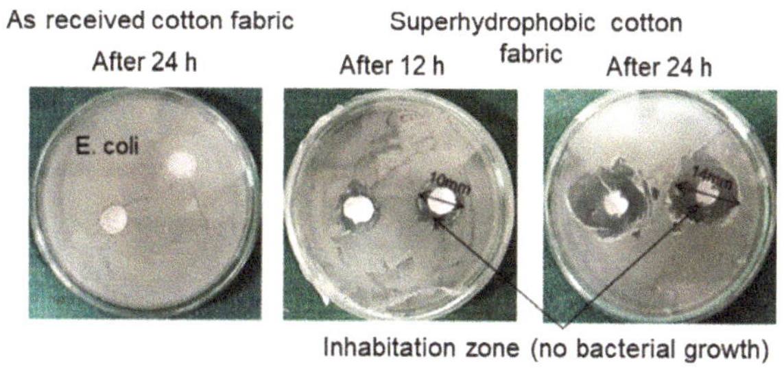

Figure 2.33 The antibacterial activities of superhydrophobic cotton fabric (middle and right) in contrast to normal cotton fabrics (left) (Chauhan et al., 2019).

2.5.7 Controlling Liquid Splash

The deposition of liquid droplets on solid surfaces is of great importance to many fundamental scientific principles and technological applications, such as spraying, coating, and printing. It is challenging to control liquid splash in some fields. Here, with a specific example of pesticide spraying during agricultural work, the problems are that more than 50% of agrochemicals are lost in the pesticide-spraying process because of the undesired bouncing and splashing behaviors on hydrophobic or superhydrophobic leaves that have "waterproof" properties (Bergeron et al., 2000; Galliker et al., 2012; de Ruiter et al., 2014). If pesticides cannot be properly deposited on crops, then pests might not be controlled, and plant injury might occur (Bergeron et al., 2000). Excessive pesticide application might contribute to soil, air, or water pollution and might negatively affect human health.

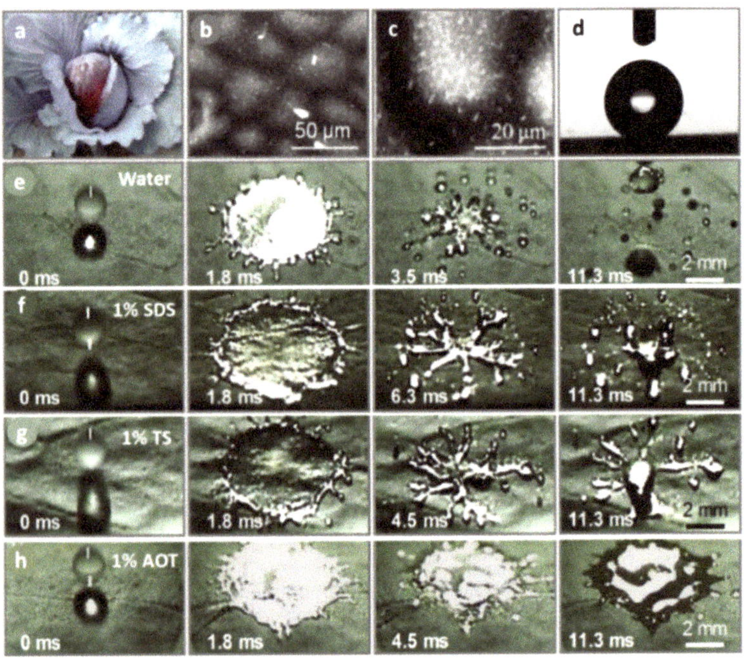

Figure 2.34 (a–d) Photograph of a *Brassica oleracea* L. leaf (a) with micro- and nanostructures (b–c) for superhydrophobicity (d). (e–h) The behaviors of the impacting of various liquid droplets, i.e., water (e), 1% SDS (f), 1% TS (g), and 1% AOT (h), on a *B. oleracea* L. leaf. The 1% AOT liquid droplet was regulated well, with little loss of liquid after the droplet impacted the surface (g), in contrast to the other formulations (e–f) (Song et al., 2017b).

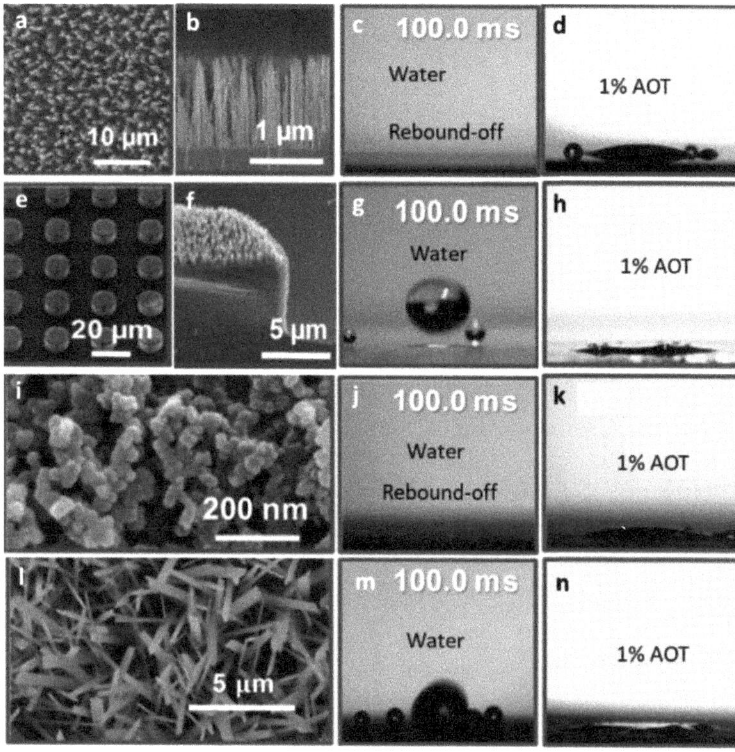

Figure 2.35 Controlling the behaviors of impacting liquid (pure water or 1% AOT) droplets on various structured surfaces. (a–d) A silicon nanowire arrayed surface. (e–h) The silicon micropillar arrayed surface, along with nanowires covering the tops of micropillars. (i) SiO_2 nanoparticle surface. (l) CuO nanowire surface (Song et al., 2017b).

Figure 2.34a presents the photograph of a *Brassica oleracea* L. leaf, whose surface contains micro-/nanostructures (Fig. 2.34b,c), with a water contact angle of ~156.2° (Fig. 2.34d) indicating superhydrophobicity. Figure 2.34e,f shows the impacts of the following droplets on a *B. oleracea* L. leaf: pure water (Fig. 2.34e) and aqueous solutions (1%) containing the micelle surfactant sodium dodecyl sulfate (SDS) (Fig. 2.34f), trisiloxane molecules (TSs) (Fig. 2.34g), and the vesicle surfactant sodium bis(2-ethylhexyl) sulfosuccinate [Aerosol OT (AOT)] (Fig. 2.34h). The splashing of liquid occurs in different receding stages, and the 1% AOT surfactant generates a substantially strong depression without receding behavior. The impacting droplet containing 1% AOT achieves complete spread on the leaf surface during the contact process, but the impacting water exhibits complete rebounding, and 1% SDS and 1% TS result in partial rebounding. For pesticide spraying, the formulation with 1% AOT is desired (Song et al., 2017b).

To achieve the desired result (e.g., water repellency or spreading) of a spray, micro-/nanostructures on the surface can be designed, as shown in Fig. 2.35. The presented silicon nanowire arrayed surfaces display water repellency and 1% AOT spreading (Fig. 2.35a–d). The surfaces with arrayed silicon micropillars and nanowires also exhibit water repellency and 1% AOT spreading (Fig. 2.35e–h). Moreover, the surfaces with micro-/nanostructured superhydrophobic SiO_2 (Fig. 2.35i–k) and the superhydrophobic CuO surface with nanoneedles all achieve water repellency and 1% AOT spreading (Fig. 2.35l–n).

2.5.8 Fluorine-Free Superwetting Systems

It is considerably significant to adopt a simple, substrate-independent, completely fluorine-free, and environmentally friendly way to fabricate superhydrophilic, superhydrophobic, and slippery surfaces. For instance, by taking advantage of polyglycerol as a second-generation mussel-inspired adhesive, highly hierarchical structures are easily prepared in a two-step coating approach (Wei et al., 2014a; Schlaich et al., 2016a). Mussel-inspired dendritic polyglycerol (MI-dPG) is synthesized by an optimized reaction protocol (Wei et al., 2014b). Starting from the polyglycerol amine ($dPG-NH_2$, Mn = 12 000 g/mol, Mw = 18 000 g/mol), 40% of the terminal amine groups of the scaffold are functionalized with catechol moieties using 3,4-dihydroxyhydrocinnamic acid (DHHA) via the 1-ethyl-3-(3-dimethylaminopropyl) carbodiimide (EDC) coupling strategy (Fig. 2.36a) under slightly acidic conditions (Schlaich et al., 2016a). The resulting polymeric architecture (Fig. 2.36b) effectively mimics mussel foot protein-5 (mfp-5) regarding the functional groups, molecular structure, and molecular weight (Wei et al., 2014b). As a result, MI-dPG can serve as a remarkable surface adhesive and generate roughness at the same time by controlling the degree of crosslinking. Figure 2.36c–e shows the different cases of a superhydrophilic (SHL) surface with a contact angle of <10° (c), superhydrophobic (SHP) surface with a contact angle of 168° (d), and environmentally friendly slippery surface (eSLIPS) with a water contact angle of 43° (e) on coated glass.

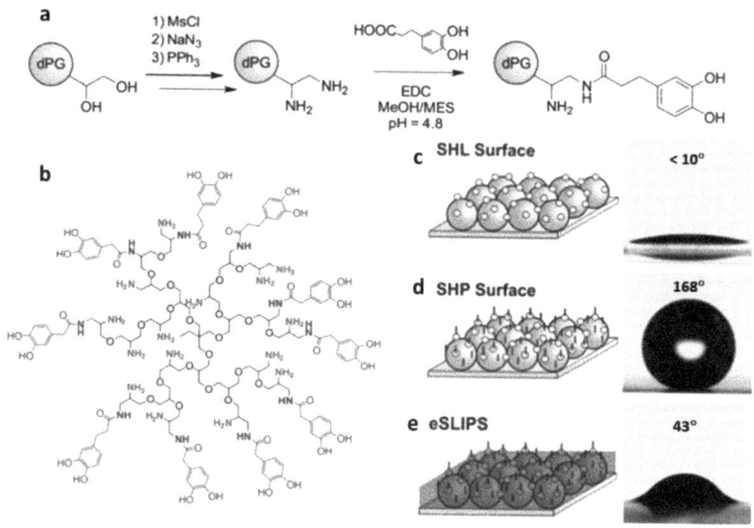

Figure 2.36 Design of surfaces and wettability. (a) Synthetic approach for mussel-inspired dendritic polyglycerol (MI-dPG) and (b) the resulting chemical structure. (c–e) Preparation of a superhydrophilic (SHL) surface (c), superhydrophobic (SHP) surface (d), and environmentally friendly slippery surface (eSLIPS) (e) with water contact angles of <10°, 168°, and 43°, respectively, on coated glasses (right, optical images) (Schlaich et al., 2016a).

To achieve a "lotus-like" hierarchical structure, a simple two-step coating approach is performed, which achieves a highly hierarchical MI-dPG-coated surface with two-tier roughness at both the micro- and nanolevels, as shown in Fig. 2.37a,b. The self-cleaning property is shown in Fig. 2.37c, where dust can be removed by the sweeping of a water droplet (Frame II of Fig. 2.37c). Furthermore, the surface can achieve self-cleaning via the water droplet sliding off after the surface is contaminated by oil (Fig. 2.37d).

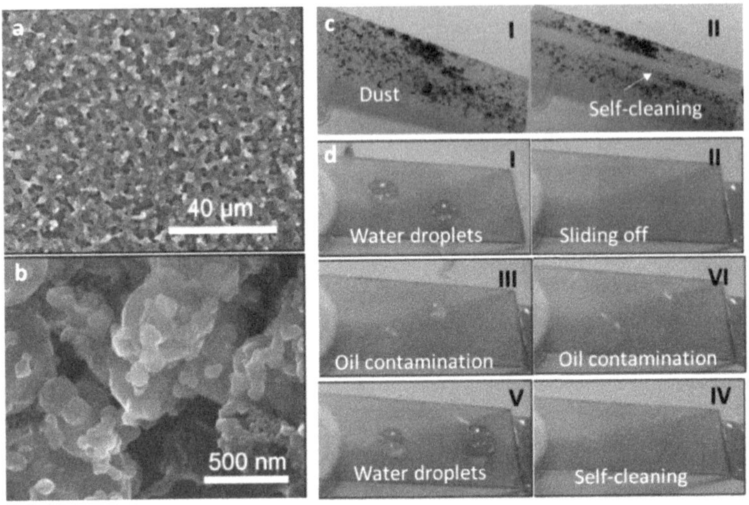

Figure 2.37 (a–b) SEM images of the MI-dPG surface observed on a large scale (a) and magnified observation at the nanoscale (b). (c–d) The self-cleaning property of the hierarchical MI-dPG-coated surface for dust removal (c) and oil contamination removal (d) (Schlaich et al., 2016b).

2.5.9 High Stress-Resistant Super-Repellent Materials

Superhydrophobic particles and coatings with high mechanical resistance have received widespread attention due to their promising applications in antiwetting, anti-icing, anticorrosion, antibacterial, moisture protection, and so on (Farhadi et al., 2011; Si and Guo, 2015; Ramakrishna et al., 2017). Figure 2.38a displays a schematic representation of the two-step method followed to synthesize siloxane-grafted silica (SOS) particles. To prepare the superhydrophobic coating, a 20 wt.% solution of poly(vinylidenefluoride-co-hexafluoropropylene) (PVDF-co-HFP) was mixed in dimethylacetamide (DMAc) by ultrasonication at 60°C for 2 h. 3-Aminopropyltriethoxysilane (APTES, 12 mL) was then added to the colloidal solution under an inert atmosphere, and the reaction was performed for 18 h at 100°C. Finally, aminopropyl-terminated silica nanoparticles (ASPs) were obtained, which were superhydrophilic in nature. In the next step, oxirane end-functionalized siloxane oligomers and the proper ASPs were ultrasonicated in triethylamine at 60°C for 1 h, with an ASP:siloxane oligomer:triethylamine ratio of 1 (w):4 (w):40 (v). With a catalyst loading of 1 wt.% 2,4,6-tris(dimethylaminomethyl)phenol, the reaction was carried out at 50°C for 8 h. Finally, a yellowish-white superhydrophobic powder (SOS) was obtained after washing the particles and drying under vacuum at 100°C for 8 h, as shown in the inset (photograph) of Fig. 2.38a. Furthermore, the different weight percentages of SOS particles were incorporated into the polymer solution by mixing for 10 min in the presence of 45% of SOS particles, with the product called 45FPS.

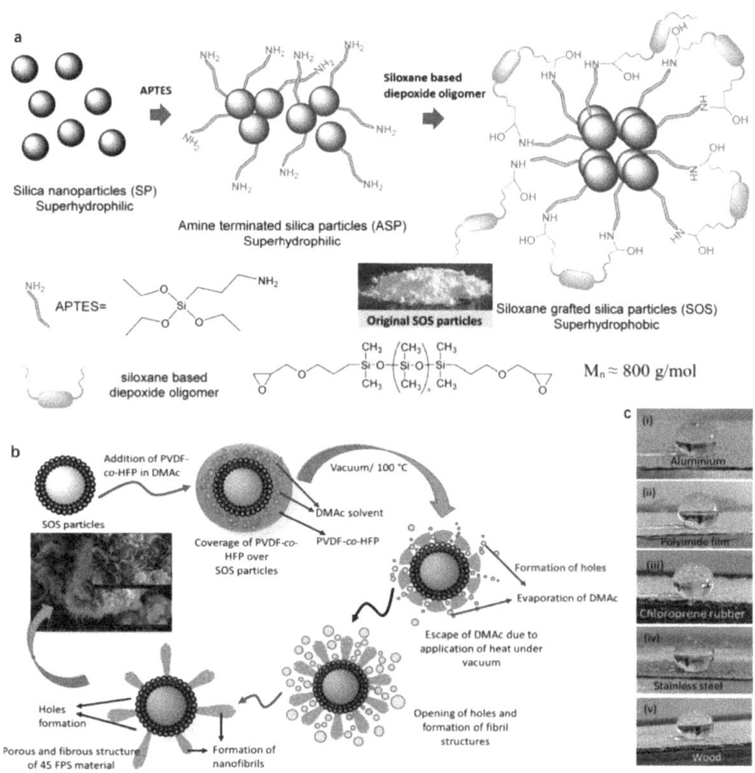

Figure 2.38 (a) Illustration of the synthesis of siloxane-grafted silica (SOS) particles from pristine silica nanoparticles and (b) illustration of the fabrication and SEM image of the porous–fibrous morphology on 45 FPS materials (inset). (c) Photographs of the self-cleaning property of 45 FPS materials coated on different substrates, including aluminum (Frame i), polyimide film (Frame ii), chloroprene rubber (Frame iii), stainless steel (Frame iv), and wood (Frame v) (Ramakrishna et al., 2017).

Figure 2.38b shows the process of SOS particles mixed with PVDF-co-HFP in DMAc to form a colloidal solution, where the DMAc solvent is distributed uniformly in the matrix. By heating at 100°C under vacuum conditions, the DMAc solvent is slowly evaporated (the boiling point is 165°C), the process of which creates holes in the copolymer mass, accordingly forming a fibrous structure wrapped over the SOS particles, i.e., the porous–fibrous morphology on the 45FPS superhydrophobic surface.

The 45FPS materials can be coated on various substrates such as (i) glass, (ii) aluminum, (iii) polyimide film, (iv) chloroprene rubber, (v) stainless steel, and (vi) wood, where the liquid (pH = 13, pH = 7, pH = 1, glycerol, honey, and ethylene glycol, etc.) droplets adopt high contact angles, in addition to self-cleaning properties (Fig. 2.38c) (Ramakrishna et al., 2017).

2.5.10 Coatings from Identical Precursors

Coatings to repel water-borne pollutants are highly desirable, for example, for stain formation inhibition, hazardous cleaning of skyscraper windows, car windshields on rainy days, repellence of oil-borne contaminants, inhibition of graffiti, grease and fingerprint residues, and so on (Hu et al., 2019).

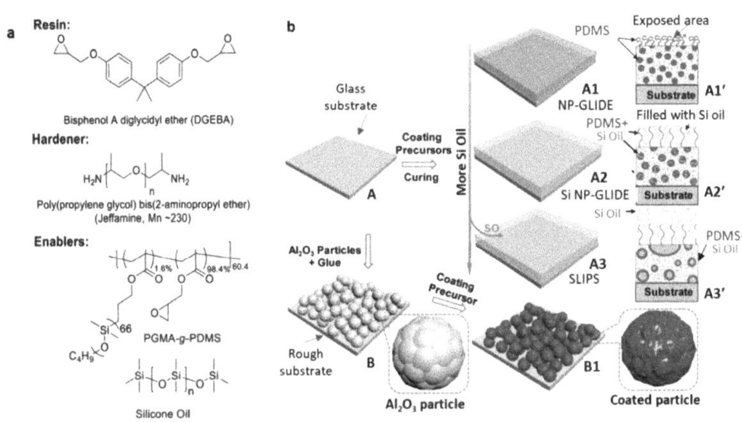

Figure 2.39 (a) Chemical structures of DGEBA (resin), Jeffamine (hardener), PGMA-*g*-PDMA, and silicone oil. (b) Preparation of NP-GLIDE (A→ A1), silicone-infused NP-GLIDE (A → A2), SLIPS (A → A3), and superhydrophobic (A → B → B1) coatings. A1′, A2′, and A3′ represent cross-sectional structures of the NP-GLIDE, SINP-GLIDE, and SLIPS coatings, respectively (Hu et al., 2019).

The coatings can be prepared by using chemical components, including bisphenol A diglycidyl ether (DGEBA), oligo(propylene glycol) bis(2-aminopropyl ether) (Jeffamine), poly(glycidyl methacrylate)-graft-poly(dimethylsiloxane) (PGMA-g-PDMS), and silicone oil (Si-oil) (Fig. 2.39a). Figure 2.39b illustrates the preparation process of the coatings, where the DGEBA, Jeffamine, and PGMA-g-PDMS are cast on glass plates to generate the NP-GLIDE coating (A → A1, side view in A1′). The silicone-infused NP-GLIDE or SI NP-GLIDE coatings can also be obtained (A → A2, side view in A2′). The pre-SLIPS coatings can be prepared, where the SLIPS coatings are obtained by using pre-SLIPS with additional silicone (A → A3, side view in A3′). Superhydrophobic coatings were obtained by casting a pre-SLIPS formulation onto a glass plate decorated with alumina particles (A → B → B1).

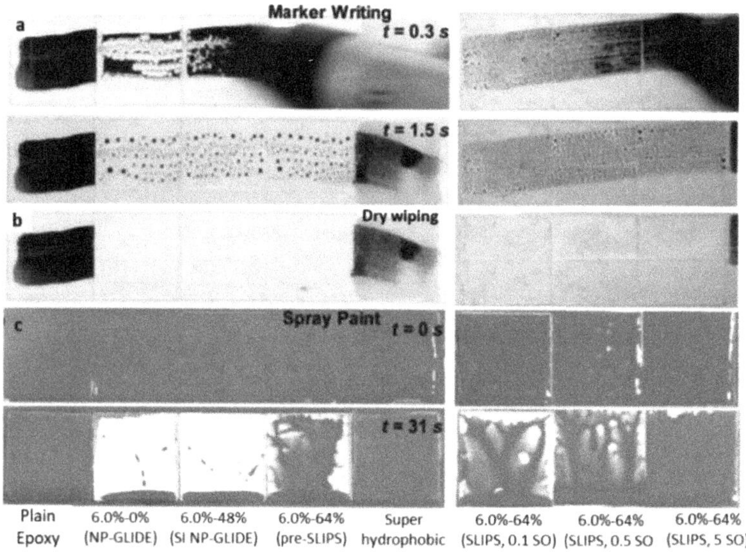

Figure 2.40 The different effects of inks (marker writing (a), marker writing after dry wiping (b)) and spray paint (c) on different coated surfaces (plain epoxy, NP-GLIDE, SI NP-GLIDE, pre-SLIPS, hydrophobic, SLIPS 0.1 SO, SLIPS 0.5 SO, and SLIPS 5 SO) (Hu et al., 2019).

Figure 2.40 shows the different effects of inks on different coated surfaces. Figure 2.40a shows the cases in which the ink contracted on the smooth 6.0–0% (NP-GLIDE), 6.0–48% (SI NP-GLIDE),

6.0–64% (pre-SLIPS) coatings, and the 6.0–64% SLIPS coating with 0.10, 0.50, and 5.0 mg of silicone oil, where the ink films contracted faster on the SLIPS coatings than on the NP-GLIDE, SI NP-GLIDE, and pre-SLIPS coatings, and where no contraction was visible on the base epoxy coating. Figure 2.40b indicates that the contracted ink patches are readily removed from the smooth coatings that initially contracted the ink. Figure 2.40c shows that the NP-GLIDE and SI NP-GLIDE coatings are able to contract the paint. This figure indicates that the best contraction capabilities of the NP-GLIDE and SINP-GLIDE coatings are for oil-based ink and paint, respectively. The NP-GLIDE coatings are better candidates for anti-graffiti and anti-smudge applications. Moreover, SLIPS coatings are better suited for providing water repellency. This research provides valuable insight into the development of the best-performing self-cleaning coatings for certain applications.

2.5.11 Anti-Fingerprint Coatings

It is known that when a finger touches a glass surface, a fingerprint contamination layer remains. Such undesired residue is predominantly water but also contains skin oils, sweat, and cosmetics, resulting in an oily liquid (Thomas and Reynoldson, 1975), which is deposited on a transparent surface, such as the touchscreen of a mobile phone, iPad, or pocket computer; accordingly, this residue decreases the optical transmittance and impairs visibility (Uhlmann et al., 2007). Therefore, developing transparent hydrophobic and oleophobic coatings as anti-fingerprint materials has become an urgent and important problem (Siriviryanun and Imae, 2014) for applications such as electronic devices and wearable electronics (Scruton et al., 1975; Choi et al., 2017).

For industrial applications, the anti-fingerprint coating must have the characteristics of smoothness, rigidity, transparency, and thinness (usually 2–3 μm). To obtain the desired amphiphobic properties for such anti-fingerprint properties, a certain kind of coating is designed, i.e., using intermediate perfluoro-1H,1H,2H,2H-decyl/octyl and trimethoxysilylethylene cyclotetrasiloxane (FVPS) and nano-SiO_2 bonded with perfluorodecyl/octyl oligosiloxane (FSi@SiO_2) resin, which is synthesized to achieve strong hydrophobic and oleophobic properties as an anti-fingerprint coating (An et al., 2018). Figure 2.41a shows an illustration of the fabrication of the anti-fingerprint coating. Figure 2.41b shows a transmission electron microscopy (TEM) image of the FVPS resin; FSi@SiO_2 is a nanohybrid resin (Fig. 2.41c). Figure 2.41d shows the solution of the coating materials. The nanoparticles are 200–400 nm in size and 321 nm on average (Fig. 2.41e). After the glass surface of the mobile phone screen is coated, it shows excellent transparency and strong hydrophobicity, where the presence of a large amount of C–Fs cooperating with nanoparticles, as well as their aggregates, leads to a strong hydro- and oleophobic coating prepared from the FSi@SiO_2 resin (Fig. 2.41f). The anti-fingerprint ability of the FSi@SiO_2 coating surface (top, Frame (a)) is better than that of the bare glass (bottom, Frame (b)) (Fig. 2.41g). This study is useful to protect the cleanliness of surfaces, including displays, phones, glass doors, glass buildings, etc.

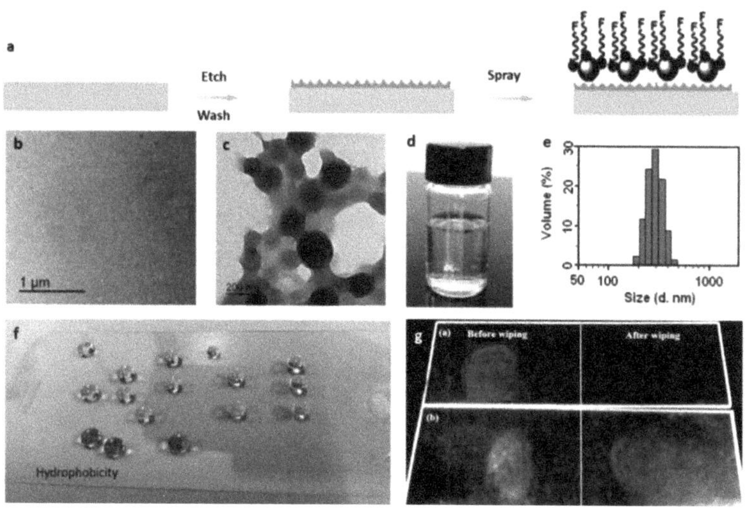

Figure 2.41 (a) Illustration of the fabrication of anti-fingerprint coatings. (b) SEM image of the surface on the coating. (c) TEM image of nanoparticles in coating materials. (d) Solution of coating materials. (e) Analysis of nanoparticle size. (f) Superhydrophobicity of the coating. (g) The excellent anti-fingerprint property of the FSi@SiO_2 coating surface (tops, Frame (a)) in contrast to that of bare glass (bottoms, Frame (b)) (An et al., 2018).

Thinking Questions

1. What is water repellency? Can you give examples?
2. What are bioinspired surfaces and materials with water repellency? Can you give examples?
3. Can you cite new literature about some materials and surfaces with water repellency?

References

Adamson A. W. and A. P. Gast. *The Physical Chemistry of Surfaces*, Wiley, New York, 1997.

An Q., Z. Lyu, W. Shangguan, B. Qiao, and P. Qin. The synthesis and morphology of a perfluoroalkyl oligosiloxane@SiO_2 resin and its performance in anti-fingerprint coating. *Coatings* **8**, 100 (2018).

Bergeron V., D. Bonn, J. Y. Martin, and L. Vovelle. Controlling droplet deposition with polymer additives. *Nature* **405**, 772 (2000).

Bird J. C., R. Dhiman, H.-M. Kwon, and K. K. Varanasi. Reducing the contact time of a bouncing drop. *Nature* **503**, 385 (2013).

Blossey R. Self-cleaning surfaces: Virtual realities. *Nat. Mater.* **2**, 301 (2003).

Caponigro M. and C. Erilsen. Surface film locomotion by the water strider, Gerris remigis Say. *Am. Midland Nat.* **95**, 268 (1976).

Chaudhury M. K. and G. M. Whitesides. How to make water run uphill. *Science* **256**, 1539 (1992).

Chauhan P., A. Kumar, and B. Bhushan. Self-cleaning, stain-resistant and anti-bacterial superhydrophobic cotton fabric prepared by simple immersion technique. *J. Colloid Interface Sci.* **535**, 66 (2019).

Chen P., L. Chen, D. Han, J. Zhai, Y. Zheng, and L. Jiang. Wetting behavior at micro-/nanoscales: Direct imaging of a microscopic water/air/solid three-phase interface. *Small* **5**, 908 (2009).

Chen S., Y. Song, and F. Xu. Highly transparent and hazy cellulose nanopaper simultaneously with a self-cleaning superhydrophobic surface. *ACS Sustainable Chem. Eng.* **6**, 5173 (2018).

Cheng Y. T. and D. E. Rodak. Is the lotus leaf superhydrophobic? *Appl. Phys. Lett.* **86**, 144101 (2005).

Cheng T., D. E. Rodak, C. A. Wong, and C. A. Hayden. Effects of micro- and nano-structures on the self-cleaning behaviour of lotus leaves. *Nanotechnology* **17**, 1359 (2006).

Choi H., K. C. Park, H. Lee, T. Crouzier, M. F. Rubner, R. E. Cohen, G. Barbastathis, and G. H. Mckinley. Superoleophilic titania nanoparticle coatings with fast fingerprint decomposition and transparency. *ACS Appl. Mater. Interfaces* **9**, 8354 (2017).

Davies G. B., T. Krüger, P. V. Coveney, J. Harting, and F. Bresme. Assembling ellipsoidal particles at fluid interfaces using switchable dipolar capillary interactions. *Adv. Mater.* **26**, 6715 (2014).

de Ruiter J., R. Lagraauw, D. van den Ende, and F. Mugele. Wettability-independent bouncing on flat surfaces mediated by thin air films. *Nat. Phys.* **11**, 48 (2014).

Decher G. Fuzzy nanoassemblies: Toward layered polymeric multicomposites. *Science* **277**, 1232 (1997).

Deng T., K. K. Varanasi, M. Hsu, N. Bhate, C. Keimel, J. Stein, and M. Blohm. Nonwetting of impinging droplets on textured surfaces. *Appl. Phys. Lett.* **94**, 133109 (2009).

Deng X., L. Mammen, H. Butt, and D. Vollmer. Candle soot as a template for a transparent robust superamphiphobic coating. *Science* **335**, 67 (2012).

Dickinson M. How to walk on water. *Nature* **424**, 621 (2003).

Ellinas K., M. Chatzipetrou, I. Zergioti, A. Tserepi, and E. Gogolides. Superamphiphobic polymeric surfaces sustaining ultrahigh impact pressures of aqueous high- and low-surface-tension mixtures, tested with laser-induced forward transfer of drops. *Adv. Mater.* **27**, 2231 (2015).

Farhadi S., M. Farzaneh, and S. Kulinich. Anti-icing performance of superhydrophobic surfaces. *Appl. Surf. Sci.*, **257**, 6264 (2011).

Feng L., S. Li, Y. Li, H. Li, L. Zhang, J. Zhai, Y. Song, B. Liu, L. Jiang, and D. Zhu. Super-hydrophobic surfaces: From natural to artificial. *Adv. Mater.* **14**, 1857 (2002).

Fox J., J. J. Wie, B. W. Greenland, S. Burattini, W. Hayes, H. M. Colquhoun, M. E. Mackay, and S. J. Rowan. High-strength, healable, supramolecular polymer nanocomposites. *J. Am. Chem. Soc.* **134**, 5362 (2012).

Galliker P., J. Schneider, H. Eghlidi, S. Kress, V. Sandoghdar, and D. Poulikakos. Direct printing of nanostructures by electrostatic autofocussing of ink nanodroplets. *Nat. Commun.* **3**, 890 (2012).

Gao X. and L. Jiang. Water-repellent legs of water striders. *Nature* **432**, 36 (2004).

Graeber G., O. B. M. Kieliger, T. M. Schutzius, and D. Poulikakos. 3D-printed surface architecture enhancing superhydrophobicity and viscous droplet repellency. *ACS Appl. Mater. Interfaces* **10**, 43275 (2018).

Gueymard C. A., D. Myers, and K. Emery. Proposed reference irradiance spectra for solar energy systems testing. *Sol. Energy* **73**, 443 (2002).

Guo P., Y. Zheng, C. Liu, J. Ju, and L. Jiang. Directional shedding-off water on natural/bio-mimetic taper-ratchet array surfaces. *Soft Matter* **8**, 1770 (2012).

Hashimoto K., H. Irie, and A. Fujishima. TiO_2 photocatalysis: A historical overview and future prospects. *Jpn. J. Appl. Phys.* **44**, 8269 (2005).

He M., Q. L. Zhang, X. P. Zeng, D. Cui, J. Chen, H. L. Li, J. J. Wang, and Y. L. Song. Hierarchical porous surface for efficiently controlling microdroplets' self-removal. *Adv. Mater.* **25**, 2291 (2013).

He M., X. Zhou, X. P. Zeng, D. P. Cui, Q. L. Zhang, J. Chen, H. L. Li, J. J. Wang, Z. C. Cao, Y. L. Song, and L. Jiang. Hierarchically structured porous aluminum surfaces for high-efficient removal of condensed water. *Soft Matter* **8**, 6680 (2012).

He M., Y. Ding, J. Chen, and Y. Song. Spontaneous uphill movement and self-removal of condensates on hierarchical tower-like arrays. *ACS Nano* **10**, 9456 (2016).

Hu D., B. Chan, and J. Bush. The hydrodynamics of water strider locomotion. *Nature* **424**, 663 (2003).

Hu H., G. Liu, and J. Wang. Preparation and comparison of NP-GLIDE, SLIPS, superhydrophobic, and other coatings from identical precursors at different mixing ratios. *J. Mater. Chem. A*, **7**, 1519 (2019).

Hwang G. B., A. Patir, E. Allan, S. P. Nair, and I. P. Parkin. Superhydrophobic and white light-activated bactericidal surface through a simple coating. *ACS Appl. Mater. Interfaces* **9**, 29002 (2017).

Jeong H., J. Heo, B. Son, D. Choi, T. H. Park, M. Chang, and J. Hong. Intrinsic hydrophobic cairnlike multilayer films for antibacterial effect with enhanced durability. *ACS Appl. Mater. Interfaces* **7**, 26117 (2015).

Jeong H., Y.-A. Cho, Y. Cho, E. Kang, H.-W. Ahn, and J. Hong. Durable urushiol-based nanofilm with water repellency for clear overlay appliances in dentistry. *ACS Biomater. Sci. Eng.* **2**, 344 (2016).

Jiang G., L. Luo, L. Tan, J. Wang, S. Zhang, F. Zhang, and J. Jin. Microsphere-fiber interpenetrated superhydrophobic PVDF microporous membranes with improved waterproof and breathable performance. *ACS Appl. Mater. Interfaces* **10**, 28210 (2018).

Jung Y. C. and B. Bhushan. *Langmuir* **24**, 6262 (2008).

Kim S. H., M. Kim, J. H. Lee, and S.-J. Lee. Self-cleaning transparent heat mirror with a plasma polymer fluorocarbon thin film fabricated by a continuous roll-to-roll sputtering process. *ACS Appl. Mater. Interfaces* **10**, 10454 (2018).

Kim T.-H., S.-H. Ha, N.-S. Jang, J. Kim, J. H. Kim, J.-K. Park, D.-W. Lee, J. Lee, S.-H. Kim, and J.-M. Kim. Simple and cost-effective fabrication of highly flexible, transparent superhydrophobic films with hierarchical surface design. *ACS Appl. Mater. Interfaces* **7**, 5289 (2015).

Kreder M. J., J. Alvarenga, P. Kim, and J. Aizenberg. Design of anti-icing surfaces: Smooth, textured or slippery? *Nat. Rev. Mater.* **1**, 15003 (2016).

Li B. and J. Zhang. Durable and self-healing superamphiphobic coatings repellent even to hot liquids. *Chem. Commun.* **52**, 2744 (2016).

Li H., W. Fang, Y. Li, Q. Yang, M. Li, Q. Li, X.-Q. Feng, and Y. Song. Spontaneous droplets gyrating via asymmetric self-splitting on heterogeneous surfaces. *Nat. Commun.* **10**, 950 (2019).

Li J., Y. Hou, Y. Liu, C. Hao, M. Li, M. K. Chaudhury, S. Yao, and Z. Wang. Directional transport of high-temperature Janus droplets mediated by structural topography. *Nat. Phys.* **12**, 606 (2016).

Liu Y., L. Moevius, X. Xu, T. Qian, J. M. Yeomans, and Z. Wang. Pancake bouncing on superhydrophobic surfaces. *Nat. Phys.* **10**, 515 (2014).

Liu Y., X. Chen, and J. H. Xin. Can superhydrophobic surfaces repel hot water? *J. Mater. Chem.* **19**, 5602 (2009).

Lu Y., S. Sathasivam, J. Song, C. R. Crick, C. J. Carmalt, and I. P. Parkin. Robust self-cleaning surfaces that function when exposed to either air or oil. *Science* **347**, 1132 (2015).

Lv C., P. Hao, Z. Yao, and F. Niu. Departure of condensation droplets on superhydrophobic surfaces. *Langmuir* **31**, 2414 (2015).

McGovern T. W. and T. M. Barkley. Botanical dermatology. *Int. J. Dermatol.* **37**, 321 (1998).

Mei H., D. Luo, P. Guo, C. Song, C. Liu, C. Y. Zheng, and L. Jiang. Multi-level micro-/nanostructures of butterfly wings adapt at low temperature to water repellency. *Soft Matter* **7**, 10569 (2011).

Mertaniemi H., R. Forchheimer, O. Ikkala, and R. H. A. Ras. Rebounding droplet-droplet collisions on superhydrophobic surfaces: From the phenomenon to droplet logic. *Adv. Mater.* **24**, 5738 (2012).

Minemawari H., T. Yamada, H. Matsui, J. Tsutsumi, S. Haas, R. Chiba, R. Kumai, and T. Hasegawa. Inkjet printing of single-crystal films. *Nature* **475**, 364 (2011).

Mishchenko L., B. Hatton, V. Bahadur, J. A. Taylor, T. Krupenkin, and J. Aizenberg. Design of ice-free nanostructured surfaces based on repulsion of impacting water droplets. *ACS Nano* **4**, 7699 (2010).

Naik A. J. T., S. Ismail, C. Kay, M. Wilson, and I. P. Parkin. Antimicrobial activity of polyurethane embedded with methylene blue, toluidene blue and gold nanoparticles against *Staphylococcus aureus*; illuminated with white light. *Mater. Chem. Phys.* **129**, 446 (2011).

Nishimura R., K. Hyodo, H. Sawaguchi, Y. Yamamoto, Y. Nonomura, H. Mayama, S. Yokojima, S. Nakamura, and K. Uchida. Fractal surfaces of molecular crystals mimicking lotus leaf with phototunable double roughness structures. *J. AM. Chem. Soc.* **138**, 10299 (2016).

Patankar N. Mimicking the lotus effect: Influence of double roughness structures and slender pillars. *Langmuir* **20**, 8209 (2004).

Ramakrishna S., D. Mathew, and K. S. Santhosh Kumar. Self-assembled and elastomeric arm decorated surfaces for high stress resistant super-repellent materials. *J. Mater. Chem. A*, **5**, 699 (2017).

Richard D., C. Clanet, and D. Quéré. Contact time of a bouncing drop. *Nature* **417**, 811 (2002).

Saito T., S. Kimura, Y. Nishiyama, and A. Isogai. Cellulose nanofibers prepared by TEMPO-mediated oxidation of native cellulose. *Biomacromolecules* **8**, 2485 (2007).

Schlaich C., L. Cuellar Camacho, L. Yu, K. Achazi, Q. Wei, and R. Haag. Surface-independent hierarchical coatings with superamphiphobic properties. *ACS Appl. Mater. Interfaces* **8**, 29117 (2016a).

Schlaich C., L. Yu, L. C. Camacho, Q. Wei, and R. Haag. Fluorine-free superwetting systems: Construction of environmentally friendly superhydrophilic, superhydrophobic, and slippery surfaces on various substrates. *Polym. Chem.* **7**, 7446 (2016b).

Scruton B., B. W. Robins, and B. H. Blott. The deposition of fingerprint films. *Appl. Phys.* **8**, 714 (1975).

Sehmi S. K., S. Noimark, J. C. Bear, W. J. Peveler, M. Bovis, E. Allan, A. J. MacRobert, and I. P. Parkin. Lethal photosensitisation of *Staphylococcus aureus* and *Escherichia coli* using crystal violet and zinc oxide-encapsulated polyurethane. *J. Mater. Chem. B* **3**, 6490 (2015).

Shen Y., J. Tao, Z. Chen, C. Zhu, G. Wang, H. Chen, and S. Liu. Rational design of the nanostructure features on superhydrophobic surfaces for enhanced dynamic water repellency. *ACS Sustainable Chem. Eng.* **6**, 9958 (2018).

Shen Y., Y. Xie, J. Tao, H. Chen, C. Zhu, M. Jin, and Y. Lu. Rationally designed nanostructure features on superhydrophobic surfaces for enhancing self-propelling dynamics of condensed droplets. *ACS Sustainable Chem. Eng.* **7**, 2702 (2019).

Si Y. and Z. Guo. Superhydrophobic nanocoatings: from materials to fabrications and to applications. *Nanoscale* **7**, 5922 (2015).

Siriviryanun A. and T. Imae. Anti-fingerprint properties of non-fluorinated organosiloxane self-assembled monolayer-coated glass surface. *Chem. Eng. J.* **246**, 254 (2014).

Song J., M. Gao, C. Zhao, Y. Lu, L. Huang, X. Liu, C. J. Carmalt, X. Deng, and I. P. Parkin. Large-area fabrication of droplet pancake bouncing surface and control of bouncing state. *ACS Nano* **11**, 9259 (2017a).

Song M., J. Ju, S. Luo, Y. Han, Z. Dong, Y. Wang, Z. Gu, L. Zhang, R. Hao, and L. Jiang. Controlling liquid splash on superhydrophobic surfaces by a vesicle surfactant. *Sci. Adv.* **3**, e1602188 (2017b).

Tadanaga K., N. Katata, and T. Minami. Super-water-repellent Al_2O_3 coating films with high transparency. *J. Am. Ceram. Soc.* **80**, 1040 (1997).

Thomas G. L. and T. E. Reynoldson. Some observation on fingerprint deposits. *J. Phys. D Appl. Phys.* **8**, 724 (1975).

Togasawa R., M. Tenjimbayashi, T. Matsubayashi, T. Moriya, K. Manabe, and S. Shiratori. A fluorine-free slippery surface with hot water repellency and improved stability against boiling. *ACS Appl. Mater. Interfaces* **10**, 4198 (2018).

Tuteja A., W. Choi, M. Ma, J. M. Mabry, S. A. Mazzella, G. C. Rutledge, G. H. McKinley, and R. E. Cohen. Designing superoleophobic surfaces. *Science* **318**, 1618 (2007).

Uchida K., N. Nishikawa, N. Izumi, S. Yamazoe, H. Mayama, Y. Kojima, S. Yokojima, S. Nakamura, K. Tsujii, and M. Irie. Photoinduced reversible formation of microfibrils on a photochromic diarylethene microcrystalline surface. *Angew. Chem. Int. Ed.* **49**, 5942 (2010); *Angew. Chem.* **122**, 6078 (2010).

Uhlmann P., R. Frenzel, B. Voit, U. Mock, B. Szyszka, B. Schmidt, D. Ondratschek, J. Gochermann, and K. Roths. Research agenda surface technology: Future demands for research in the field of coatings materials. *Process Org. Coat.* **58**, 122 (2007).

Urata C., B. Masheder, D. F. Cheng, and A. Hozumi. A thermally stable, durable and temperature-dependent oleophobic surface of a polymethylsilsesquioxane film. *Chem. Commun.* **49**, 3318 (2013).

Vasileiou T., J. Gerber, J. Prautzsch, T. M. Schutzius, and D. Poulikakos. Superhydrophobicity enhancement through substrate flexibility. *Proc. Natl. Acad. Sci.* USA **113**, 13307 (2016).

Wang W., K. Lockwood, L. M. Boyd, M. D. Davidson, S. Movafaghi, H. Vahabi, S. R. Khetani, and A. K. Kota. Superhydrophobic coatings with edible materials. *ACS Appl. Mater. Interfaces* **8**, 18664 (2016).

Watanabe H., A. Fujimoto, and A. Takahara. Characterization of catechol-containing natural thermosetting polymer "Urushiol" thin film. *J. Polym. Sci., Part A: Polym. Chem.* **51**, 3688 (2013).

Wei J., G. Zhang, J. Dong, H. Wang, Y. Guo, X. Zhuo, C. Li, H. Liang, S. Gu, C. Li, X. Dong, and Y. Li. Facile, scalable spray-coating of stable emulsion for transparent self-cleaning surface of cellulose-based materials. *ACS Sustainable Chem. Eng.* **6**, 11335 (2018).

Wei Q., K. Achazi, H. Liebe, A. Schulz, P. L. Noeske, I. Grunwald, and R. Haag. Mussel-inspired dendritic polymers as universal multifunctional coatings. *Angew. Chem. Int. Ed.* **53**, 11650 (2014a).

Wei Q., T. Becherer, P.-L. M. Noeske, I. Grunwald, and R. Haag. A universal approach to crosslinked hierarchical polymer multilayers as stable and highly efficient antifouling coatings. *Adv. Mater.* **26**, 2688 (2014b).

Wu L., Z. Dong, M. Kuang, Y. Li, F. Li, L. Jiang, and Y. Song. Printing patterned fine 3D structures by manipulating the three-phase contact line. *Adv. Funct. Mater.* **25**, 2237 (2015).

Xu L., X. Yao, and Y. Zheng. Direction-dependent adhesion of water strider's legs for water-walking. *Solid State Sciences* **14**, 1146 (2012).

Zheng X., J. Weng, Q. Huang, B. Hu, T. Qiao, and P. Deng. Fabrication of a stable poly(vinylpyrrolidone)/poly-(urushiol) multilayer ultrathin film through layer-by-layer assembly and photo-induced polymerization. *Colloids Surf. A* **337**, 15 (2009).

Zheng Y. *Bio-inspired Wettability Surfaces: Developments in Micro- and Nanostructures*, Pan Stanford Publishing, 2015, 0-216.

Zheng Y., D. Han, J. Zhai, and L. Jiang. In situ investigation on dynamic suspending of microdroplet on lotus leaf and gradient of wettable micro- and nanostructure from water condensation. *Appl. Phys. Lett.* **92**, 084106 (2008).

Zheng Y., X. Gao, and L. Jiang. Directional adhesion of superhydrophobic butterfly wings. *Soft Matter* **3**, 178 (2007).

Zhu T., C. Cai, J. Guo, R. Wang, N. Zhao, and J. Xu. Ultra water repellent polypropylene surfaces with tunable water adhesion. *ACS Appl. Mater. Interfaces* **9**, 10224 (2017).

Chapter 3

Water-Collecting Surfaces

In this chapter, biological water-collecting surfaces in nature are introduced based on mechanism, for instance the cooperation of roughness and curvature on wetted spider silk, superhydrophobic/hydrophilic patterns on beetle backs, the geometric gradient of cactus spines, and hierarchical microchannels for water transport on pitcher plants. First, bioinspired spindle-knot structured fibers are introduced, including methods of spindle-knot fiber fabrication, large droplets by spindle-knots, water collection by spindle-knots, flexible spindle-knots for droplet capture and release, gradient-sized spindle-knots on fibers, large-scale cavity microfibers with spindle-knots, wet assembly of spindle-knots, stimuli-responsive fibers, tunable-wettability interfaces, and digital control coded microfibers. Second, bioinspired cactus spines for fog-water harvesting are introduced, including heterogeneous rough conical wires, conical spines with micro-/nanostructures, cactus-stem-inspired cone-arrayed surfaces, and nanocones on 3D fiber networks. Third, a bioinspired beetle back for fog collection is introduced, including hydrophilic–superhydrophobic patterned surfaces, star-shaped wettable patterns, hybrid membranes with asymmetric microtopology, slippery asymmetric bumps, and inkjet printing for direct micropatterns. Finally, water collection engineering and applications are introduced, including metal-foam

Bioinspired Materials Surfaces: Nature-Inspired Micro-/Nanostructures
Yongmei Zheng

ISBN 978-981-5129-34-2 (Hardcover), 978-1-003-51388-9 (eBook)
www.jennystanford.com

structured surfaces, cooling tower fog harvesting in power plants, metal–organic framework (MOF) materials, electrostatic fog-water collection, and patterned nanobrushes.

3.1 Biological Water-Collecting Surfaces in Nature

Water harvesting by novel materials and surfaces is significant for the lives of human beings (Vince, 2010; Domen et al., 2016). According to the United Nations (UN), more than 700 million people in the world suffer from water scarcity due to the barren regions in some countries. By 2025, this number will almost triple. In 2005, the UN launched a 10-year action named "Water for Life," aimed at solving the problem of limited water access in numerous regions worldwide. One of the goals of this action is to "strengthen international cooperation in the fields of technologies for enhanced water productivity, financing opportunities, and an improved environment." With large regions having little rainfall and a changing global climate, looking for new routes to collect water is one of the most important global challenges (Domen et al., 2016; Pinchasik et al., 2016). In the absence of sufficient precipitation, collecting water from fog or oversaturated vapor is a promising route due to its low cost and energy loss. Natural fog or oversaturated vapor includes larger water molecules and water droplets with high humidity (relative humidity (RH) of 90~100%) but has not yet been used effectively. How can water be harvested from fog or oversaturated vapor? Could novel materials or surfaces be used to achieve this task?

Interestingly, biological surfaces in nature have evolved functions for harvesting water from air, such as spider silks (Zheng et al., 2010), beetle backs (Parker and Lawrence, 2001), cactus spines (Ju et al., 2012), and pitcher plants (Sarracenia) (Chen et al., 2018). Spider silk is composed of random nanofibrils and aligned nanofibrils when wetted for water collection (Fig. 3.1a–c). The beetle back has a wax-free peak and microhumps surrounded by a superhydrophobic region (Fig. 3.1d–f). Cactus spines have overlapping scales and conical geometry (Fig. 3.1g–i). Pitcher plants have spines with multilevel micro-/nanogrooves (Fig. 3.1j–l). These biological surfaces have inspired the design of novel materials and surfaces with promise for future use in water engineering. Bioinspired surfaces and materials have been achieved at the micro- and nanolevel, as well as one-, two-, and three-dimensional materials in recent years. Three design aspects are considered to contribute to efficient fog harvesting: (1) enhanced condensation, (2) control of lateral adhesion, and (3) guided transport. Thus, the chemistry and physics of material surfaces should be scientifically designed to integrate properties and functional control.

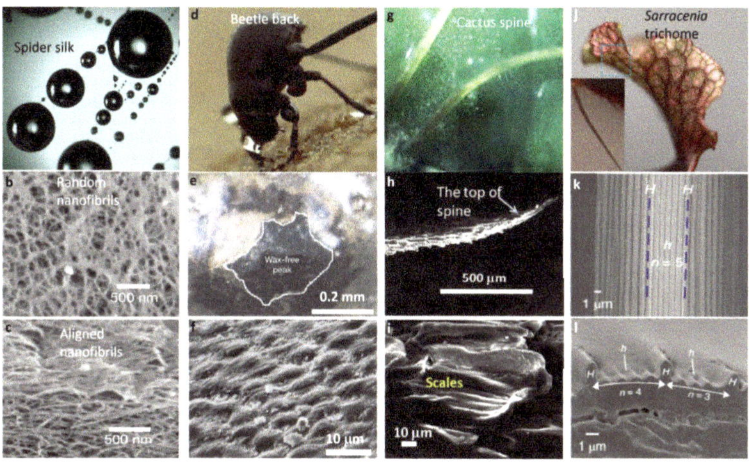

Figure 3.1 Multistructured biological surfaces with water-harvesting abilities. (a–c) Spider silk composed of random and aligned nanofibrils (Zheng et al., 2010). (d–f) Beetle back with wax-free peak and microhumps (Parker and Lawrence, 2001). (g–i) Cactus spine with overlapping scales and conical geometry (Ju et al., 2012). (j–l) Pitcher plant spine with multilevel micro-/nanogroove (Chen et al., 2018).

3.1.1 Cooperation of Roughness and Curvature

Cooperation between roughness and curvature is necessary to propel water transport during water harvesting. *Uloborus walckenaerius* capture silk has attracted much attention for its wet-rebuilt structure and directional water collection (Zheng et al., 2010). The wet-rebuilt structures consist of random nanofibrils (Fig. 3.1b) and aligned nanofibrils (Fig. 3.1c) on the surface of wetted spider silk. Such wet-rebuilt structures generate cooperation between roughness and curvature, which reveals an important mechanism of directional water collection.

In this mechanism of water collection, both the surface energy gradient and Laplace pressure can generate a cooperative driving force (Fig. 3.2) to act on the water droplet for directional movement. In detail, on wetted spider silk, the spindle-knots and joints are composed of random nanofibrils and aligned nanofibrils that can form discontinuous and continuous three-phase contact lines (TCLs), respectively. Continuous TCLs along aligned nanofibrils (at the joint) will facilitate drop spreading and movement, whereas discontinuous TCLs along random nanofibrils (at the spindle-knot) halt the drops (Fig. 3.2a). The wettability difference between the spindle-knot and joint generates a surface energy gradient resulting in an apparent surface energy gradient along the silk.

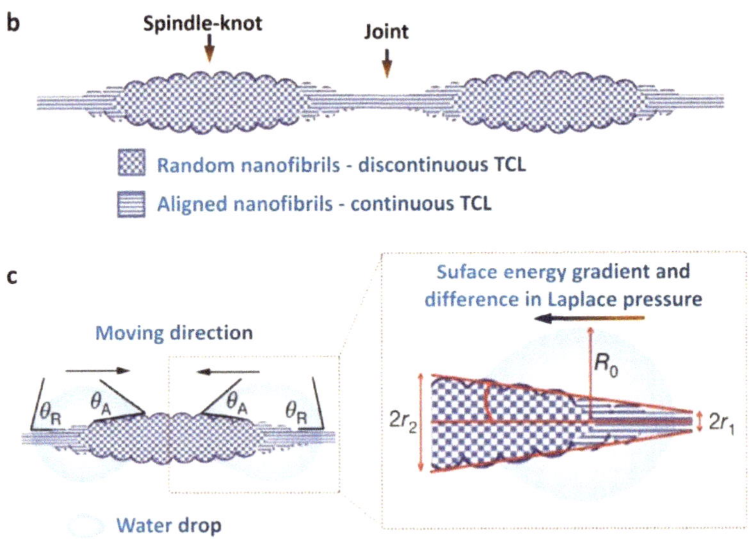

Figure 3.2 Mechanism of water collection on wetted spider silk, along with cooperative driving forces including two components: (a) force from the surface energy gradient between the spindle-knot and the joint and (b) force from Laplace pressure produced by the curvature difference between the spindle-knot and the joint (Zheng et al., 2010).

The roughness difference between the spindle-knot and joint leads to a surface energy gradient, driving droplets to move from joint to spindle-knot (more hydrophilic). The driving force dF can be described as follows (Chaudhury and Whitesides, 1992; Daniel et al., 2001):

$$dF = \int_{L_j}^{L_k} \gamma(\cos\theta_A - \cos\theta_R)dl \tag{3.1}$$

where γ is the surface tension of water and θ_A and θ_R represent the advancing and receding water contact angles on spider silk. The surface energy gradient resulting from roughness will drive water from the less hydrophilic region (joint) to the more hydrophilic region (spindle-knot).

The second driving force from the "joint" toward the "spindle-knot" is the gradient of Laplace pressure. As illustrated in Fig. 3.2b, a fiber with spindle-knots can be treated as two oppositely curved and joined conical objects. Each conical object induces a difference in Laplace pressure (ΔP) functioning on the water drop as follows (Lorenceau and Quéré, 2004):

$$\Delta P_{\text{curvature}} = \int_{r_1}^{r_2} \frac{2\gamma}{(R+R_0)^2} \sin\beta dz \tag{3.2}$$

where R is the local radius of the fiber, R_0 is the drop radius ($R_0 = (3\text{ V}/4\pi)^{1/3}$), V is the drop volume, β is the half apex angle of the spindle-knot, and z is the integrating variable along the diameter of the spindle-knot. The Laplace pressure on the high-curvature position (the "joint" with local radius r_1) is greater than that on the low-curvature position (the "spindle-knot" with local radius r_2) because r_1 is smaller than r_2, which results in a difference in the Laplace pressure acting to move the water drop from the "joint" toward the "spindle-knot." It is noted that the Laplace pressure and surface gradient act cooperatively as driving forces for directional water movement from joint to spindle-knot along spider silk.

3.1.2 Superhydrophobic/Hydrophilic Patterns

Superhydrophobic/hydrophilic patterns are another model of water harvesting, observed on beetle backs. Some beetles in the Namib Desert collect drinking water from fog-laden wind on their backs (Parker and Lawrence, 2001), which consist of alternating hydrophobic, wax-coated and hydrophilic, nonwaxy regions (Hamilton and Seely, 1976). For example, the Namib Desert in southern Africa supports a unique sand-dune fauna (Konh, 1962) and experiences high winds, extreme daytime temperatures, and dense early morning fog (Roberts et al., 1991), but rainfall is little (Louw, 1972). The tenebrionid beetle *Stenocara* sp. tilts its body to face the wind to collect water (Watanable, 2000). Interestingly, there exists a unique control mechanism to collect the fog-water droplets and avoid loss (Machin, 1970) by means of alternating high/low surface energy regions of the elytra. Figure 3.1d shows a photo of the desert beetle *Stenocara* sp., whose elytra are covered with bumps 0.5–1.5 mm in height and ~0.5 mm in diameter. The peaks of these bumps are smooth in the wax-free region, but the valleys and sides of the humps are covered with a waxy microstructure consisting of flattened hemispheres (Fig. 3.1e,f). The hydrophilic (i.e., wax-free) region gathers a water droplet, and the superhydrophobic region repels the droplet. The cooperation of these contrasting hydrophilic/superhydrophobic regions and the wind velocity allows the surface of the beetle elytra to capture fog water for drinking (as illustrated in Fig. 3.3).

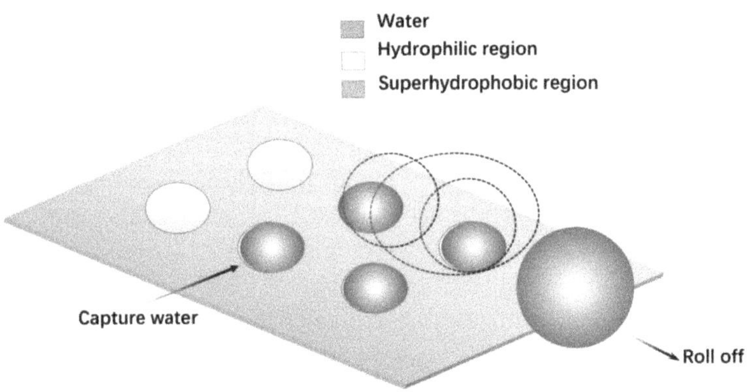

Figure 3.3 Illustration of the mechanism of the desert beetle elytra with hydrophobic/hydrophilic patterns for fog harvesting.

3.1.3 Geometric Gradient of the Cactus Spine

The geometric gradient of the conical spine is the third model of water harvesting, which is observed in natural cactus spines. Cactus species in deserts have many needle-like spines that enable fog-water harvesting. For instance, the cactus *Opuntia microdasys* in the Chihuahua Desert has an integrated multifunctional system that facilitates efficient fog collection [Ju et al., 2012]. This unique fog-collection system consists of integrated multilevel surface structures, including the geometric cone of the spine, ratchets along the surface of the spine, and hydrophilic trichomes (Fig. 3.4a–c). Scanning electron microscopy (SEM) of a single spine shows the tip containing oriented barbs, the middle containing gradient grooves, and the base containing belt-structured trichomes (Fig. 3.4d). The spine has an apex angle (2α) of ~12.3° (Fig. 3.4e). Such multistructures, including typical geometric gradients, play an important role in fog harvesting.

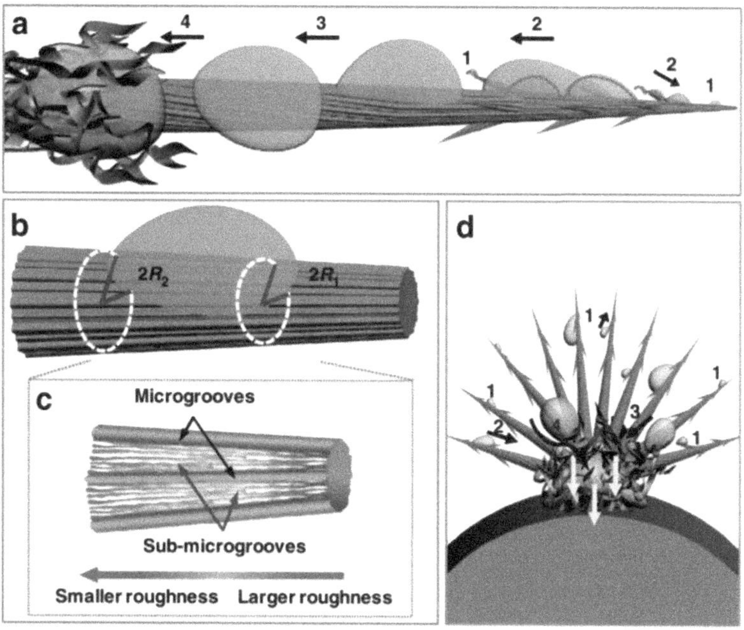

Figure 3.4 (a) Illustration of cactus spine with a fog-harvesting system via cooperation of multiconical geometries, where the droplets coalesce and transport as follows: droplet 1 → 2 → 3 → 4. (b–c) The curvature radii are different at different locations along the conical grooves (b), along with the roughness gradient. (c) Fog-harvesting system of cactus spines, including trichomes (Ju et al., 2012).

Based on the multilevel geometric structures, the driving forces result from the gradient of Laplace pressure and the gradient of surface free energy, which can act on condensed water droplets during water harvesting. The curvature of the conical shape generates a Laplace pressure difference (ΔP) between the two opposite sides of the droplet (Lorenceau and Quéré, 2004), which can be expressed as follows:

$$\Delta P = -\int_{R_1}^{R_2} \frac{2\gamma}{\left(R+R_0\right)^2} \sin\alpha dz \tag{3.3}$$

where R is the local radius of the spine (R_1 and R_2 are the local radii of the spine at the two opposite sides of the droplet), γ is the surface tension of water, R_0 is the drop radius, α is the half apex angle of the conical spine, and dz is the incremental radius of the spine.

In addition, the surface energy gradient and gradient groove contribute to fog harvesting. By cooperation among the multiple spines, the multiple trichomes, and the spine trichomes (Fig. 3.4f), the water droplets undergo processes of (1) deposition, (2) collection, and (3) transportation and then subsequently are quickly absorbed into the cactus stem by the trichomes (4, absorption). This water-harvesting system composed of spine trichomes inspired the fabrication of a novel device for application in fog-water engineering.

3.1.4 Hierarchical Microchannels for Water Transport

Water transport on a surface plays an important role in gathering water for collection. The surface of a Sarracenia trichome provides ultrafast water transport for droplets (Chen et al., 2018). The high velocity of water transport is attributed to the unique hierarchical microchannel organization of the trichome. There are two types of ribs with different heights regularly distributed around the trichome cone, where two neighboring high ribs form a large channel that contains 1–5 low ribs (defined as smaller base channels). This configuration results in two successive but distinct modes of water transport. Initially, a thin film of water forms rapidly inside the base channels, which is followed by ultrafast water sliding on top of the thin film.

Such a mechanism can inspire the design of microchannels for microfluidic applications. As shown in Fig. 3.5a, the Sarracenia trichome can be depicted as a conical spine, along with an extended structure region marked with colored square and the letters b, c, d, e from top to bottom. The sectional views are marked $b_\perp$, $c_\perp$, $d_\perp$, $e_\perp$, and the corresponding SEM images are at the right, as shown in Fig. 3.5b–e, where the special high–low channels form along the spine of the Sarracenia trichome from top to bottom.

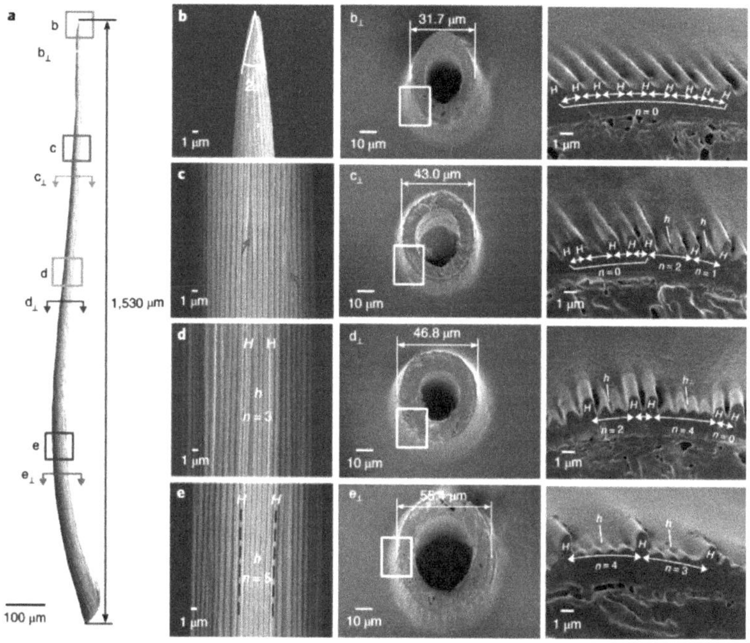

Figure 3.5 (a) Optical image of the surface of a spine of Sarracenia trichome. (b–e) Appearance and surface structure of the Sarracenia trichome, with multilevel microgrooves along the conical geometry of the spine (Chen et al., 2018).

3.2 Bioinspired Spindle-Knot Structured Fibers

3.2.1 Methods of Spindle-Knot Fiber Fabrication

Spider silk generates integrated wet-built nanofibril structures for collecting water, and the roughness and curvature structure are illustrated by the spindle-knot and joint on wetted spider silk, which form a cooperative driving force (e.g., differences in the Laplace and surface energy gradients, both from curvature and roughness between the spindle-knot and joint) to act on droplets. This indicates that the spindle-knot and joint are necessary for the task of water collection.

Recently, some methods have reported the design of biomimetic multilevel spindle-knot structures on fibers by mimicking rough curvature features, such as spindle-knot, hump, and bead on string. Various methods and techniques have been developed to fabricate rough spindle-knots similar to those of spider silk, such as dip-coating (Fig. 3.6a), wet assembly (Fig. 3.6b), fluid coating (Fig. 3.6c), electrospinning (Fig. 3.6d), and microfluidics (Fig. 3.6e) (Chen and Zheng, 2014).

Figure 3.6a illustrates the dip-coating method for the fabrication of spindle-knot fibers. As a nylon fiber is drawn out fast and horizontally from a polymer solution, the polymer film is formed on the fiber and subsequently breaks up into polymer droplets. After the drying of the polymer droplets, spindle-knots form along the fiber. Figure 3.6b illustrates the wet-assembly method for humped structured fabrication. In this method, the formation of fog, including nanoparticles, in the humidity of the environment should be well controlled to induce water condensation, including nanoparticles, on thin uniform fibers (e.g., submicrofibers). With the formation of the as-condensed droplets, the nanoparticles in droplets assemble into periodic humps. After drying, the humps form spindle-knots along the fibers. Figure 3.6c illustrates the fluid-coating method, which can continuously fabricate spindle-knot fibers by controlling the coating velocity. Figure 3.6d illustrates the coaxial electrospinning approach, including a spinnable inner fluid with high viscosity and a sprayable outer fluid with low viscosity. Thus, heterostructured bead-on-string fibers can be fabricated. The electrospinning method was developed for the one-step fabrication of bead-on-string fibers. A microfluidic method can be used to produce spindle-knot structures by designing multiple components to mimic the spinning process of spiders by controlling the fluid valve, as shown in Fig. 3.6e.

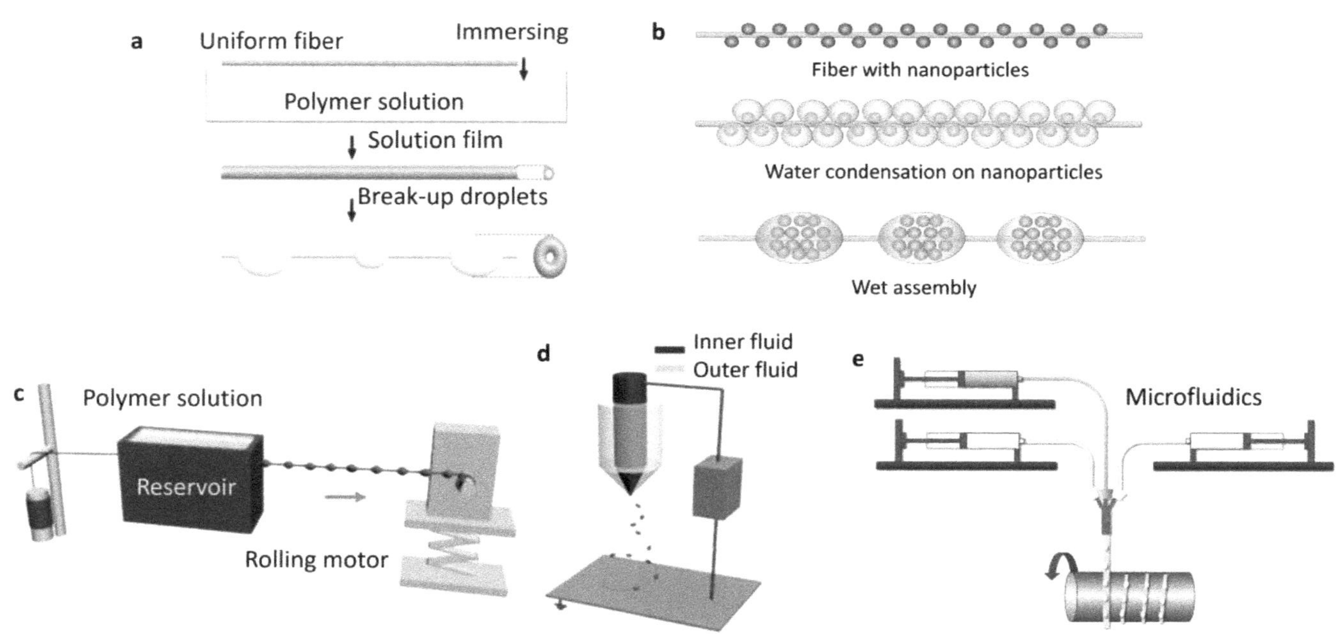

Figure 3.6 Illustration of methods to fabricate spindle-knot fibers: (a) dip-coating, (b) wet assembly, (c) fluid coating, (d) electrospinning, (e) microfluidics (Chen and Zheng, 2014).

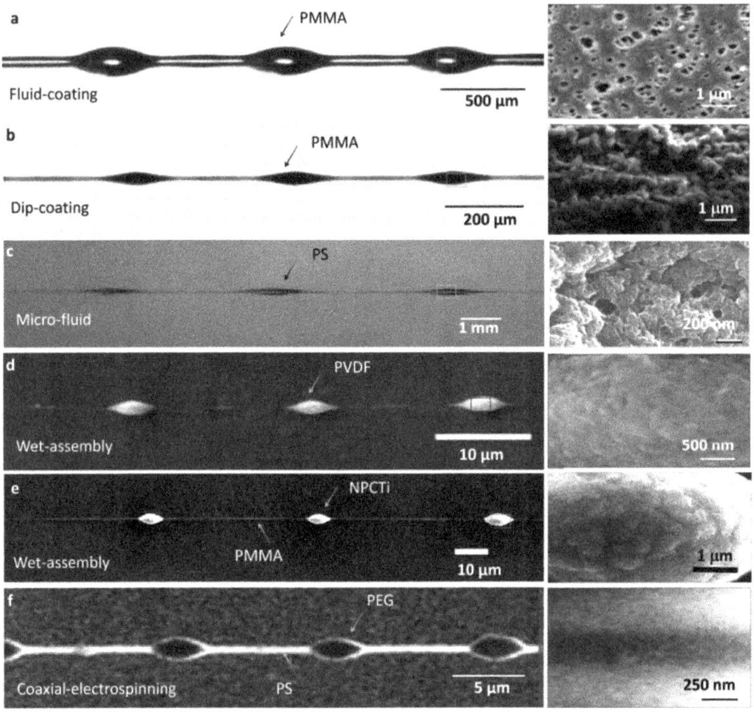

Figure 3.7 Various-sized rough spindle-knot fibers fabricated from polymers (PMMA, PS, PVDF, PEG, etc.) via the following methods: (a) fluid coating, (b) dip-coating, (c) microfluidics, (d–e) wet assembly, (f) coaxial electrospinning.

Various bioinspired spindle-knot fibers at the micro- and nanolevels can be fabricated, as shown in Fig. 3.7. In fluid coating, a polymethyl methacrylate (PMMA) solution is used to obtain polymer droplets. After drying and phase separation, the polymer droplets are shrunk into spindle-knots with a porous surface (Fig. 3.7a). The dip-coating forms the spindle-knot fiber (Fig. 3.7b). The small diameter of the main fiber results in a small spindle-knot, with a smaller diameter than the main fiber. The microfluidics method can form spindle-knot fibers by using polystyrene (PS) solution (Fig. 3.7c). The wet-assembly method can be used to assemble various nanoparticles together to form humps, such as NPCTi from hydrolysis of CCl_4, PMMA or polyvinylidene fluoride (PVDF) nanoscale lamellae in wet assembly during water condensation from humidity (Fig. 3.7d,e). The coaxial electrospinning method can produce a hydrophilic bead and hydrophobic string, e.g., polyethylene glycol (PEG) bead-on-PS string (Fig. 3.7f). After fabrication, the rough curvature gradient feature forms a joint and spindle-knot (Bai et al., 2012), which can be observed by atomic force microscopy (AFM), as shown in Fig. 3.8. More roughness is formed at the spindle-knot site, and less roughness is formed at the joint site (Fig. 3.8e–g), which is consistent with the SEM image (Fig. 3.8a–c).

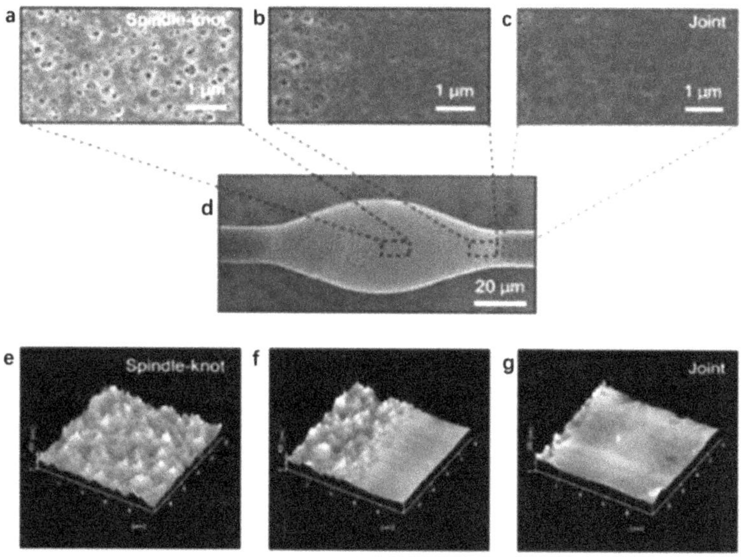

Figure 3.8 (a–d) SEM images of spindle-knot with gradient roughness along the surface from spindle-knots to joints. (e–g) AFM image of the surface from spindle-knot to joint, distinctly showing the roughness gradient (Bai et al., 2012).

3.2.2 Large Droplet Formation by Spindle-Knots

A larger droplet can be hung on a spindle-knot fiber than on a uniform fiber. The uniform fiber allows relatively small droplets to detach easily. Interestingly, the spindle-knot generates a slope and curvature effect (Tian et al., 2011) and enlarges the solid–liquid–gas TCL to pin a relatively large droplet on the spindle-knot fiber. The curvature of the spindle-knot controls the droplet receding angle (θ_r) (Fig. 3.9a, right) and is related to the height (H) and semi-apex angle

(α) of the spindle-knots, usually defined as the "curvature effect" in droplet pinning. The TCL on the spindle-knot fiber is stable, while the TCL on the uniform fiber is unstable (Fig. 3.9b); thus, the droplet is hung in different ways.

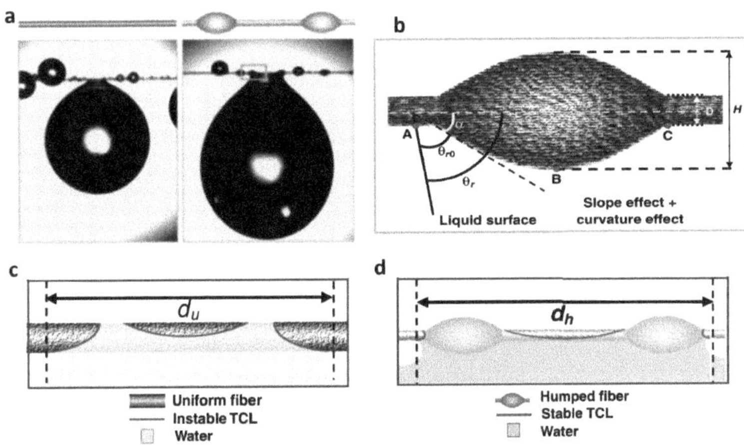

Figure 3.9 (a) Comparison of hanging droplets on uniform fiber (left) and spindle-knot fiber (middle). The geometry of the spindle-knot generates slope and curvature effects (right) to achieve a larger hanging droplet (middle). (b) Different three-phase contact lengths for droplets on uniform fibers (left) with unstable TCLs and spindle-knot fibers (right) with stable TCLs (Tian et al., 2011).

Figure 3.10 shows the in situ observation of the spindle-knot pinning of large-scale hanging droplets. The growing droplet is stably pinned by the spindle-knot, and the TCL is stable during droplet growth (Fig. 3.10a). However, the TCL is unstable and moves along the uniform fiber during droplet growth, and the droplet can easily detach even at a small volume (Fig. 3.10b). To describe the pinning cases at solid–liquid contact interfaces between fiber and droplet, the contact lengths of solid–liquid interfaces can be defined as L_B for the spindle-knot fiber and L_N for the uniform fiber. The spindle-knot structure on the fiber enlarges the contact length for the pinning of the droplet, which is greater than that of the uniform droplet, i.e., $L_B > L_N$; as a result, the spindle-knot fiber hangs a large droplet with greater volume than that of the uniform fiber (Chen et al., 2014).

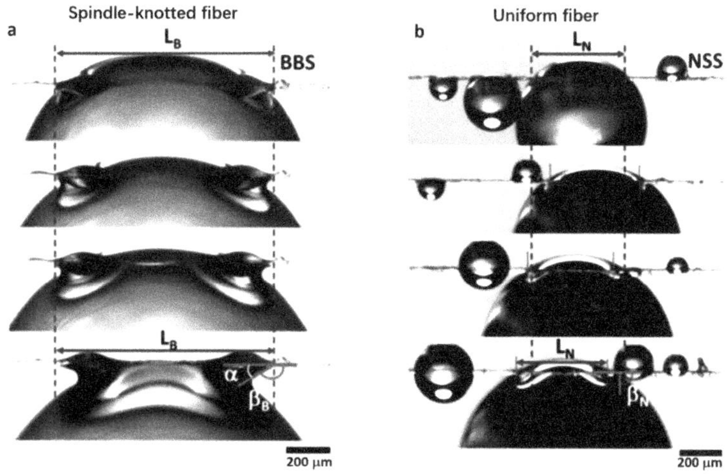

Figure 3.10 Difference in hanging droplets during droplet growth. (a) Stable TCL of spindle-knot fiber for a large hanging droplet with contact length L_B. (b) Unstable TCL of uniform fiber during droplet growth, with contact length L_N (Chen et al., 2014).

3.2.3 Water Collection by Spindle-Knots

The effectiveness of water collection is related to the sizes of spindle-knots. The different sizes of spindle-knots can be used to collect water under different fog flow rates, as shown in Fig. 3.11. The spindle-knots can be selected at volumes (V_{knot}) of 0.12 nL, 0.08 nL, 0.03 nL, and smooth (inset of Fig. 3.11a). For the large spindle-knot, the fiber would achieve the highest volume of as-collected water at a fog flow velocity of ~75 cm/s (Bai et al., 2011). The maximum volume of water at a fog flow velocity of ~75 cm/s is ~35 nl, achieved in 12 s. Different fog flow velocities, such as low (~25 cm/s), middle (~75 cm/s), and high (~100 cm/s), can be used to achieve different as-collected volumes of water droplets (Fig. 3.11b). The maximum volume of the water droplet on the spindle-knot can reach ~0.83 nL within ~7.2 s at low fog flow velocity. The water volume follows the relationship $V \sim t^3$. The volume of the collected water droplet is approximately 15 times that of the spindle-knot itself.

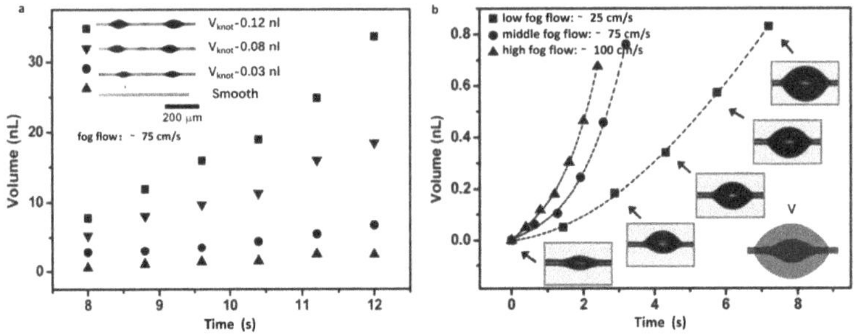

Figure 3.11 (a) Relationship of water collection in volume versus time for different sized spindle-knots (V_{knot} = 0.12 nl, 0.08 nl, 0.03 nl, and smooth). (b) Relationship of water collection in volume versus times for different flow velocity rates (V = 25, 75, 100 cm/s) (Bai et al., 2011).

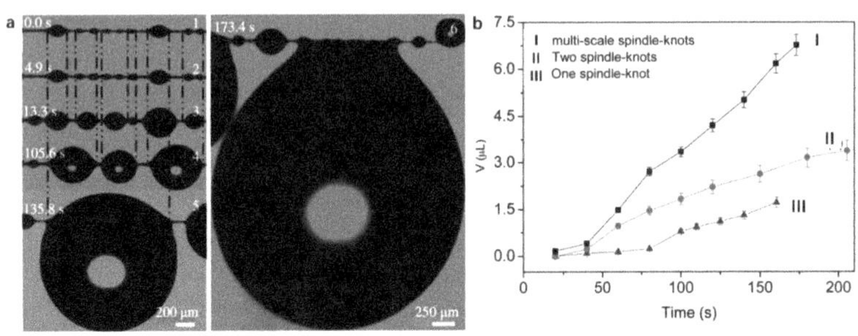

Figure 3.12 (a) Multilevel spindle-knots for strong water droplet hanging ability. (b) Water volumes collected by multiscale spindle-knots (Frame I), two spindle-knots (Frame II), and one spindle-knot (Frame III) (Hou et al., 2012).

Multilevel spindle-knots on fiber introduce stronger water-collection ability. Figure 3.12a shows optical images from in situ observations of water collection via multiscale spindle-knots on bioinspired fibers. Water condensation appears on different spindle-knots, along with droplet growth and coalescence. The droplets on small spindle-knots coalesce toward the large spindle-knots. Although the distance between the large spindle-knots is long, the small spindle-knot offers a bridge for transport to larger spindle-knots. Finally, the water drop detaches from two large spindle-knots in 174.3 s, forming a much larger hanging droplet. Figure 3.12b compares the water droplet hanging ability of different numbers of spindle-knots (fiber length of 1 mm, humidity of 90–95%). The multiscale spindle-knots on fiber achieve the maximal hanging drop volume of ~6.76 μL, which takes 173 s before detachment (Frame I). Two spindle-knots on a fiber can achieve a maximal hanging drop volume of ~3.38 μL with more than 205 s before detachment. One spindle-knot on a fiber may have a maximal hanging drop volume of ~1.71 μL with ~160 s before detachment (Hou et al., 2012). This study indicates that fibers with multiscale spindle-knots display more efficient water collection and a larger hanging ability.

3.2.4 Flexible Spindle-Knot for Droplet Capture and Release

Elastic bioinspired fibers (EBFs) are interesting because they can release as-pinned droplets from fibers via force effects. For example, bioinspired fibers with a polydimethylsiloxane (PDMS) spindle-knot were prepared by immersing elastic nylon fiber specimens in PDMS/curing agent/methylbenzene solution (10:1:2, PDMS:curing agent:methylbenzene, mass proportion) (Wang et al., 2015a) and then drawing them out horizontally at a velocity of 150 mm/min. A thin polymer solution film coated on the original nylon fiber broke into a series of polymer drops as a result of Rayleigh instability (Rayleigh, 1878). Thus, periodic spindle-knots formed to achieve EBFs after the polymer drops solidified at 70°C for 12 h.

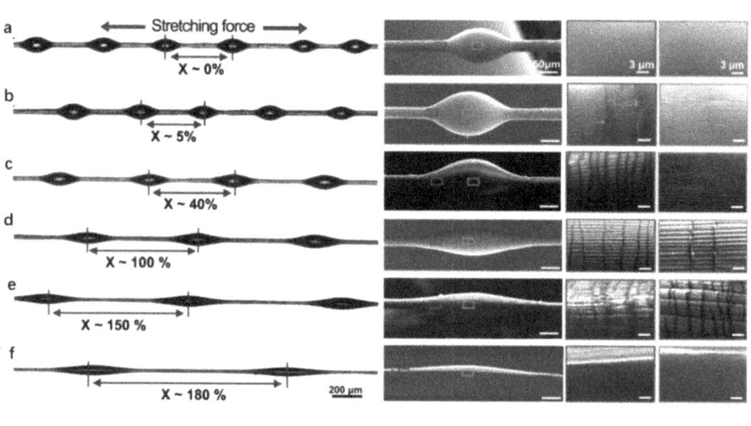

Figure 3.13 The change in surface roughness due to fiber stretching: X = 0% (a), 5% (b), 40% (c), 100% (d), 150% (e), and 180% (f). The right insets are SEM images of the corresponding structural change with different stretching ratios (Wang et al., 2015a).

EBFs at different stretching ratios are made by applying a certain tensile force. The stretching ratio X is calculated as follows: X = (m/m_0) – 1 (where m denotes the periodicity length of spindle-knots after stretching, and m_0 denotes the original periodicity length of the spindle-knots). Figure 3.13 shows surface changes in the stretching ratio (X) of 0%, 5%, 40%, 100%, 150% to 180%, and the height, length, and periodicity of the spindle-knot are ≈90, ≈240, ≈500 to ≈75, ≈510, and ≈1380 μm, respectively. The half apex angles change from 29° to 11° (Wang et al., 2015a).

The roughness of the spindle-knot surface affects the hanging of droplets on the fibers. As shown in Fig. 3.14a, the different values of X alter the maximum hanging droplet formed in different times: for X = 0, 30, 50, and 105%, the largest droplet is formed in 170 s, 184 s, 194 s, and 217 s, respectively, along with the tendency of increasing surface roughness. As X changes from high to low, the as-attached droplet can be released to detach from the surface of the fiber. As shown in Fig. 3.14b, the droplet can be released from the fiber after a change in X: from 105% to 44% for 4.0 μL, to 58% for 4.5 μL, to 61% for 5.5 μL, and to 76% for 6.5 μL.

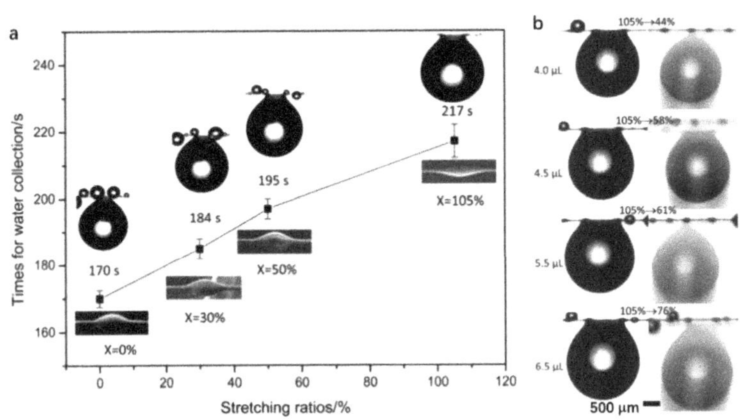

Figure 3.14 (a) Relationship of water collection and stretching ratios versus time. (b) Optical images showing stretching change for hanging-droplet attachment and detachment (Wang et al., 2015a).

3.2.5 Gradient-Sized Spindle-Knots on Fibers

The gradient is an important concept as the origin of the driving force. The existence of gradients on fibers generates the driving force, such as gradient-sized spindle-knots (from small to large arrangement), in addition to the roughness and curvature of every spindle-knot. A large spindle-knot increases the pinning force for a water droplet compared to that of a small spindle-knot.

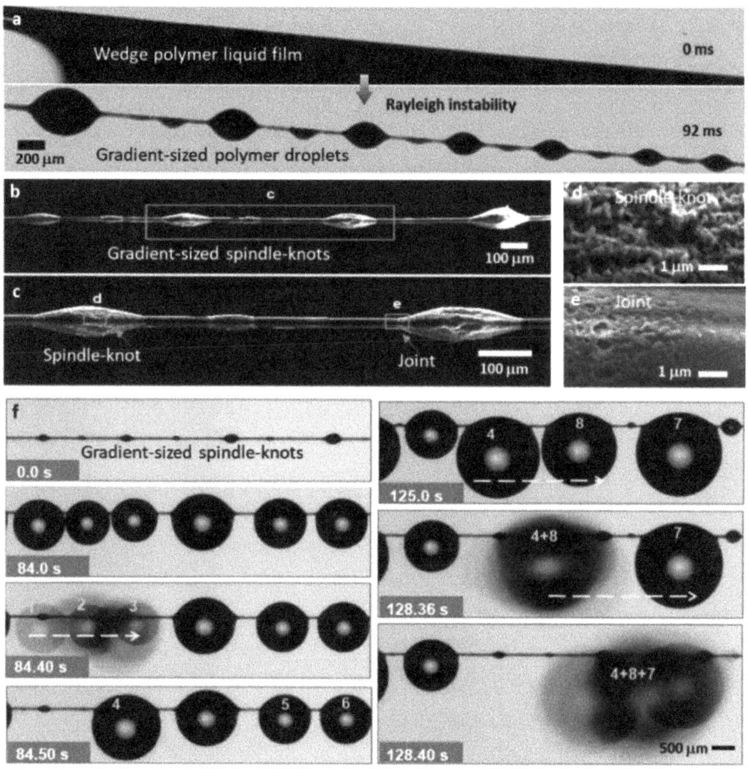

Figure 3.15 (a) Optical images of the fabrication process from wedge polymer liquid film to gradient-sized polymer droplets via Rayleigh instability. (b–e) SEM images of gradient-sized spindle-knots (b–c) and gradient roughness between spindle-knot (d) and joint (e). (f) Water droplets coalesce and transport directionally (see yellow arrows) (Chen et al., 2013).

Such gradient-sized spindle-knots can be achieved on fibers via fabrication methods such as tilt-angle dip-coating (Chen et al., 2013) and changeable velocity in fluid coating (Xue et al., 2014). For example, fabricated fibers can transport droplets along fibers in the direction of the gradient. In a gradient spindle-knot fiber formed by tilt-angle dip-coating, the main fiber is a uniform nylon fiber (~18 mm), and the polymer solution is polyvinylidene fluoride (PVDF, Mw

55 30 000, Aldrich) dissolved in N,N-dimethylformamide (DMF). The main component of the spindle-knot is PVDF, and the surface has gradient porosity from the center to the joint of every spindle-knot (Fig. 3.15a–e). The resulting fiber exhibits microdrop transport over a long distance (Fig. 3.15f); for example, a microdrop of ~0.25 μL travels a distance of ~5.00 mm with a velocity of 0.10–0.22 m/s. This is attributed to the multilevel cooperation of the release energy of drop coalescence along the gradient spindle-knots, in addition to capillary adhesion force and the continuous difference in Laplace pressure; accordingly, water drops are driven to move rapidly and directionally over a long distance (Chen et al., 2013; Xue et al., 2014).

A changeable velocity in the fluid-coating method can also form gradient-sized spindle-knots on the fiber. Changing the fluid-coating velocity can produce different modes of gradients, such as unidirectional increasing size (from small to large) and asymmetrically increasing size (large or small size occurs on either sides). Figure 3.16 shows the unidirectional gradient mode (spindle-knot size from small to large, marked s1, s2, s3, s4, and s5 with gradient length L_{gra}). With water condensation, the droplets (marked 1, 2, 3, 4, and 5) grow and coalesce on every spindle-knot. Subsequently, the as-grown droplet further coalesces (1 + 2 + 3 + 4 → 1 + 2 + 3 + 4 + 5) along the gradient-sized spindle-knots. Finally, a large droplet forms and targets a large spindle-knot (Fig. 3.16b) for water collection. Such transport results in two main aspects that form the driving force (as illustrated in Fig. 3.16c,d): One is the force (f_l) from the difference in Laplace pressure (Chen et al., 2013; Xue et al., 2014) observed along the gradient of the spindle-knots because $r_{1(\text{small knot})} = r_{1(\text{large knot})}$, $r_{2(\text{small knot})} < r_{2(\text{large knot})}$, then $f_{l(\text{small knot})} > f_{l(\text{big knot})}$, and the nonequilibrium Laplace pressure will propel the water drop from the small knot to the big knot.

Second, the capillary adhesion force (f_c) is enhanced by the increasing geometric features of the spindle-knots along the fiber. The overall result is that droplets directionally coalesce from the joint to the spindle-knot, and a continuous coalescence induces droplet transport from the small spindle-knot to the large spindle-knot. The target of droplet transport is controlled by the largest spindle-knot. Figure 3.16e shows a photograph of droplet transport to the target at a large spindle-knot along the gradient direction (Xue et al., 2014).

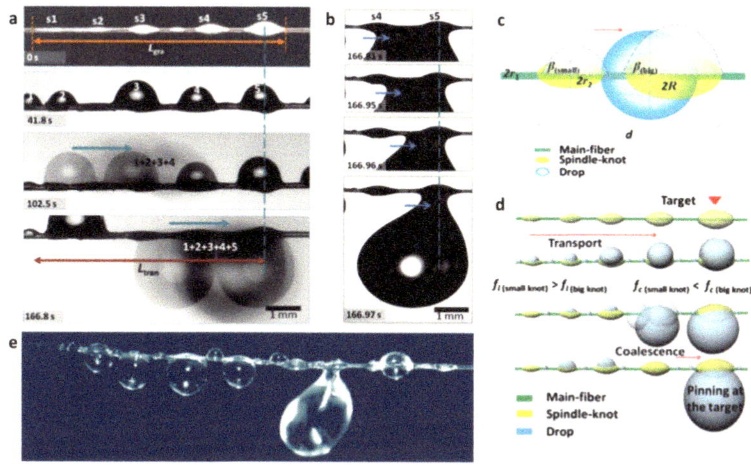

Figure 3.16 (a–b) Gradient-sized spindle-knot fiber for directional water collection. (c–d) Illustration of the mechanism of droplet transport on gradient spindle-knot fibers with gradient adhesion, where a larger spindle-knot has a higher adhesion, and gradient spindle-knots generate Laplace pressure as the driving force to act on the as-coalesced droplet. (e) Photo of droplet transport to the target at a large spindle-knot along the gradient direction (Xue et al., 2014).

3.2.6 Large-Scale Cavity Microfibers with Spindle-Knots

A simple gas-in-water microfluidic method can be used to fabricate large-scale cavity-spindle-knot microfibers (Fig. 3.17). In this fabrication, a hybrid solution is employed as the continuous phase, which is composed of polyethylene glycol (PEG), polyvinyl alcohol (PVA), and sodium alginate. Nitrogen gas is used as the dispersed phase.

In addition, the high-precision syringe pump and homemade gas pressure controller are used to inject the hybrid solution and nitrogen gas, respectively. The resultant jet wrapping the generated bubbles is collected in $CaCl_2$ solution to form a solidified cavity microfiber. The periodicity and size of spindle-knots can be well controlled by different flow rates and gas pressures, such as 0.6 mL/h at 22.40 kPa (Fig. 3.17b), 0.6 mL/h at 20.96 kPa (Fig. 3.17c), 1 mL/h at 26.04 kPa (Fig. 3.17d), and 1 mL/h at 24.07 kPa (Fig. 3.17e). The appropriate flow rate and gas pressure generate the desired periodicity and size of the spindle-knot (Tian et al., 2017).

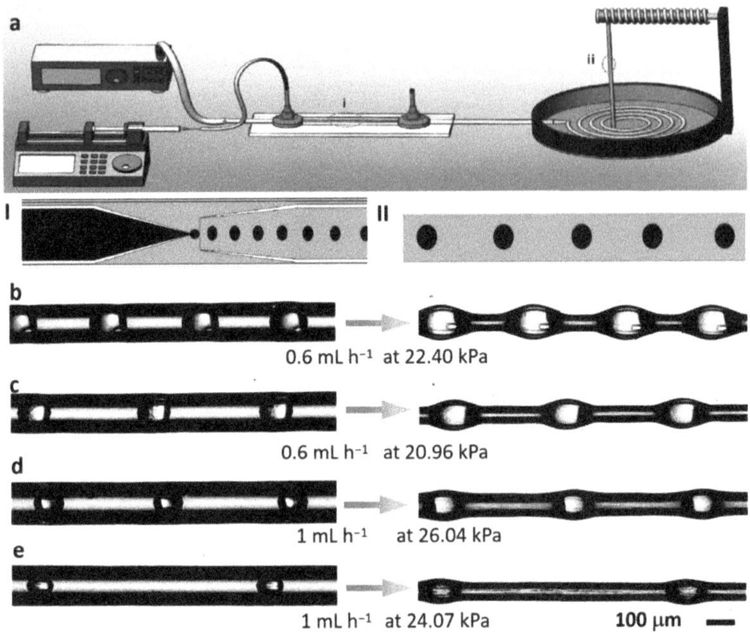

Figure 3.17 (a) Illustration of the microfluidic method to fabricate the cavity microfiber. (b–e) Optical images of controllable microfluidic methods for the formation of cavity spindle-knots with different periodicities (Tian et al., 2017).

The cavity microfiber is endowed with unique surface roughness, mechanical strength, and long-term durability due to the design of the cavity as well as polymer composition, as shown in Fig. 3.18a,b. The optical image can be seen in Fig. 3.18c. There are uniform periodic spindle-knots distributed on the fiber, which can be fabricated on a large scale (Fig. 3.18d). Outstanding water-collection performance can be observed on human-made web (Fig. 3.18e); it is estimated that the maximum water volume collected on a single knot is almost 495 times that of the knot on the cavity microfiber. Figure 3.18f shows the relationship between the volume of water collected and different humidities. With increasing humidity, the as-collected water volume increases. At a humidity of 84%, the collected water volume reaches more than 15 μL in 300 s.

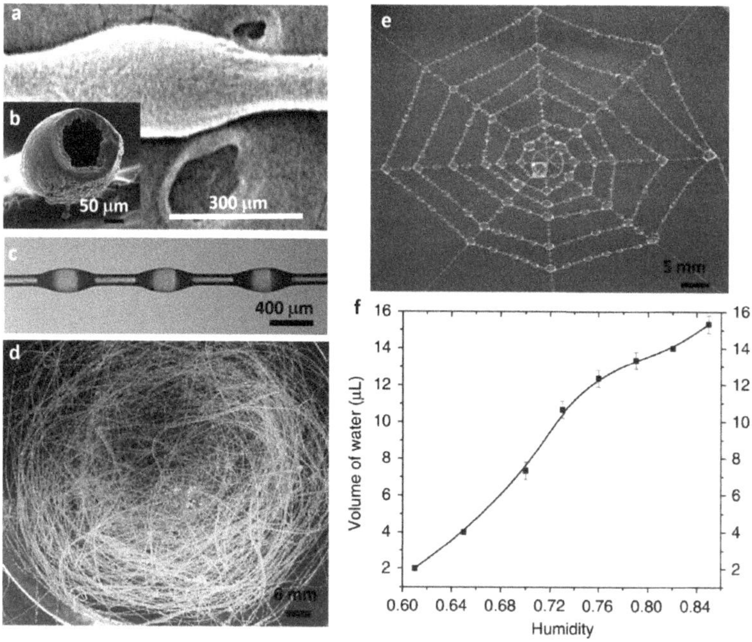

Figure 3.18 (a) SEM image of spindle-knot. (b) Cavity of spindle-knot. (c) Optical image of the cavity microfiber. (d) The cavity microfiber on a large scale. (e) Optical image of water collection on the web of the cavity microfiber. (f) Relationship of as-collected water volume versus humidity (Tian et al., 2017).

3.2.7 Wet Assembly for Spindle-Knots

A bioinspired heterostructured bead-on-string fiber composed of PMMA and titanium tetrachloride ($TiCl_4$)-hydrolyzed nanoparticles is prepared by an integrated wet-assembly system, including PMMA electrospinning, fog with nanoparticles, and multistage water coalescence. The wet assembly is regulated by the difference in surface energy and Laplace pressure in humid environments (Song et al., 2014; Zhao et al., 2014). Figure 3.19a illustrates the wet-assembly process. In fog with nanoparticles, due to water condensation, the fog droplet with nanoparticles is formed on the fiber, and continuous condensation causes droplet growth, including additional nanoparticles in the water droplet. The coalescence

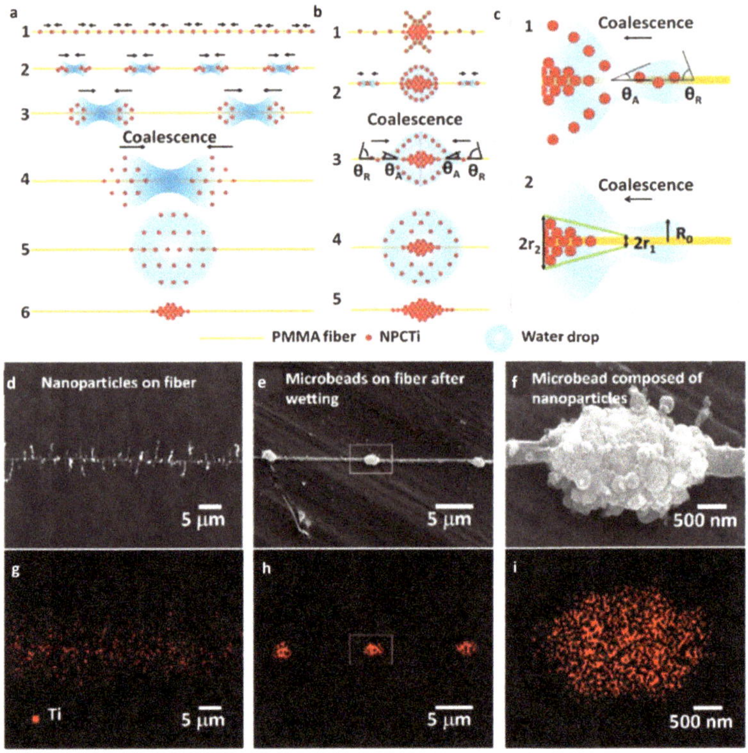

Figure 3.19 (a) The mechanism of the wet assembly of nanoparticles into spindle-knot formation under controllable humidity and time of water condensation. (d–f) SEM images of spindle-knots after wet assembly of nanoparticles. (g–i) EDS mapping analysis of PMMA-Ti on the spindle-knot (Zhao et al., 2014).

of the as-grown droplet makes the nanoparticles accumulate for self-assembly in humidity. In wet assembly, a driving force (df) is expressed as $\mathrm{df} = \int_{x_{joint}}^{x_{hump}} \gamma(\cos\theta_A - \cos\theta_R)\mathrm{dx}$, where γ is the surface tension of water, x is the integration variable along the axis of the fiber from joint to hump, and θ_A and θ_R are the advancing and receding angles of droplets with higher and lower NPCTi concentrations ($\theta_A < \theta_R$), respectively. In addition, the Laplace pressure (ΔP) acts on the droplet, i.e., $\Delta P = \int_{r_1}^{r_2} \frac{\gamma}{(r+R_0)^2} \sin\alpha \mathrm{dz}$, where r is the local radius of the fiber, R_0 is the droplet radius ($R_0 = (3V/4\pi)^{1/3}$ and V is the droplet volume), α is the half apex angle of the bead, and z is the integration variable along the diameter of the bead. The Laplace pressure on the joint with the high curvature (with a local radius r_1) is larger than that on the bead with low curvature (with a local radius r_2; $r_1 < r_2$). Thus, beads are formed, corresponding to larger droplets on the fiber after drying. The humidity and condensation times can cooperatively control the periods or sizes of spindle-knots distributed on fibers. Figure 3.19d–i shows the result of wet assembly to achieve spindle-knots, where the spindle-knots are composed of PMMA-Ti components, along with rough surfaces of spindle-knots that were observed by SEM and energy dispersive spectroscopy (EDS) mapping analysis, respectively (Zhao et al., 2014).

Another bioinspired fiber is fabricated by employing the water-coalescence-induced self-assembly of nanofragments attached to the fibers during water condensation. The sizes and distances of the humps on the fiber are easily controlled by regulating the coalescence time and RH. Figure 3.20a–d shows an SEM image of the "fragment-on-string" fiber. A local magnified SEM image shows the fragment of PVDF (11%) during electrospinning. The fiber is 200 nm in diameter, and the fragments are approximately 100–200 nm in diameter. After wet assembly of nanofragments (RH = 80%, in time of 30 s), the microhumps on the nanofiber form in a highly ordered arrangement, with distances of ~15.52 μm. The surfaces of the microhumps are rather bumpy, consisting of a number of nanopapillae with diameters of ~150 nm. Fog collection is observed on fibers with a hump periodicity of 61.65 μm (Fig. 3.20f, Frames 1–5). At ~9.1 s, tiny water drops (~21.18 μm in diameter) begin to condense at the location of the hump (Frame 2, Fig. 3.20f). At ~16 s, those tiny drops grow into larger drops (~39.12 μm in diameter) on the fiber (Frame 3, Fig. 3.20f). At ~92.4 s, some small water drops coalesce into two large droplets (Frame 4, Fig. 3.20f). The largest droplet is achieved at 333.8 s, with a diameter greater than 1300 μm and a volume of ~1.32 mm^3, which is approximately 200 million times the volume of the wetted region on the fiber that holds the hanging droplet. Estimated from frame 5 in Fig. 3.20f, the wetted length (defined as S) is ~80 μm (Song et al., 2014). This indicates a strong hanging-droplet ability on this wet-assembly humped fiber.

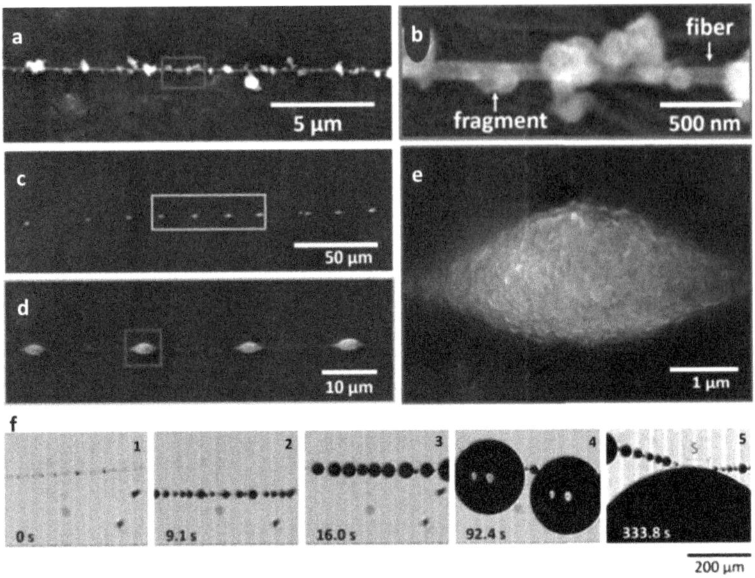

Figure 3.20 (a–c) SEM images of the "fragment-on-string" fiber before (a–b) and after wet assembly (c–e). (f) Optical images of the process in water collection (Frame 1 to Frame 5); the much larger droplet can be hung by thin fiber with short contact length S, observed in time of 333.8 s (Song et al., 2014).

3.2.8 Stimuli-Responsive Fiber

Graphene oxide (GO) fibers with a spindle-knot structure and photothermally responsive phase-transition behaviors are fabricated by combining spinning and emulsification procedures in microfluidics. The resultant GO fibers enable fog capture and near-infrared (NIR)-triggered water-collection applications. Figure 3.21a illustrates the fabrication process of GO fibers with spindle-knots. By employing a capillary microfluidic device with a sequential coflow structure, alginate microfibers, droplets containing GO suspension, and nisopropylacrylamide (NIPAM) pregel solutions are simultaneously generated through ionic gelation and shear-induced emulsification.

GO/NIPAM composite fibers with anisotropic structures are generated continuously. The as-prepared fibers are hydrophilic and anisotropic along the axial direction, which endows them with biomimetic fog-capture capacity. Figure 3.21b shows the SEM images of the GO fiber, including three spindle-knots in a periodicity of ~1000 μm. Figure 3.21c shows the SEM image of a joint with an aligned structure. Figure 3.21d,e shows SEM images of a spindle-knot with porous structure at different magnifications (Shang et al., 2017). This material design gives the fibers a photothermally responsive water-collection ability because of the NIR absorption of GO and the thermally responsive shape transition of the NIPAM hydrogel. Figure 3.21f illustrates the stimuli-responsive behaviors of the fibers. Figure 3.21g,h shows the fog capture of the fibers before and after exposure to a foggy environment. Figure 3.21i–k shows the photothermally responsive water collection of the fibers. The initial state of the fibers with fog captured on the spindle-knots. During NIR irradiation, the spindle-knots shrink, and water droplets are released from the fibers for water collection. When the NIR is switched off, the spindle-knots swell again.

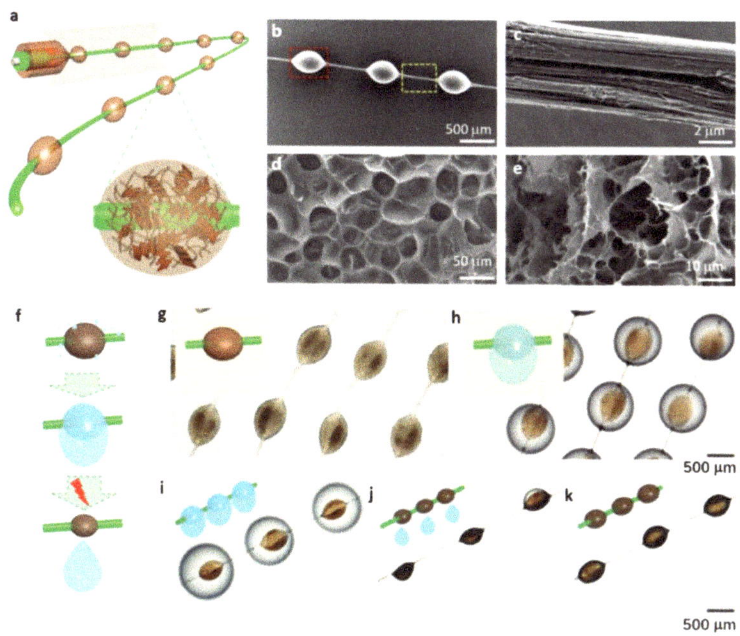

Figure 3.21 (a) Illustration of the microfluidic method for spindle-knots. (b–e) SEM images of surfaces of spindle-knots (b) with porous structures (d–e) and joints with aligned structures (c). (f) Illustration of stimuli-responsive behaviors of fibers. (g–h) Fog capture of the fibers before (g) and after being exposed to a foggy environment (h). (i–k) Photothermally responsive water collection of the fibers. During NIR irradiation, the spindle-knots shrink, and water droplets are released from the fibers for water collection (Shang et al., 2017).

3.2.9 Tunable-Wettability Interface

A bioinspired fiber can be achieved by the dip-coating method using azobenzene (Azo) polymer materials, including gradients of roughness and curvature similar to the spindle-knots of wetted spider silk, as shown in Fig. 3.22a–d. Figure 3.22e shows the relationship between the diameter (D) of condensation droplets and condensation time (T), along with surface wettability properties that are changed by irradiation with Vis and UV. Along with alternative visual (Vis) or ultraviolet (UV) irradiation on spindle-knots, photoresponsive wettability plays a role in water gathering, in conjunction with the cooperation between roughness and curvature, which together effectively regulate the separation of water droplets from the spindle-knots or coalescence toward the spindle-knots, respectively. After Vis irradiation, there is a linear relationship, i.e., $\lg D_{Vis} = 0.444 + 0.337 \lg T$. After UV irradiation, there is a different linear relationship, i.e., $\lg D_{UV} = 0.617 + 0.320 \lg T$. At a given T, $D_{UV} > D_{Vis}$, which indicates a higher efficiency of water gathering on the bioinspired fibers after UV irradiation (Feng et al., 2013).

Another kind of smart bioinspired spindle-knot fiber can be fabricated by using the N-isopropylacrylamide (NIPAAm) polymer. The bioinspired spindle-knot fiber would be temperature responsive, in conjunction with roughness and curvature features similar to those of wetted spider silk (Fig. 3.23a–f). In this as-designed spindle-knot interface, the reversible formation of intermolecular hydrogen bonding between PNIPAAm chains and water molecules and intramolecular hydrogen bonding between C=O and N–H groups in PNIPAAm chains below and above the lower critical solution temperature (LCST) was observed. The direction of motion of tiny water droplets can be manipulated reversibly by the effective cooperation of multiple gradients such as roughness, curvature, and temperature-responsive wettability.

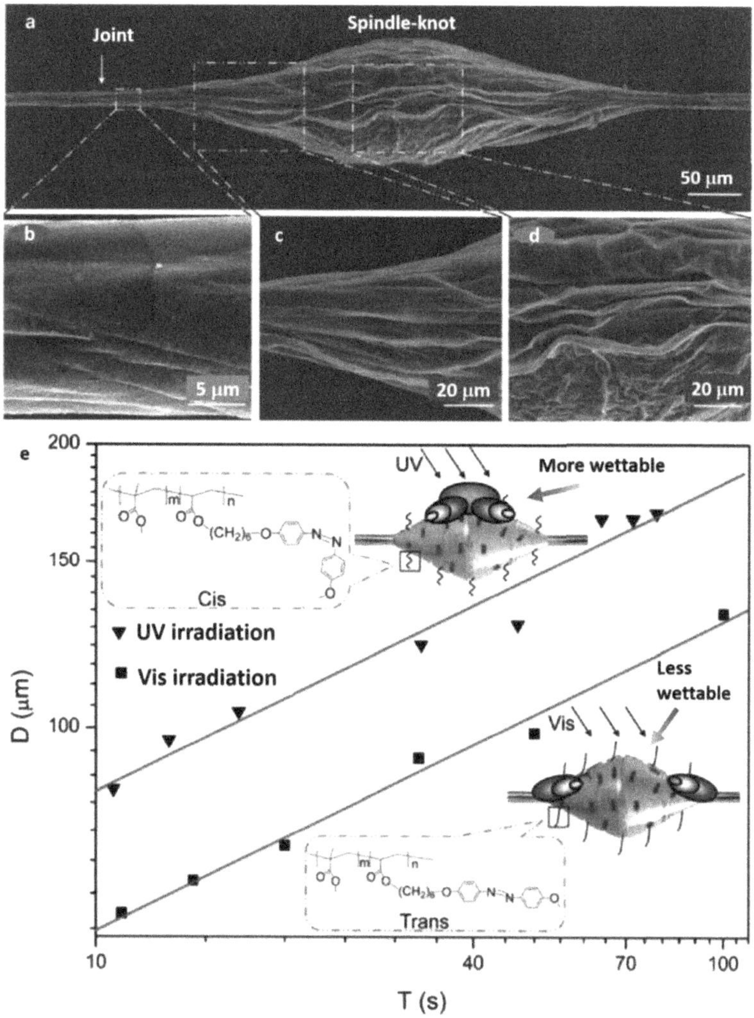

Figure 3.22 (a–d) SEM image of photoresponsive wettable surface on spindle-knot and joint (a) with gradient roughness from joint (b) to spindle-knot (c–d). (e) The relationship between droplet diameter (D) and coalescence time (T) versus the Vis/UV irradiation for the directional control of droplet coalescence. Condensed water droplets coalesce and move toward the joint from the spindle-knot under Vis irradiation. Condensed water droplets coalesce and move toward the spindle-knot from the joint under UV irradiation (Feng et al., 2013).

As shown in Fig. 3.23g,h, during the switching between heating up and cooling down, the direction of increasing wettability can be changed and marked with L (less wettable) and M (more wettable) along the spindle-knot, along with the contact angle (CA) change. A more wettable surface induces a lower CA, and a less wettable one induces a higher CA. At T < LCST, the droplet moves toward the spindle-knot (Fig. 3.23i,k,m). In contrast, at T > LCST, droplets move away from the spindle-knot (Fig. 3.23j,l,n), although the spindle-knots have a high apex angle (the red arrows indicate the droplet movement direction) (Hou et al., 2013).

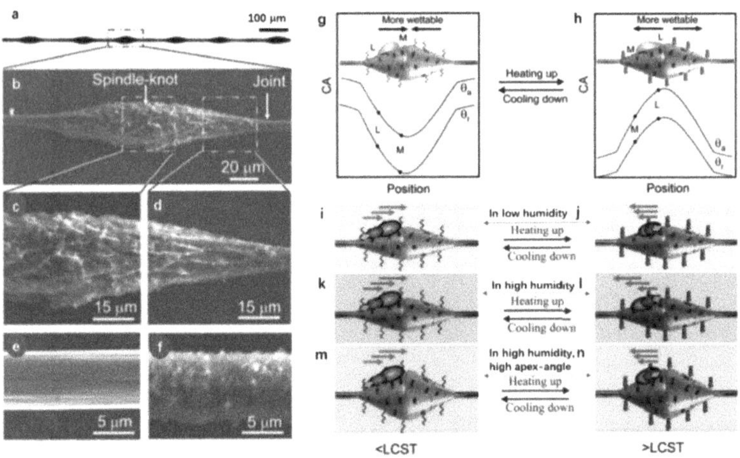

Figure 3.23 (a) Photo of a bioinspired spindle-knot fiber with a temperature-responsive wettable surface. (b–f) SEM images of bioinspired spindle-knot fiber (b), including gradient structure between spindle-knot (c–d) and joint (e–f). (g–h) Illustration of droplet movement toward a more wettable surface under heating up and cooling down. (i–m) Illustration of directional droplet movement under different humidity and different temperatures (e.g., heating up/cooling down). At temperatures lower than LCST, the droplets move toward the spindle-knot (see red arrows) (left), and at temperatures greater than LCST, the droplets move toward the joint (see red arrows) (right) (Hou et al., 2013).

3.2.10 Digital Control Coded Microfiber

Inspired by the anatomy of the silk-spinning system of spiders (Fig. 3.24a)—i.e., the silk gland produces several proteins, the spinning duct focuses and solidifies the protein solution, and the

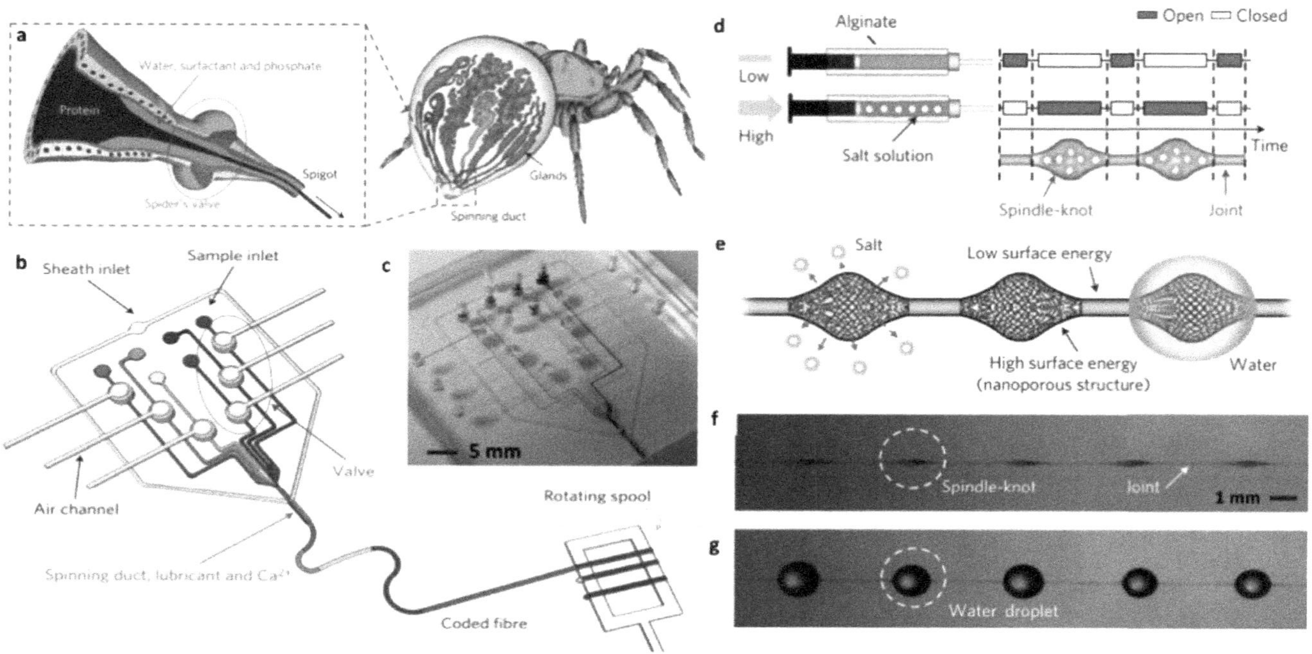

Figure 3.24 (a) Illustration of the silk-spinning system of spiders. (b) Illustration of the design of a bioinspired microfluidics system. (c) Photo of the bioinspired microfluidics system. (d) Illustration of the method to fabricate the spindle-knot and joint while controlling the open or closed modes over time. (e) Illustration of spindle-knot fiber with high surface energy (composed of salt and nanoporous structure) on the spindle-knot and low surface energy on the joint. (f–g) Optical images of spindle-knot silk (f) for water collection (g) (Kang et al., 2011).

valve controls the flow of the protein—a bioinspired microfluidic system has been designed by using multiple channels (Fig. 3.24b). In this design, the microfluidic spinning chip can be used to control the fluid to open or close modes in periodicity (Fig. 3.24c).

In this design, different components with different high/low surface energies, such as alginate and salt, can be regulated in different channels (Fig. 3.24d). By time control, the spindle-knot is formed with the salt solution, and the joint is formed with alginate solution, which can be periodically distributed on the fiber (Fig. 3.24e). The spindle-knot has high surface energy (i.e., hydrophilicity due to the salt component), along with a porous structure after salt release, and the joint has low surface energy (i.e., hydrophobicity due to the alginate component). Figure 3.24f shows the as-fabricated coding fiber with spindle-knots and joints. Water collection occurs on the spindle-knots (Fig. 3.24g) due to the high surface energy (hydrophilicity) (Kang et al., 2011).

3.3 Bioinspired Cactus Spines for Fog-Water Harvesting

Since the cactus spines offer another surface with water-collection ability, including conical geometry and grooves along the cone, they have inspired the fabrication of biomimetic surface and geometry features at the micro- and nanolevel for use in water collection.

3.3.1 Heterogeneous Rough Conical Wires

An artificial periodic roughness-gradient conical copper wire (PCCW) was designed with inspiration from cactus spines and wet spider silk, as shown in Fig. 3.25a, and fabricated by controlled electrochemical corrosion methods with periodic current gradients. The periodic roughness (Ra) formed along the cone wire (Fig. 3.25b) is observed by AFM; Ra = 80.4 nm, 154 nm, and 63.1 nm correspond to regions 1, 2, and 3, respectively, in Fig. 3.25a.

PCCW can harvest fog on periodic points of the conical surface from air and transport the drops for a long distance without external force (Fig. 3.25c–e). Optical images from in situ observation show fast droplet transport on the PCCW in 0.22 s (Fig. 3.25c), and the coalesced force also induces fast droplet transport in less than 0.1 s. This result is attributed to the dynamic as-released energy generated from drop deformation in drop coalescence on the PCCW, in addition to gradients of both geometric curve and periodic roughness, where the geometric curve induces differences in Laplace pressure, and periodic roughness induces the surface energy gradient. These gradients generate cooperative driving forces for high-efficiency droplet transport (Xu et al., 2016).

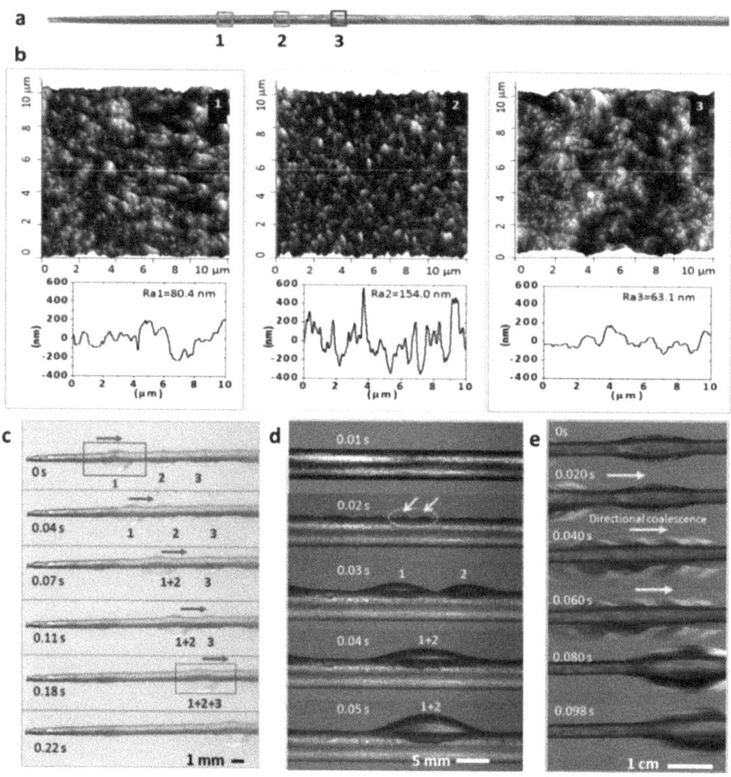

Figure 3.25 (a) Photo of periodic roughness-gradient conical copper wire (PCCW). (b) AFM images of continuous regions (marked with numbers 1, 2, and 3 in (a) on the PCCW). (c) Optical images of in situ observations of droplet transport in 0.22 s. (d–e) Fast droplet transport in less than 0.1 s via coalescence of droplets for fog harvesting (Xu et al., 2016).

A fog collector with an array system of PCCWs achieves a continuous process of efficient water collection better than previous results. Figure 3.26a illustrates a PCCW array fixed on a square frame with a groove to be connected to the ends of the PCCWs, and the bottom of the frame stores the water. The arrows indicate the movement directions of water drops. It is estimated that water-collection rates are different when fog flows to impact the PCCWs at different velocities (e.g., 0.8, 1.2, 1.6, 2.0, and 2.4 m/s) and at different impact angles (30°, 45°, 60°, 75°, and 90°) (Fig. 3.26b). The maximal water-collection rate can reach up to 0.618 g/cm^2/h at an impact angle of 90° and fog velocity of 2.4 m/s at an RH of 90% and a temperature (T) of 15°C.

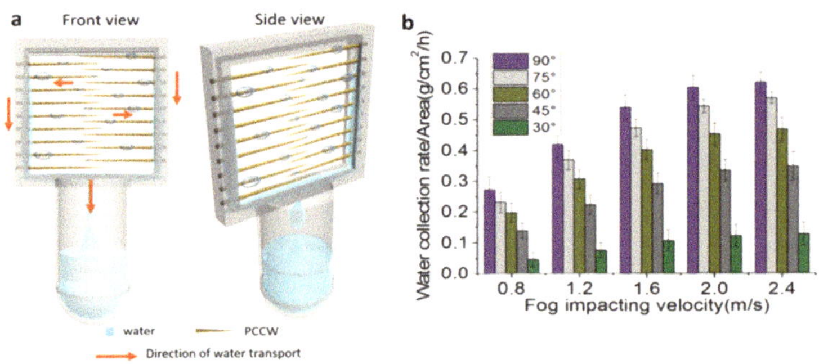

Figure 3.26 (a) Illustration of a water collector composed of PCCW arrays. The arrows indicate the directions of water transport. (b) Comparison of water-collection rates versus fog impact velocities on PCCW collector with different tilted-up angles: 90°, 75°, 60°, 45°, 30° (Xu et al., 2016).

3.3.2 Conical Spine with Micro-/Nanostructures

Cactus spines and trichomes form an integrated fog-collection system. Inspired by these, a micro-/nanostructured conical spine and Janus membrane integrated system (MNCS+JM) has been designed. In this design, the surface of the conical spine is covered with rough micro- and nanostructures (MNCS) (Zhou et al., 2018). MNCS can be fabricated by using aluminum (Al) wire via an integrated electrochemical corrosion and hydrothermal method.

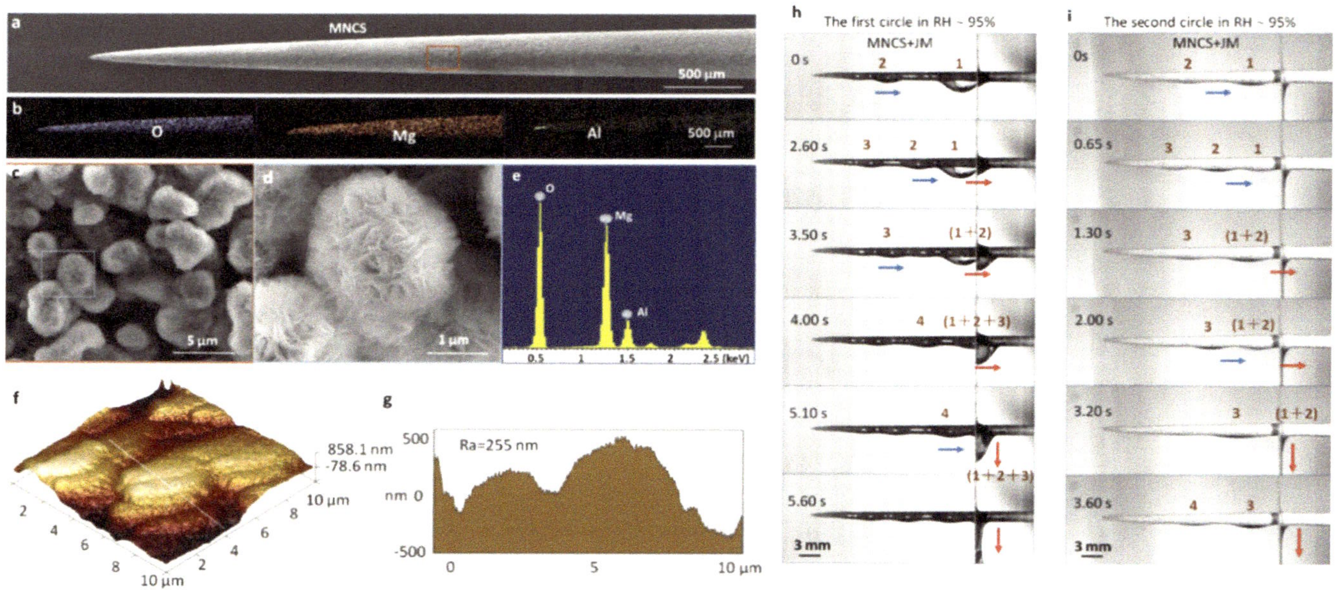

Figure 3.27 (a) SEM image of a micro-/nanostructured conical spine (MNCS). (b, e) EDS analysis on component elements of the spine surface, predominantly oxygen (O)–, magnesium (Mg)–, and Al. (c–d, f–g) SEM images and AFM images of rough micro- and nanostructured conical spines. (h–i) Optical images from in situ observations of directional water transport by MNCS+JM for water collection (Zhou et al., 2018).

Figure 3.27a shows an SEM image of MNCS with conical geometry. Figure 3.27b shows SEM images with EDS analysis of component elements of the spine surface, predominantly oxygen (O)–, magnesium (Mg)–, and Al, indicated in blue (left), red (middle), and green (right), respectively. Figure 3.27c,d shows microscale papillae structures with an average diameter of 3 μm. Each micropapilla is stacked with nanoscale lamellae with a thickness of 20 nm and a length of 500 nm. Quantitative EDS analysis spectroscopy reveals the mass percentages of O (63.31%) and Mg (28.19%) for a certain area of the micro-/nanostructures. This indicates the components of magnesium hydroxide ($Mg(OH)_2$), which is hydrophilic. The surface roughness was analyzed by AFM, and Ra = 255 nm on average (Fig. 3.27f,g). These properties indicate that MNCS can form effective water-harvesting materials due to hydrophilic, conical, and roughness properties, including gradients.

MNCS would be a robust material for fog collection, as observed under a fog flow velocity of ≈0.5 m/s and RH of ≈95%. The tiny fog droplets can be captured, coalesced, and transported on MNCS. For efficient water collection, a Janus membrane (JM) can be fabricated by inspiration from cactus trichomes, hydrophobic and hydrophilic properties on two surfaces of a membrane. The JM was fixed to cross the MNCS inside the hydrophobic surface and outside the hydrophilic surface. Once the transported droplet contacts the JM, a force results from the two-surface (hydrophobic/hydrophilic) wettability difference. Thus, the droplet is further driven to transport from the hydrophobic surface side to the hydrophilic surface side of the JM. Thus, the MNCS+JM system would effectively propel the droplet to achieve continuous transport–coalescence–transport in a circle for collection. As shown in Fig. 3.27h, the first circle of droplet transport through JM would be completed in 5.60 s. However, the second circle of droplet transport through JM would be in 3.60 s, which is faster. This implies that the wetted MNCS would be formed along the surface of the spine, which favors rapid droplet transport after the first circle.

The MNCS+JM system can be optimized by the spine length, JM position at the spine, number of MNCS, hydrophobicity/hydrophilicity of JM, etc. (Fig. 3.28). Thus, there is a higher fog-collection rate on average (e.g., ≈55 g/h) for MNCS30+JM than that of MNCS15+JM (≈31 g/h) at the same observation time (e.g., 60 min) under an RH of 95%. This indicates that the increasing water-collection rate is directly related to the number of spines, which provides higher efficiency. The mechanism is attributed to a cooperative effect between the difference in Laplace pressure and the released surface energy in droplet coalescence, in addition to the wettability force of the superhydrophobic–hydrophilic difference in the JM.

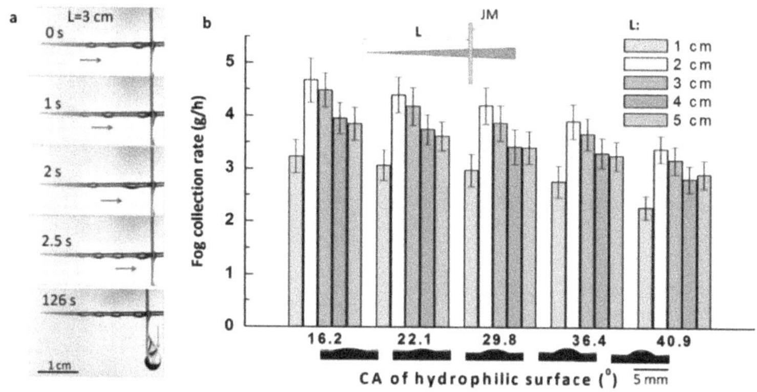

Figure 3.28 (a) Optical images of the MNCS+JM system (spine length L = 3 cm) for droplet transport in 126 s. (b) Fog-collection rates versus spine length and different hydrophilic surfaces (CA: 16.2°, 22.1°, 29.8°, 36.4°, 40.9°) in JM. Optimal system for efficient water collection (Zhou et al., 2018).

3.3.3 Cactus-Stem-Inspired Cone-Arrayed Surfaces

The cactus *O. microdasys* has an integrated fog-collection system arising from evenly distributed clusters of spines and trichomes on the cactus stem. The intersite of the clusters on the cactus stem is densely covered with cones (Fig. 3.29a,b), which are also capable of collecting water from fog efficiently (Fig. 3.29c). Inspired by these cones, using a simple method combining mechanical perforating and template replica technology, PDMS cone arrays have been fabricated with different arrangements. Interestingly, one hexagonal arrangement proves to be more efficient due to the more turbulent flow field around the staggered cones, which induces the rapid directional movement of water drops along each cone (Fig. 3.29d,e). In the mist, water drops deposited randomly on the cones move directly toward the base of the cone (drop 1 in Fig. 3.29d and drops 1, 2 in Fig. 3.29e) or merge with adjacent drops first and then move toward the base of the cone (drop 2+3 in Fig. 3.29d5, at 267 s), refreshing the original sites for the next cycle of deposition, as indicated by the newly appearing drop 1′ and drop 2′ (Fig. 3.29 d5, and 29e4–5, respectively) (Ju et al., 2014).

Figure 3.29f shows a comparison of the mass of the water collected on four types of surfaces (cactus stem, hexagonal arrangement, tetragonal arrangement, smooth) along deposition time under the same fog condition. At a fixed time, the surface with hexagonally arranged cones shows higher fog-collection efficiency (with mass up to ~600 mg in 200 s) than the natural cactus stem and collects the most water, while the smooth surface collects the least water.

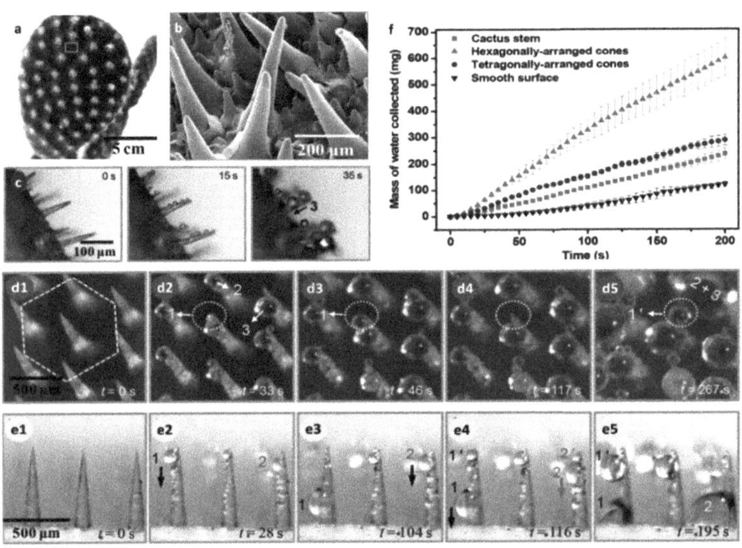

Figure 3.29 (a–b) Photo of cactus (a) and an SEM image of surface on cactus stem with cone array (b). (c) Optical images from in situ observations of water collection on cones. (d–e) Bioinspired cone array with hexagonal arrangement. The droplets move onto tops of cones in the top view (d) and side view (e). (f) Comparison between different arrangements of cones on the surface for water collection (Ju et al., 2014).

3.3.4 Nanocone on 3D Fiber Network

A bioinspired nanocone-decorated three-dimensional fiber network (N3D) has been fabricated, where an original 3D web was inspired by some new research findings on spider webs and decorated with hydrophilic zinc oxide (ZnO) nanocones inspired by cactus spines (Li et al., 2019). Figure 3.30a shows an optical photograph of the N3D with a width of 27 mm, length of 37 mm, and height of 14 mm. The SEM image (Fig. 3.30b,d, Frame I) and the corresponding EDS image (Fig. 3.30b,d, Frame II) reveal that a single fiber from N3D has a diameter of ~150 μm and is covered by dense ZnO nanocones with an apex angle of ~15° on average (Fig. 3.30d). The EDS spectrogram of single fiber from N3D exhibits atomic percentages of 17.86% C, 49.43% O, and 32.71% Zn (Fig. 3.30c).

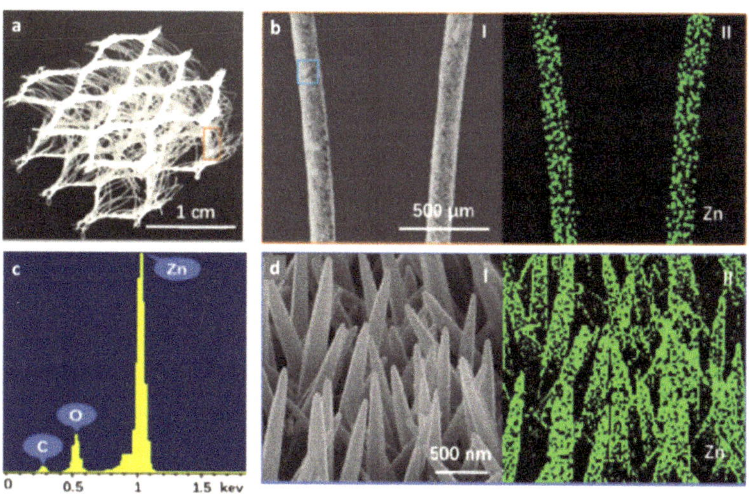

Figure 3.30 (a) Photo of the bioinspired nanocone-decorated three-dimensional fiber network (N3D). (b, d) The SEM image (Frame I) and the corresponding EDS image (Frame II) of nanocones on fiber in N3D (Li et al., 2019). (c) EDS spectrogram of single fiber from N3D shows atomic percentages of 17.86% C, 49.43% O, and 32.71% Zn.

The multilevel high specific surface area exposure on fiber together with the hydrophilic decoration enables this structure to be especially attractive to water molecules. These nanocones can capture fog droplets, generate coalesced droplets, and accordingly perform efficient droplet transport because of the Laplace pressure difference (Fig. 3.31a).

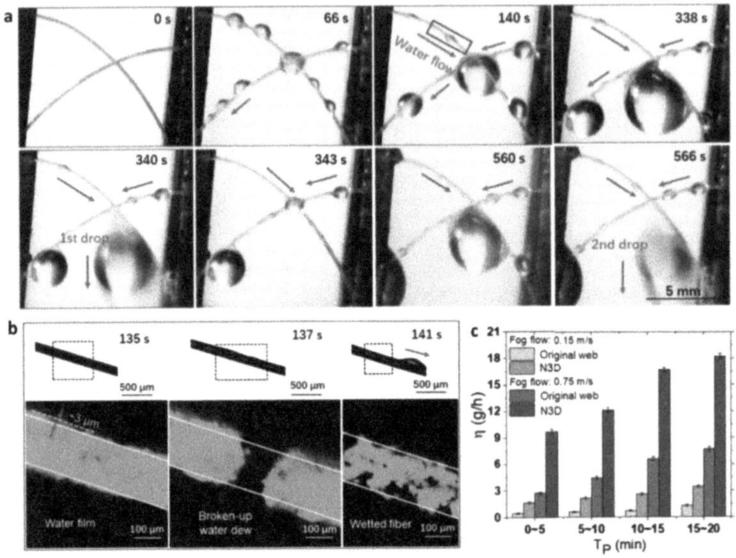

Figure 3.31 (a) Optical images from in situ observations of highly efficient water harvesting on N3D. (b) Rayleigh instability of the water film for fast water transport and collection. (c) Efficiency of water collection (η) versus periodic time (T_p) via the original web, N3D, under fog flows of 0.15 m/s and 0.75 m/s (Li et al., 2019).

In particular, a mechanism revealed that after the nanocone-decorated fiber was wetted, a water film formed and immediately broke up into droplets, owing to the force relating to Rayleigh instability (Fig. 3.31b). Consequent lower retention surfaces realize the formation of fast continuous water flow rather than the traditional intermittent course. Thus, outstanding fog-harvesting efficiency is achieved on N3D and is estimated to reach 865.1 kg/m^2/day, where the mass of water collected within 2 h can be 240 times higher than the weight of an original 3D web without nanocones (Fig. 3.31c).

3.4 Bioinspired Beetle Back for Fog Collection

Beetle backs with hydrophilic and hydrophobic regions for fog harvesting have inspired researchers to design patterned surfaces using various methods.

3.4.1 Hydrophilic–Superhydrophobic Patterned Surfaces

A facile, easy-to-operate, and low-cost method has been developed for the fabrication of a hydrophilic–superhydrophobic patterned hybrid surface for highly efficient fog-water collection. The essence of the method is incorporating a (super)hydrophobically modified metal-based gauze onto the surface of a hydrophilic polystyrene (PS) flat sheet by a simple lab-oven-based thermal pressing procedure (Fig. 3.32a). The resulting hybrid patterned surfaces consisted of PS patches sitting within the holes of the metal gauzes. The method allows easy control over the pattern's dimensions (e.g., patch size) by varying the gauze mesh size (mesh numbers: 50#, 60#, 80#, 100#) and the thermal pressing temperature (e.g., 130°C), which is translated into easy optimization of the ultimate fog-water collection efficiency (Fig. 3.32b–e, e.g., CuO-50-PFDT, CuO-60-PFDT, CuO-80-PFDT, CuO-100-PFDT, respectively).

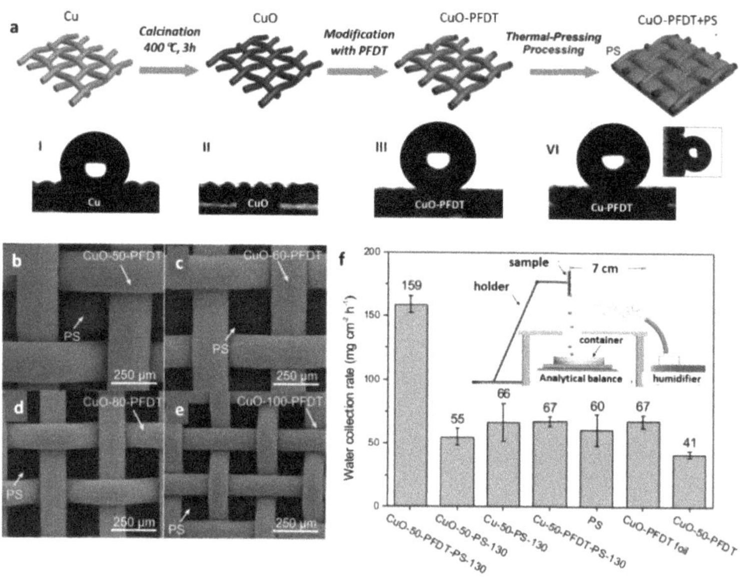

Figure 3.32 (a) Illustration of CuO-PFDT+PS mesh based on Cu mesh, along with wettability with different contact angles (see insets, optical images of droplets, Frames I–VI). (b–d) SEM images of different CuO-50/60/80/100-PFDT meshes on PS substrates, along with features of the hydrophilic–superhydrophobic patterned surfaces. (f) Comparison of water-collection rates via different mesh surfaces, along with temperature treatment of 130°C (Wang et al., 2015b).

The results showed that hydrophilic–superhydrophobic patterned hybrid surfaces with a similar pattern size to the Stenocara beetles' back pattern produced a significantly higher fog-collection efficiency than uniformly (super)hydrophilic or (super) hydrophobic surfaces (Wang et al., 2015b). Water collection on different surfaces can be examined, including a mesh number 50#, via the fog humidifier system, as shown in Fig. 3.32f. Highly efficient water collection, up to ~159 mg cm^2/h, is observed on the surface of CuO-50-PFDT-PS-130, which is far more than that on other surfaces.

3.4.2 Star-Shaped Wettability Patterns

The surface with star-shaped wettability patterns is designed with inspiration from the water-collecting strategies of desert beetles and spider silk (Bai et al., 2014). In this fabrication, first, a superhydrophilic surface can be fabricated by depositing TiO_2 slurry onto a bare glass substrate via a spin-coating method (Fig. 3.33a). Then, it is treated with heptadecafluorodecyl-trimethoxysilane (FAS) to form a superhydrophobic one (Fig. 3.33b). Photomasks with circle-shaped patterns or four-, five-, six-, and eight-pointed star-shaped patterns are used to obtain the shape features of wettability patterns via irradiation with UV light (Fig. 3.33c) due to the photocatalytic decomposition of the FAS monolayer.

By integrating a surface energy gradient and Laplace pressure gradient, surfaces with star-shaped wettability patterns can quickly drive tiny water droplets toward more wettable regions (Fig. 3.33d,e). A driving force ($F_{shape\text{-}wet\text{-}grad}$) arises from the shape gradient to move the droplet directionally toward the center of the pattern. A star-patterned surface might have a higher water-collection efficiency than a circle-patterned surface because a larger driving force ($F_{shape\text{-}wet\text{-}grad} > F_{wet\text{-}grad}$) might move the tiny droplets faster toward the collecting site.

Water collection can be compared on various patterned surfaces, such as superhydrophilic, superhydrophobic, circular, and four-, five-, six-, and eight-pointed star-shaped patterns. Figure 3.34a illustrates the fog-harvesting system. The water-collection results are shown in Fig. 3.34b. The highest efficiency of water collection is on a surface with a five-pointed star pattern at up to 2.75 $g/cm^2 \cdot h$. Generally, star-shaped patterned surfaces achieve more efficient water collection than circle-patterned, superhydrophilic, and superhydrophobic surfaces.

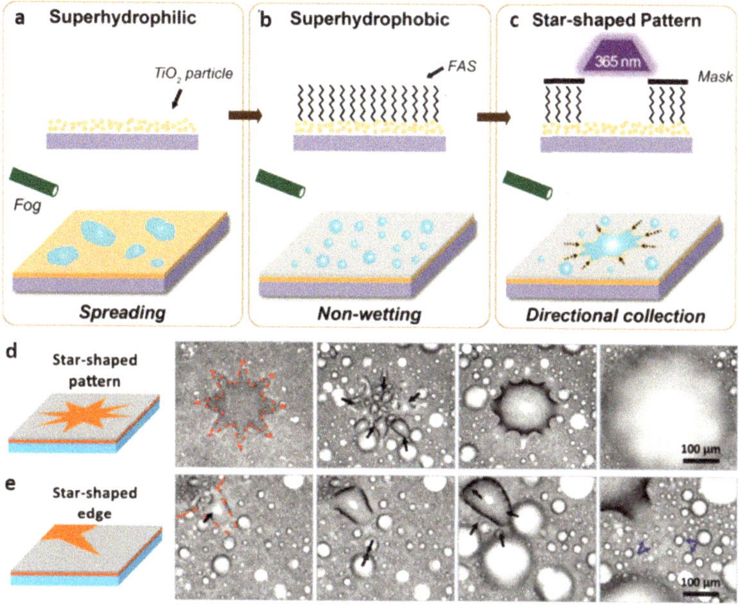

Figure 3.33 (a–c) Illustration of the fabrication process from nonpatterned (a) to star-shaped wettable pattern (b) and directional fog collection (c). (d–e) Illustration (left) and optical images for water-collection direction and efficiency on a star-shaped pattern to regulate the fog collection (right) (Bai et al., 2014).

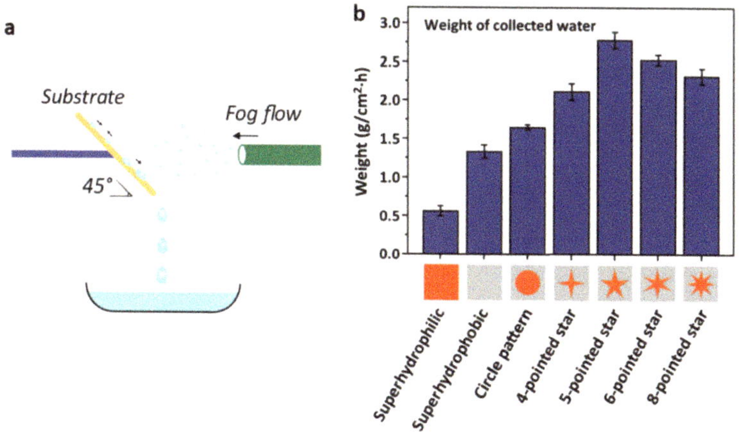

Figure 3.34 (a) Illustration of the water-collection system. (b) Comparison of water collections between various patterned surfaces (Bai et al., 2014).

3.4.3 Hybrid Membrane with Asymmetric Microtopology

Inspired by the asymmetric microgeometry of the cactus spine and hydrophobic/hydrophilic alternating micropattern of the desert beetle's back, a hybrid membrane with asymmetric microtopology and anisotropic wettability has been designed to achieve highly efficient fog collection (Hu et al., 2019).

Figure 3.35a shows the fabrication scheme of the hybrid membrane with anisotropic wettability and topology. First, a layer of PVDF-HFP nanofibers (PNFs) is electrospun on Cu mesh. Then, $Cu(OH)_2$ nanoneedles (CNNs) grow up from the Cu mesh and pierce out of the PNF network through anodization. Figure 3.35b–g shows SEM images of surfaces composed of nanostructured fibers. The wettabilities are the difference in the contact angle (CA) for water on PNFs (CA ≈ 141.3°), on PNFs on Cu mesh (CA ≈ 140.7°), on anodized Cu mesh (CA = 0°), and on CNNS/PNF hybrid membrane (CA = 0°).

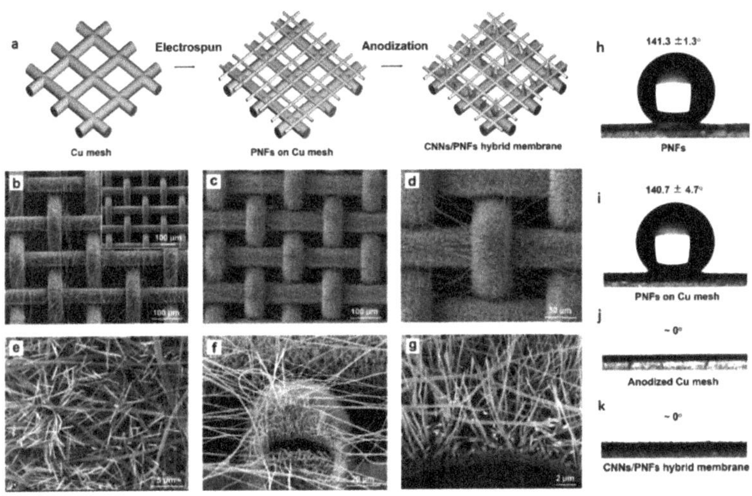

Figure 3.35 (a) Illustration of the fabrication scheme for the CNN/PNF hybrid membrane. (b–g) SEM images of membrane surfaces composed of fibers with nanostructures. (h–k) Wettability on different membranes: PNFs, PNFs on Cu mesh, anodized Cu mesh, and CNN/PNF hybrid membrane, with contact angle (CA) ~ 141.3°, 140.7°, 0°, 0°, respectively (Hu et al., 2019).

The mechanism is the cooperation of nanocones with the difference in Laplace pressure and the wettability gradient, as shown in Fig. 3.36a–f. Once the hybrid membrane is in a foggy environment, it can capture numerous fog droplets. Subsequently, the droplets grow, coalesce, and undergo directional transport from the PNF side to the CNN side due to the wettability gradient force. A water film forms on the CNN side, and the fog-collection process proceeds continuously.

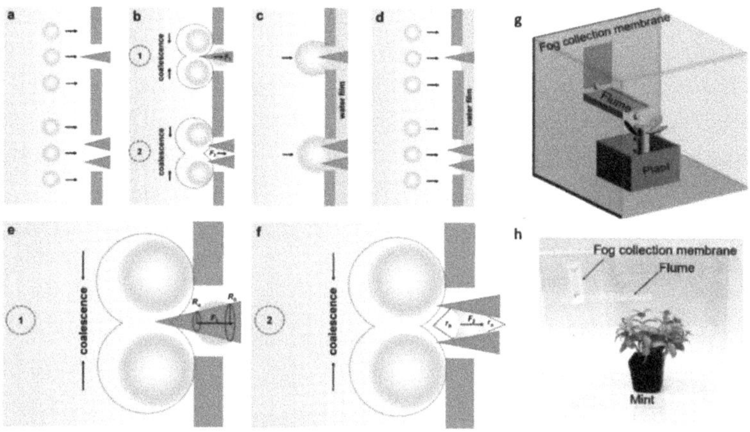

Figure 3.36 (a–f) Illustration of water collection via condensed-droplet coalescence and the cooperation of nanocones with the difference in Laplace pressure and wettability gradient. (g–h) Mechanism and application of fog harvesting (Hu et al., 2019).

Figure 3.36e,f shows the detailed process by which droplets grow and then contact either one nanoneedle or nanoneedle cluster. When the growing fog droplet (or coalesced droplet) contacts one nanoneedle, the droplet will be directly absorbed to the CNN side due to the conical geometry, which induces a Laplace pressure difference (F_1). R_a and R_b are the local radii of the nanoneedle at two opposite sides of the droplets. If the fog droplet contacts the nanoneedle cluster, the nanoneedle wedge also results in a Laplace pressure difference (F_2), where r_a and r_b represent the transverse radius at the two opposite meniscuses, respectively. Thus, fog droplets are directionally transported from the tip to the base of nanoneedles. The fog-harvesting system can be designed for application in plant growth (Fig. 3.36g,h).

3.4.4 Slippery Asymmetric Bumps

A surface has been designed by considering the principles derived from Namib Desert beetles, cacti, and pitcher plants (Park et al., 2016). The resulting surface presents an unconventional interpretation of the role of the beetle's bumpy surface geometry in promoting condensation (Fig. 3.37a,b), where the apex of convex millimetric bumps can be optimized in terms of the radius of curvature and cross-sectional shape. Thus, an asymmetric hump can be designed and fabricated with a tangential connection between the descending slope and the surrounding flat regions (Fig. 3.37e), where the surface is a nanoscale porous structure, incorporating a pitcher-plant-inspired slippery nanocoating. This design enables droplet movement opposite to the direction of gravity for water transport (Fig. 3.37d).

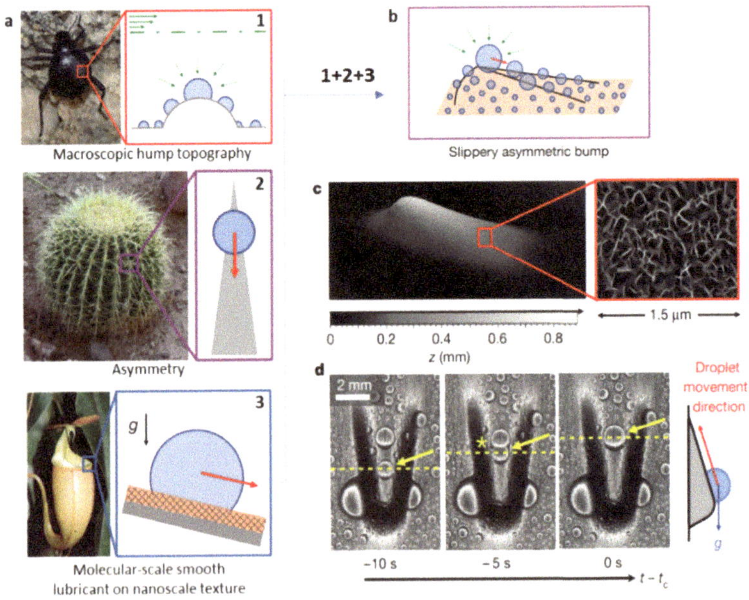

Figure 3.37 (a) Illustration of biological surfaces for droplet behaviors: beetle back with microscopic hump topography (Frame 1), cactus spine with asymmetry (Frame 2), and pitcher plant with molecular-scale smooth lubricant on a nanoscale texture (Frame 3). (b) Illustration of integrated design for the slippery asymmetric humps on the surface. (c) SEM image of asymmetric hump with covering nanostructure. (d) The directional control of droplet transport (Park et al., 2016).

The comparison between the humped surface and smooth surface is shown in Fig. 3.38. The humped surface generates an excellent water-gathered droplet for rolling-off (Fig. 3.38a), but the smooth surface pins the droplet to prevent rolling-off (Fig. 3.38b). Figure 3.38c shows water collection versus time. As the time increases from 0 to 60 min, the humped surface generates higher-efficiency water collection than the smooth surface. It is estimated that the rate is more than 30 ml/m^2 in 60 min for humped surfaces but less than 5 ml/m^2 in 60 min for smooth surfaces.

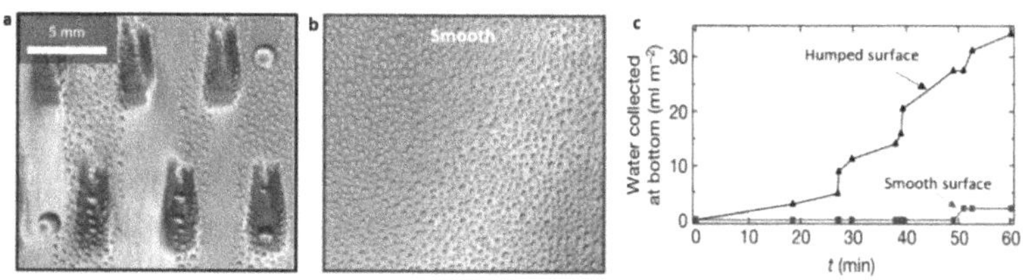

Figure 3.38 (a) The humped surface generates an excellent water-gathered droplet for rolling-off. (b) The smooth surface pins the droplet and prevents rolling-off. (c) Curve comparing the humped surface and smooth surface for water collection, where water collection on the humped surface is higher in efficiency (Park et al., 2016).

3.4.5 Inkjet Printing for Direct Micropatterns

A mussel-inspired ink consisting of an optimized solution of dopamine has been used directly in inkjet printing onto superhydrophobic surfaces. Stable Wenzel's microdroplets of the dopamine solution with well-defined micropatterns can be obtained on these surfaces. Superhydrophilic micropatterns with well-controlled dimensions are then readily achieved on the superhydrophobic surfaces by the formation of polydopamine via in situ polymerization (Fig. 3.39a–c) (Zhang et al., 2015). The micropatterned superhydrophobic surfaces show enhanced water-collection efficiency (Fig. 3.39d) compared with uniform superhydrophilic or superhydrophobic surfaces (Fig. 3.39e). Patterned surfaces of various sizes can be compared, as shown in Fig. 3.39f. The surface, along with the 500 μm pattern and pattern separation of 1000 μm, displays a high water-collection ability (~60 mg/cm^2·h on average).

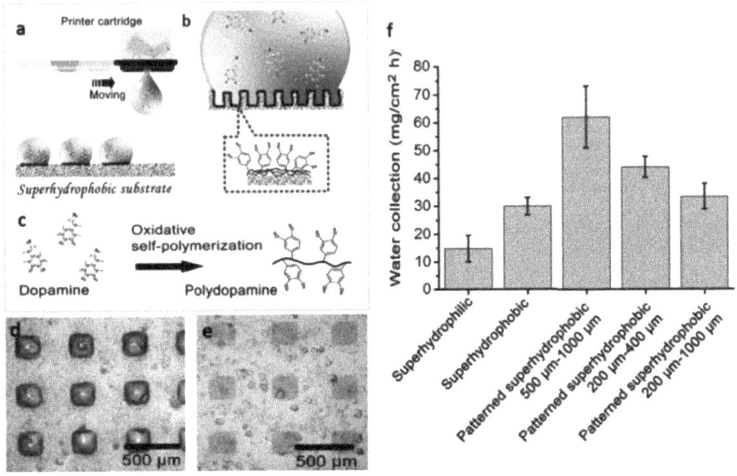

Figure 3.39 (a–c) Illustration of fabrication by inkjet printing to achieve a micropatterned superhydrophobic surface. (d–e) Excellent effect of a condensed water droplet on a micropatterned superhydrophobic surface (d), rather than a uniform superhydrophilic or superhydrophobic surface. (f) Comparison of various surfaces in terms of water-collection abilities (Zhang et al., 2015).

3.5 Water-Collection Engineering and Applications

Water harvesting from fog is a potential alternative source of good-quality fresh water with a very low-cost collection system (Schemenauer et al., 2011; Fessehaye et al., 2014), which has been developed in the coastal deserts of Peru, Sultanate of Oman, Namib Desert, Yemen, and the hills of Nepal (Shanyengana et al., 2003; Rivera and Lopez-Garcia, 2015). As shown in Fig. 3.40a, the projects have mostly been installed in regions of low precipitation and droughts (Certini et al., 2019) but face a common limitation of seasonal fogs or a limited number of fog days in different climatic and topographic locations. The standard fog collector consists of a large screen held by support structures such as poles or frames (Fig. 3.40b), perpendicular to the direction of the wind-driven fog, and is generally installed in outside environments. A mesh with fibers against a humid wind stream is used to collect water from the fog (Batisha, 2015).

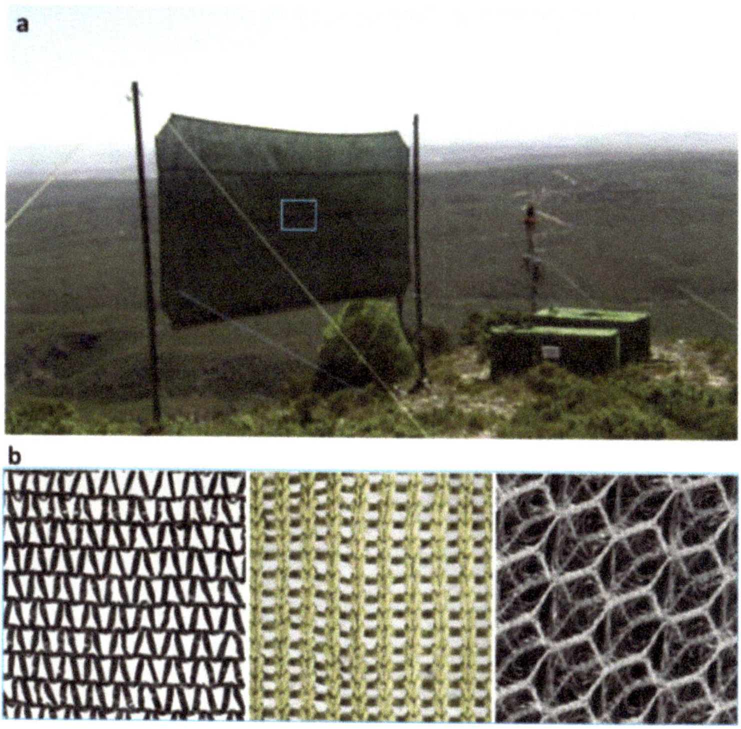

Figure 3.40 (a) Photograph of a water-collection project installed in a region of low precipitation and droughts. (b) Optical images of meshes with structures such as various poles or frames in the fog collector for water harvesting (Batisha, 2015).

3.5.1 Metal-Foam Structured Surface

Novel fog-collecting materials can be fabricated by using copper foams (Batisha, 2015) as the 3D extension of biologically inspired 1D and 2D materials used in water-harvesting applications or liquid–gas separation. The network structure of the 3D materials effectively increases the contact area and interaction time of the skeleton structure and fog compared to those of traditional 2D fog-collecting materials.

Figure 3.41a shows four kinds of copper foam materials with 10, 20, 40, and 80 pores per inch (PPI) (pore density). Figure 3.41b shows optical and SEM images of copper foam with a superhydrophilic coating, where the surface has a water contact angle of up to ~2°.

Figure 3.41c shows optical and SEM images of copper foam with a superhydrophobic coating, where the surface has a water contact angle of up to ~155°.

The fog-harvesting device is designed to generate an inertial centrifugal force by rotating the metal foams during the fog-harvesting process. The highly porous frame increases the contact area with fog droplets, and rotation detaches the water from the surface for high-efficiency collection (Fig. 3.42a,b). Figure 3.42c shows the preparation of four metal foams with 10, 20, 40, and 80 PPI for water collection, and metal foam with 20 PPI exhibits excellent water collection (Fig. 3.42c). Furthermore, different superhydrophilic and superhydrophobic surface metal foams with 20 PPI are used to collect water, and superhydrophobic copper foam (20 PPI) rotated at 1500 rpm can achieve a maximum efficiency of 86% in water collection (Fig. 3.42d).

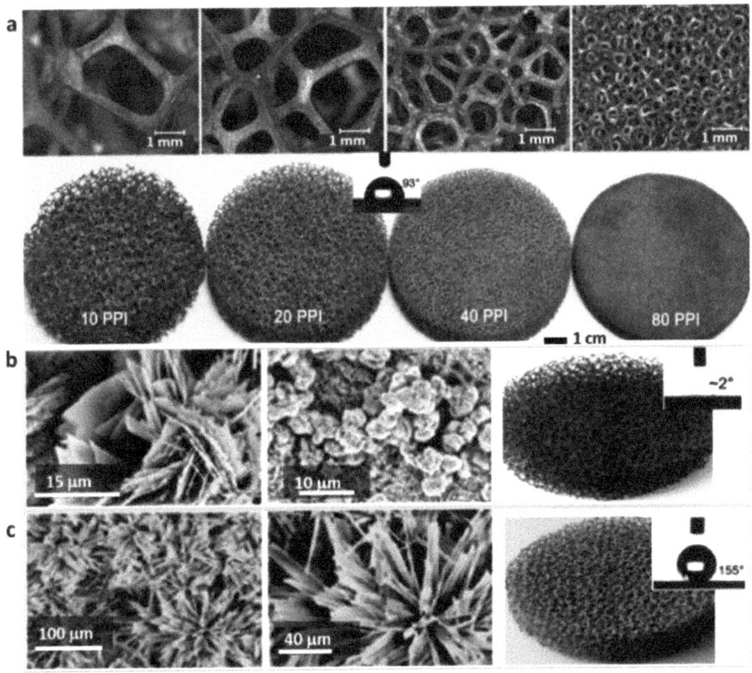

Figure 3.41 (a) Optical images of four kinds of copper foam materials with 10, 20, 40, and 80 PPI (pore density) and a contact angle of ~93°. (b–c) SEM images of copper foams with different wettabilities, i.e., ~2° (the tops, with superhydrophilic coating) and ~155° (the bottoms, with superhydrophobic coating) (Batisha, 2015).

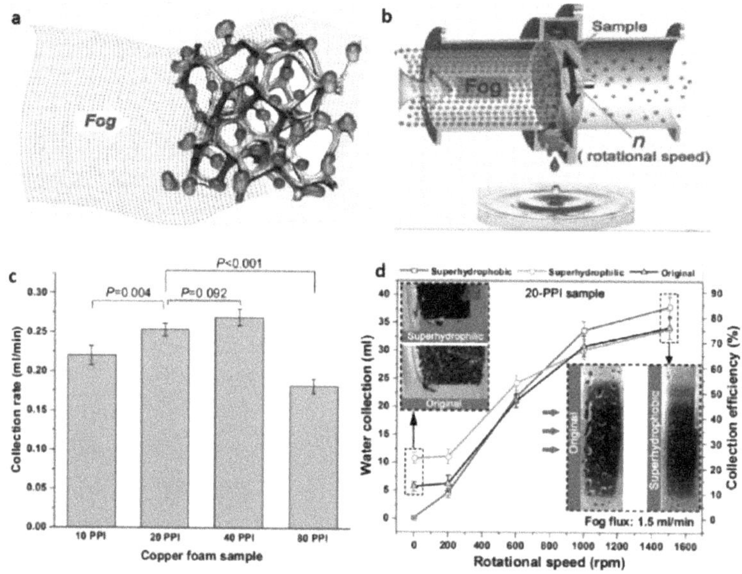

Figure 3.42 (a) Illustration of fog harvesting using copper foam. (b) Device for the theory of fog-water collection. (c–d) Optimal parameters such as PPI (c) and rotation speed (d) for efficient water collection (Ji et al., 2016).

3.5.2 Cooling Tower Fog Harvesting in Power Plants

Fog nets are installed in the cells of a cooling tower for water harvesting (Fig. 3.43a), which can be a potential source of sustainable fresh water supply (Ghosh et al., 2015). Power plant cooling towers produce dense fog plumes due to the evaporative cooling of circulating water (CW). The cooling tower fog is a technical hazard for outdoor electrical installations (Veldhuizan and Ledbetter, 1971), and a CW of 54 000 × 60 000 m^3/h is needed for a typical 500 MW power plant. However, large amounts of CW are lost due to evaporative cooling in the form of vapor or drift and blow down due to unevaporated tiny water droplets carried along with the vapor plume. The amount of water lost in a cooling tower can reach up to ~900 m^3/h for a single 500 MW unit. Generally, this huge amount of fresh water needs to be provided from the local rivers, limiting fresh water available for consumption. Innovative techniques are needed to minimize water consumption in power plants.

Inspired by the scope of this fog-harvesting opportunity, a single-layer woven wire mesh has been made of corrosion-resistant stainless steel (grade SS 304) wires having 0.3 mm diameter and procured from the local market. The mesh is fitted to a rhombus metal frame, creating a fog net with a surface area of 0.0814 m^2. A geometric group variable, termed the shade coefficient (SC), denotes the fraction of mesh projected area occupied by the mesh fiber material. Meshes with SC = 20% (designated as Type 1, in size of 34 cm × 24 cm) and SC = 33% (Type 2, in 34 cm × 24 cm) are shown in Fig. 3.43b–d. A mesh of third type (Type 3, in 24 cm × 22 cm) with SC = 33% was fitted to a triangular metal frame with a mesh area of 0.0525 m^2 (Fig. 3.43c).

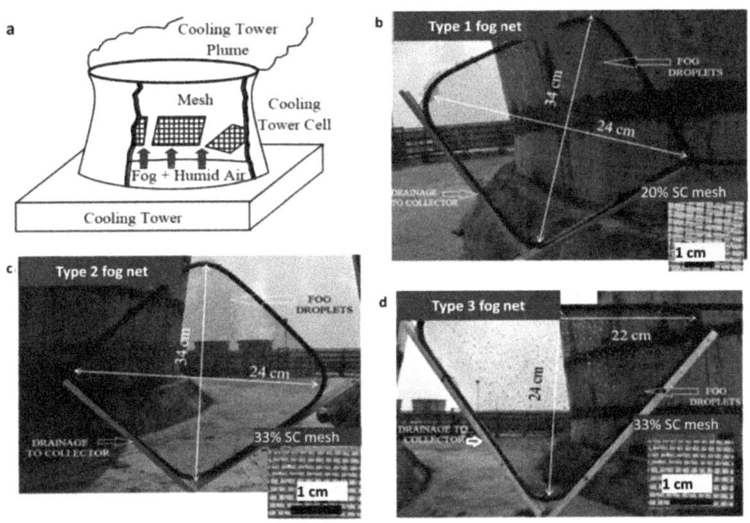

Figure 3.43 Illustration of a cooling tower (a) and a rhombus fog net with (b) 20% SC mesh (34 cm × 24 cm) and (c) 33% SC mesh (34 cm × 24 cm). (d) A triangular fog net with 33% SC mesh (24 cm × 22 cm). Inset figures show the wire mesh (Ghosh et al., 2015).

All fog-collection experiments were performed on top of the cooling tower cell at an RH of 100% and a temperature of 28 ± 0.5°C. The fog collector is positioned as illustrated. The fog velocity was –4.8 m/s. Fog nets of the three types were held at inclination angles (α) of 15, 25, 35, 45, 55, 65, and 75° to the vertical. The inclination angle can be optimized to obtain a high fog-harvesting ability. The data on fog harvesting from the cooling tower with net types 1, 2, and 3 can be compared with the collection data for sites at Cape Verde (Fessehaye et al., 2014), Oman (Abdul-Wahab et al., 2010), and elsewhere (Jagadeesh and Reddy, 2013; Holmes et al., 2014). Type 3 mesh (33% SC) had the highest fog-harvesting rate of 2.28 L/m^2 h in the cooling tower.

3.5.3 MOF Materials

Fresh water resources are currently quite limited, such as lakes on earth (Mekonnen and Hoekstra, 2016). Although some fresh water exists in vapor and atmosphere (Schneider, 1996), dewing from moist air (Shvartsev and Zektser, 2001; Muselli et al., 2002; Lee et al., 2012) and fog (Schemenauer and Cereceda, 1994; Klemm et al., 2012; Park et al., 2013), novel approaches to water harvesting still need to be explored (Kim et al., 2017; LaPotin et al., 2019). Porous materials such as MOFs can harvest water from air (Burtch et al., 2014; Canivet et al., 2014; Wang et al., 2016) via high energy consumption to release water and have been applied in catalysis (Lee et al., 2009; D'Alessandro et al., 2010; Zhou et al., 2012), heat pumps (Jeremias et al., 2014; de Lange et al., 2015), dehumidification (Seo et al., 2012), and other roles (Yaghi et al., 2003; Chu and Majumdar, 2012; Gur et al., 2012; Furukawa et al., 2013). The amount of water harvested by MOFs can be much more than others in dew-harvesting systems at RH $< 50\%$ (Kim et al., 2017). Sunlight is often abundant in arid regions with low RH (>7 kWh/m^2 per day, equivalent to 7 h of 1 sun per day) where water resources are limited, but the natural temperature changes between night for water adsorption due to cooling and day for water release due to warmth. In this case, it would be greatly desirable to realize water harvesting by means of as-designed novel materials and devices.

A recent report indicated that MOF materials were usable for water collection (Kim et al., 2017). MOF-801 powder can be synthesized and activated by heating and then infiltrated into a porous copper foam (porosity of 95%). After brazing on a copper substrate, an adsorbent layer is formed with 1.79 g of as-activated MOF-801. This water-harvesting strategy is completely passive, relying only on the high water-uptake capacity, low-grade heat requirement for desorption, and ambient temperatures (e.g.,

20% RH) to condense and collect the water (Fig. 3.44a). The structure of MOF-801 contains three symmetrically independent cavities to capture as-condensed water molecules (Fig. 3.44b). Figure 3.44c illustrates the system of water harvesting. Figure 3.44d shows the characterization of harvested water from MOF-801 in the adsorption–desorption cycle, and the experimental curve is in agreement with the simulated curve in cumulative water harvesting. This MOF-801 can effectively harvest water (2.8 L/kg of MOF per day at 20% RH) under a nonconcentrated solar flux of less than 1 sun (1 kW/m^2), which is greater than yields from conventional adsorbents (Kim et al., 2016).

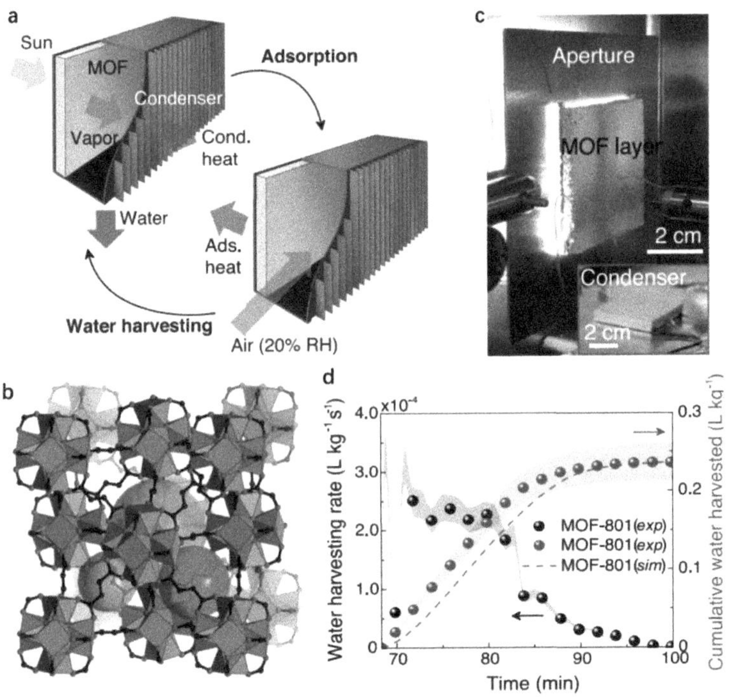

Figure 3.44 (a) Illustration of the working principle of water harvesting with MOFs. (b) Illustration of the structure of MOF-801 with three symmetrically independent cavities to capture as-condensed water molecules. (c) System of MOFs for water harvesting. (d) Comparison of the experimental data and simulated curve of MOF-801 showing the water-harvesting rate and cumulative water harvesting versus time (Kim et al., 2017).

3.5.4 Electrostatic Fog-Water Collection

Fresh water is fundamental to society: for human and cattle consumption, for irrigation purposes, and for domestic use. Current water-collection technologies harvest water from different stages of the hydrological cycle. Superficial water is extracted from lakes and run-offs, underground water is pumped from wells, and oceanic water is desalinated by reverse osmosis and distillation. In places where these sources are nonexistent, contaminated, or too difficult or expensive to exploit, atmospheric water emerges as an option (Cruzat and Jerez-Hanckes, 2018).

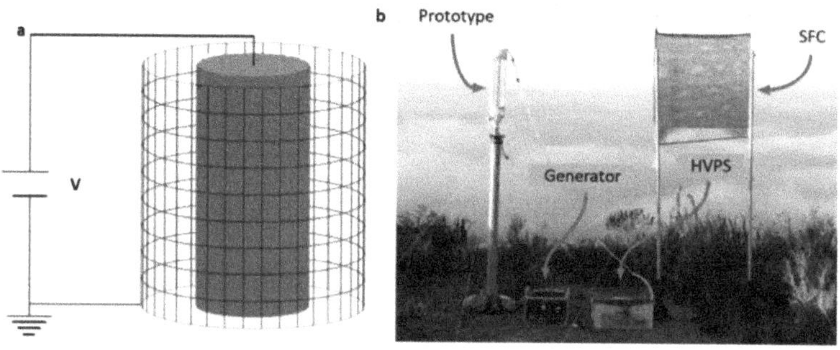

Figure 3.45 (a) Illustration of the design and (b) field experimentation setup (Cruzat and Jerez-Hanckes, 2018).

One method is electrostatic extraction of water from the atmosphere, specifically from fogs. Figure 3.45a shows a schematic of our prototype's design. There are two cylindrical coaxial electrodes of 50 cm height, one wrapping the other. The outer electrode has a radius R_{out} = 8.5 cm and is an air-permeable, electrically grounded mesh. The inner one has a radius R_{inn} = 1.4 cm, is impermeable to air, and is subjected to an applied voltage V = 40 kV. The lower and upper ends of the device are closed by electric insulators that hold the electrodes. Figure 3.45b shows the field experimentation setup, including the DC high-voltage power supply (HVPS) and a standard fog collector (SFC). The obtained experimental results prove that the collection of fog water is feasible by the proposed electrostatic method. Although the prototype is suboptimal, it achieved approximately 60% better performance than current fog collectors in terms of the water-collection rate per unit cross-sectional area (Cruzat and Jerez-Hanckes, 2018).

3.5.5 Patterned Nanobrush

An ambient ion-based method has been developed to make hydrophilic–hydrophobic patterned silver nanowires (NWs) as humidity harvesters to mimic biological surface (e.g., cactus trichomes and beetle wings) structures with water harvesting. Figure 3.46a shows an illustration of a cactus with water droplets on the trichomes. Figure 3.46b shows a schematic of the cactus showing that its sharp spines and trichomes are responsible for water capture. Figure 3.46c shows a transmission electron microscopy (TEM) image of the material that resembles the hairy structure of a cactus, although the scales shown are different. Figure 3.46d shows an illustration of a Stenocara beetle and the hydrophobic and hydrophilic patterning of the wings (Fig. 3.46e).

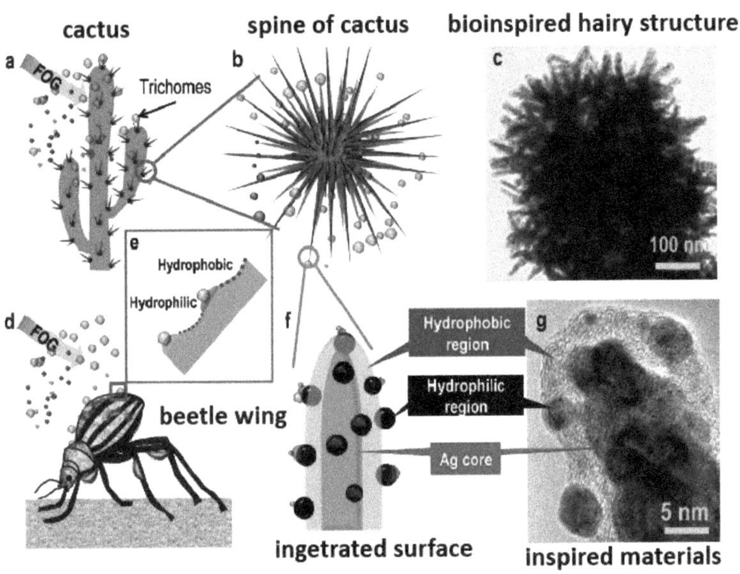

Figure 3.46 (a–c) Illustration of the spines (a) and trichomes of a cactus (b) for fog harvesting and TEM image of hairy structure materials similar to cactus spines (c). (d–f) Illustration of beetle wing (d) for fog harvesting and hydrophobic–hydrophilic patterning of beetle wing (e) and bioinspired surface integrated on the spine shape (f). (g) Bioinspired materials via fabrication of Ag nanowires and Ag nanoparticles with hydrophobic and hydrophilic regions (Sarker et al., 2018).

Figure 3.46f,g shows a schematic of the synthesized hydrophilic–hydrophobic patterned Ag nanowire (NW) and a TEM image of the same with three distinct regions: (i) the Ag core, (ii) hydrophobic coating on the Ag core, and (iii) tiny hydrophilic regions (Ag nanoparticles (NPs), see below for its synthesis) on the hydrophobic background. The Ag core gives strength to the brushes, while the fluorothiol (FT) coating creates a hydrophobic background. The subsequent electrospray deposition (ESD) of Ag on this FT-coated surface resulted in Ag NP decoration covering the NWs. These Ag NPs, embedded in the FT layer, act as hydrophilic zones responsible for the condensation of atmospheric water. A schematic illustration of atmospheric water capture by these hydrophilic–hydrophobic patterned NWs is shown in Fig. 3.46f (Sarkar et al., 2018).

These patterned NWs of ≈20 μm length and ≈200 nm width grown over a relatively large area (2 × 2 cm^2) exhibit an atmospheric water capture (AWC) efficiency of 56.6 L/m^2 d. Figure 3.47 shows the water-collection system.

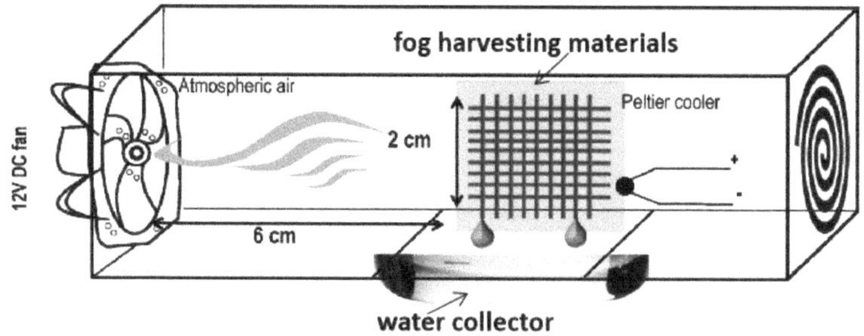

Figure 3.47 Illustration of the water-collection system (Sarker et al., 2018).

3.5.6 Super Moisture-Absorbent Gels

A recent report by the UN Intergovernmental Panel on Climate Change states that there are 3.2 billion people around the world suffering from water shortages because of the absence of water sources such as rivers, lakes, and groundwater in their living areas (Team et al., 2014). This dire warning has motivated research on a new water resource—atmospheric water, which provides ≈50 000 km^3 of water regardless of the geographical (and/or hydrologic) conditions (Deepak et al., 2013). Water-harvesting projects are significant for opening novel approaches based on material design, which has raised great scientific interest and attention in recent years (Pinchasik et al., 2016; Zhou et al., 2019).

For example, a super moisture-absorbent gel (SMAG) is designed by synergistically combining the moisture-absorbing hygroscopic polymer with the water-storing hydrophilic gel, which is substantially more robust than that of other atmospheric water-harvesting (AWH) materials (Zhao et al., 2019). SMAG is rationally designed into a hybrid gel by integration with hygroscopic chloride-doped polypyrrole (PPy-Cl) and poly(nisopropylacrylamide) (poly-NIPAM) with switchable hydrophilicity. This SMAG achieves polymeric-network-enabled moisture capture, which exhibits highly efficient AWH in a broad humidity range, rather than active-surface-based vapor adsorption by other materials.

Figure 3.48a shows a schematic of the skeleton, porous structure and interpenetrating network of poly-NIPAM and PPy-Cl clusters. SEM reveals the microstructures of a dried SMAG, showing the porous structure (Fig. 3.48b), which enhances the interaction with moisture during the adsorption process. The rough surface of the wall structure provides evidence of the inlay of PPy-Cl clusters in the poly-NIPAM matrix. The morphology comparison of pure PPy-Cl with nanoscale particles and the wall structure of the SMAG shows the interpenetration of PPy-Cl clusters and the poly-NIPAM matrix. Figure 3.48c illustrates the water harvester based on an SMAG water collector. Figure 3.48d shows a photograph of SMAG bags during water capture in the natural environment. Figure 3.48e illustrates the moisture absorption enabled by SMAGs and the express mode and normal mode for water release powered by solar energy. This study indicates a high-efficiency AWH working across a range of relative humidities via SMAGs integrated with hygroscopic PPy-Cl for moisture absorption and liquefaction, a hydrophilic network of poly-NIPAM for water storage, and hydrophilicity switching based on the poly-NIPAM for rapid water release. These synergistic functions are enabled by effective water transport along the interpenetrating networks of functional polymers. This design represents a promising strategy for developing atmospheric water management with high-capacity storage, controllable phase transformation, and tunable water transport through polymeric networks. This strategy provides insight into the synergetic effect of interpenetrating functional polymers and their potential for advanced AWH systems in the future.

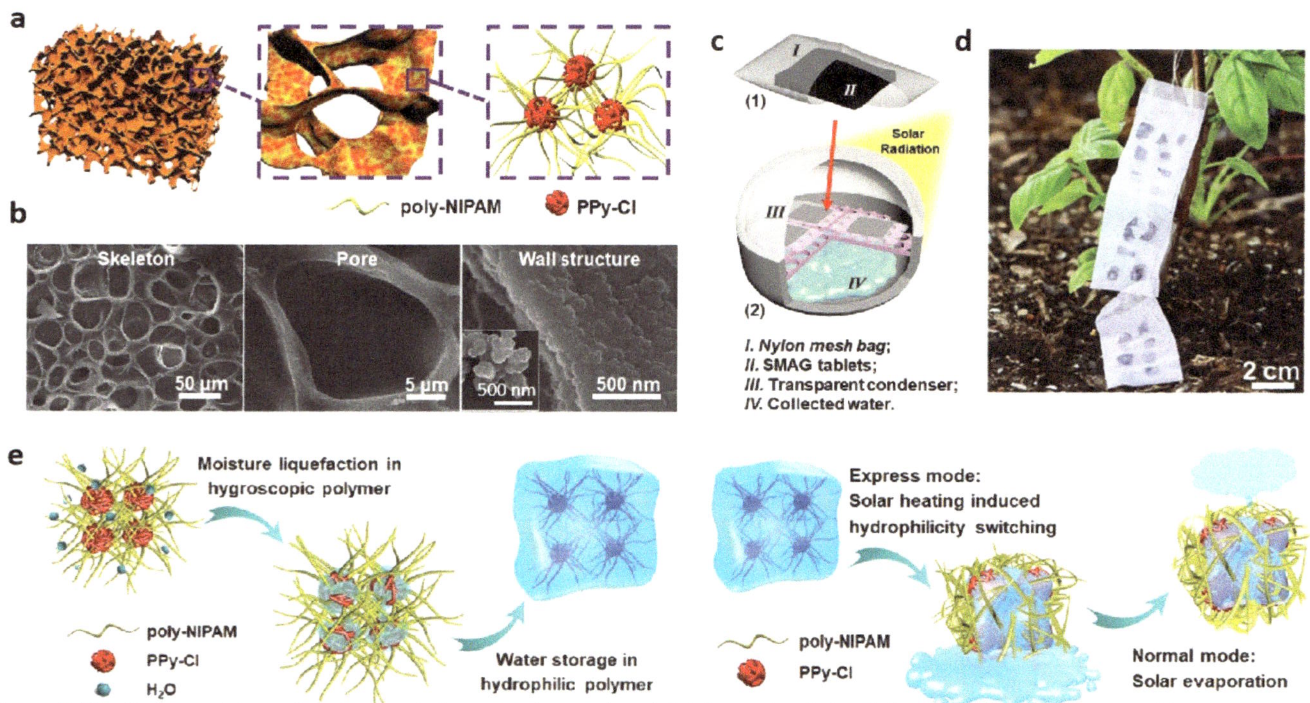

Figure 3.48 Super moisture-absorbent gels for water harvesting. (a) Schematic of the skeleton, porous structure and interpenetrating network of poly-NIPAM and PPy-Cl clusters. (b) SEM images of a dried SMAG at different magnifications, showing the porous structure. (c) Schematic illustration of the water harvester based on SMAGs for the water collector. (e) Illustration of the moisture absorption enabled by SMAGs and express mode and normal mode for the water release powered by solar energy (Zhao et al., 2019).

Taking another example, a method to sustainably harvest water from this moisture-rich zone is demonstrated by employing a nanoporous superhygroscopic hydrogel, which is capable of absorbing over 420% (highest) of its own weight in water from highly humid atmospheres (Nandakumar et al., 2019). The desorption process from the hydrogel, occurring at 55°C (lowest), is triggered by natural sunlight (A. M. 1.5), resulting in an external energy-less water-harvesting approach. The hydrogel exhibits excellent stability even after 1000 absorption–desorption cycles. Through multiple absorption–desorption cycles, it is possible to harvest over 10 L water per kilogram of hydrogel daily. This low-energy, solar-flux-powered approach harvests water from the highly humid air above sea surfaces by employing the humidification–dehumidification (HDH) approach using a superhygroscopic hydrogel. This was achieved by developing prototypes that can float over sea surfaces. This approach has significant advantages in terms of the large surface area available for utilization and efficient ability to condense the surrounding water.

The synthesized gel is an amorphous nonstoichiometric compound of zinc synthesized from its acetate salt combined with an alcohol, an amino-alcohol and water. The gel has two distinct states: dehydrated (DH) and hydrated (H). Figure 3.49a shows the simulated stable structure for a Zn:O ratio of 1:1.1 with many open pores and very few dangling bonds (blue balls represent zinc atoms; yellow balls represent oxygen atoms). There are porous networks and nanofringes on the surface of the hydrogel that offer an enormous surface area for water absorption (Fig. 3.49b,c). A device has been designed by using hydrogel materials, as illustrated in Fig. 3.49d,e, where water can be collected by the process of water condensation on the glass well of the device (Fig. 3.49f). Figure 3.49g illustrates the cross-sectional view of the prototype, corresponding to the prototype built for the in-field study of the absorption characteristics of the hydrogel (Fig. 3.49h). A potential application for water collection involves placing the devices on the sea surface (Fig. 3.50). This work demonstrates low-energy, solar-flux-triggered water harvesting from the air over the sea surface, with a facile setup demonstrated using a novel superhygroscopic hydrogel.

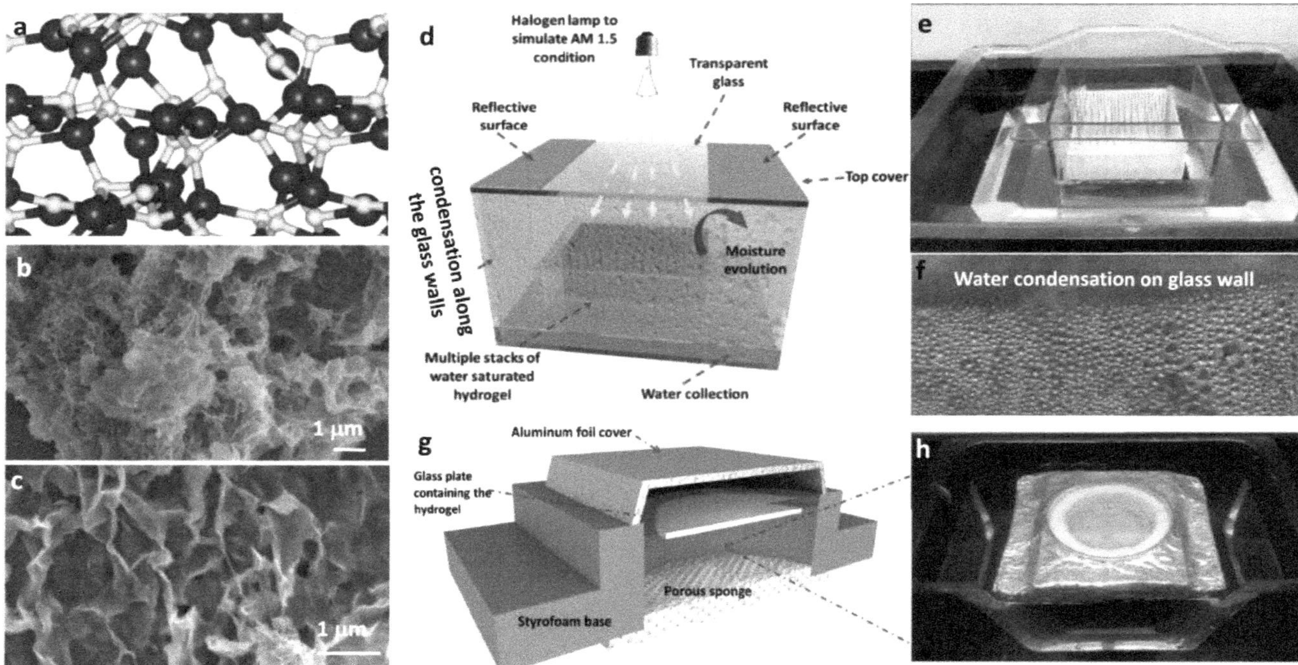

Figure 3.49 (a) Simulated stable structure for Zn:O ratio 1:1.1 (blue balls represent zinc atoms; yellow balls represent oxygen atoms). (b–c) SEM image of the hydrogel with a porous network and nanofringes on the surface for an enormous surface area. (d) Illustration of the device used for water harvesting from the saturated hydrogel. (e) Photo of the device used for water harvesting. (f) Optical image of water condensation on a glass well of the device. (g) Cross-sectional view of the prototype. (h) Prototype built for the field study of the absorption characteristics of the hydrogel (Nandakumar et al., 2019).

Figure 3.50 Photograph of the device floating on the sea surface for application (Nandakumar et al., 2019).

Recently, a highly efficient system for clean water production from a contaminated environment with a wide range of humidity was developed based on a rationally designed sodium polyacrylate (PAAS)/graphene framework (PGF). This porous framework with plentiful oxygen functional groups facilitates the sorption of water vapor in humid air and simultaneously captures impurities with van der Waals forces (Yao et al., 2020).

As illustrated in Fig. 3.51a, a lab-made prototype of this moisture purification and water harvest (MPWH) system has been designed by using PGF. The microporous PGF harvests moisture from contaminated air and desorbs the vapor under the sun to generate clean water free of impurities. The inset shows a schematic of the interaction between water molecules and PGF. The water molecules are mainly captured through the formation of hydrogen bonds with PAAS coated on graphene sheets. Figure 3.51b indicates the fabrication of PGF. In this fabrication, the mixture of GO and PAAS solutions is treated by a freeze-drying process to obtain the GO/PAAS foam. Furthermore, PGF is achieved after reduction under hydrazine hydrate ($N_2H_4 \cdot H_2O$) vapor.

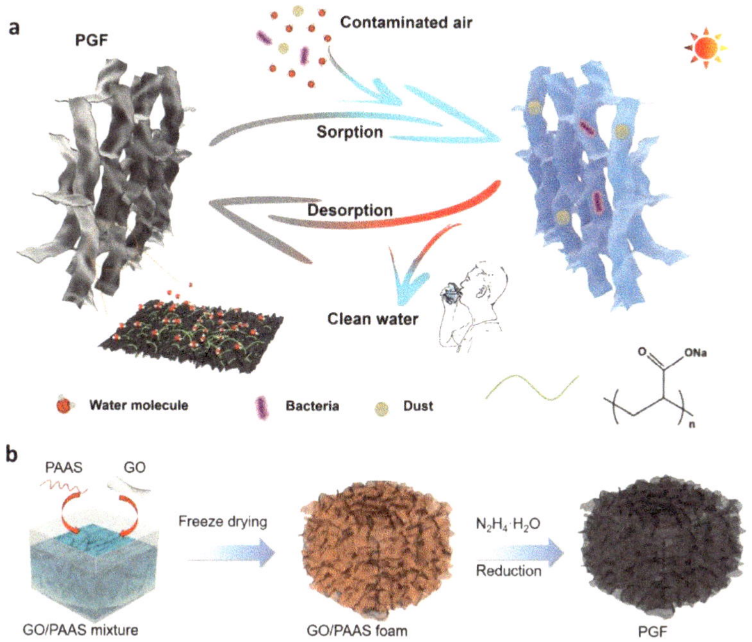

Figure 3.51 (a) The concept of the MPWH system. The microporous PGF harvests moisture from contaminated air and desorbs the vapor under the sun to generate clean water free of impurities. The inset shows a schematic of the interaction between water molecules and PGF. The water molecules are mainly captured through the formation of hydrogen bonds with PAAS coated on graphene sheets. (b) Preparation and characterization of PGF. The mixture of GO and PAAS solutions is treated by a freeze-drying process, and then the GO/PAAS foam can be obtained. Furthermore, PGF is achieved after reduction under hydrazine hydrate ($N_2H_4 \cdot H_2O$) vapor (Yao et al., 2020).

Figure 3.52 shows a photograph of PGF. Figure 3.52b,c shows SEM images of the porous microstructures of PGF. This PGF has an equilibrium water uptake of 0.14 g/g at a low RH of 15% and a superhigh uptake of 5.20 g/g at an RH of 100%, exhibiting excellent water-uptake ability over a wide range of humidities. Under solar irradiation of 1 sun (1 kW/m^2), the sorbed water can be quickly desorbed into vapor within a few minutes to generate clean water free of impurities. Thus, the PGF meets the requirements of efficient moisture capture, impurity filtration, and clean water production from natural air. A device employing the as-built MPWH system

(Fig. 3.52 c–e) can collect over 25 L of water per kilogram of PGF daily from the atmospheric environment. This PGF offers an efficient platform to capture moisture from ordinary and contaminated air for the production of high-quality clean water.

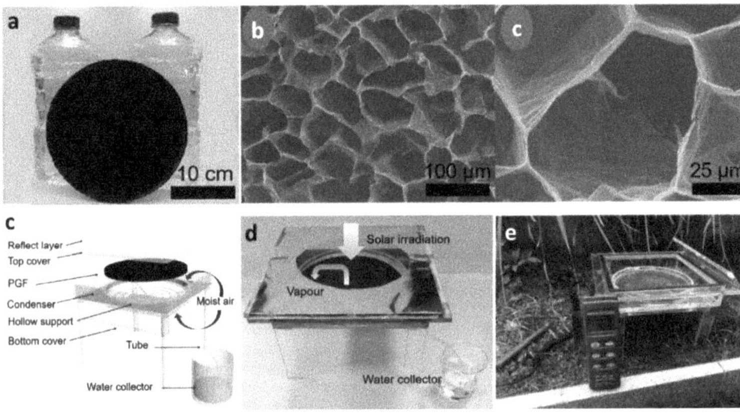

Figure 3.52 (a) Photograph of a PGF with a diameter of 20 cm. (b–c) SEM images of PGF under different magnifications, along with the cellular architecture. (c) Illustration of the device used for water sorption. (d) Photograph of the device showing water collection. (e) Photograph of the outdoor water-sorbing experiment in the natural environment (Yao et al., 2020).

Thinking Questions

1. Can you give some examples of biological surfaces and materials that have water-harvesting properties?
2. Give the key points for materials and surfaces that can harvest water from air.
3. Can you give examples of methods and techniques for the fabrication of materials and surfaces for water collection?
4. Do you know some bioinspired materials and surfaces with water-harvesting abilities?
5. What are some robust materials and surfaces for the application of fog-water harvesting?

References

Abdul-Wahab S. A., A. M. Al-Damkhi, H. Al-Hinai, K. A. Al-Najar, and M. S. Al-Kalbani. Total fog and rainwater collection in the Dhofar region of the Sultanate of Oman during the monsoon season. *Water Int.* **35**, 100–109 (2010).

Bai H., J. Ju, R. Sun, Y. Chen, Y. Zheng, and L. Jiang. Controlled fabrication and water collection ability of bioinspired artificial spider silks. *Adv. Mater.* **23**, 3708–3711 (2011).

Bai H., J. Ju, Y. Zheng, and L. Jiang. Functional fibers with unique wettability inspired by spider silks. *Adv. Mater.* **24**, 2786–2791 (2012).

Bai H., L. Wang, J. Ju, R. Sun, Y. Zheng, and L. Jiang. Efficient water collection on integrative bioinspired surfaces with star-shaped wettability patterns. *Adv. Mater.* **26**, 5025–5030 (2014).

Batisha A. F. Feasibility and sustainability of fog harvesting. *Sustainability of Water Qual Ecol.* **6**, 1–10 (2015).

Burtch N. C., H. Jasuja, and K. S. Walton. Water stability and adsorption in metal–organic frameworks. *Chem. Rev.* **114**, 10575–10612 (2014).

Canivet J., A. Fateeva, Y. Guo, B. Coasne, and D. Farrusseng. Water adsorption in MOFs: Fundamentals and applications. *Chem. Soc. Rev.* **43**, 5594–5617 (2014).

Certini G., G. Castelli, E. Bresci, G. Calamini, A. Pierguidi, L. N. V. Paredes, and F. Salbitano. Fog collection as a strategy to sequester carbon in drylands. *Sci. Total Envir.* **657**, 391–400 (2019).

Chaudhury M. K. and G. M. Whitesides. How to make water run uphill. *Science* **256**, 1539–1541 (1992).

Chen H., T. Ran, Y. Gan, J. Zhou, Y. Zhang, L. Zhang, D. Zhang, and L. Jiang. Ultrafast water harvesting and transport in hierarchical microchannels. *Nat. Mater.* **17**, 935–942 (2018).

Chen Y., J. He, L. Wang, Y. Xue, Y. Zheng, and L. Jiang. Excellent bead-on-string silkworm silk with drop capturing abilities. *J. Mater. Chem. A*, **2**, 1230 (2014).

Chen Y., L. Wang, Y. Xue, L. Jiang, and Y. Zheng. Bioinspired tilt-angle fabricated structure gradient fibers: Micro-drops fast transport in a long-distance. *Sci. Rep.* **3**, 2927 (2013).

Chen Y. and Y. Zheng. Bioinspired micro-/nanostructure fibers with a water collecting property. *Nanoscale* **6**, 7703 (2014).

Chu S. and A. Majumdar. Opportunities and challenges for a sustainable energy future. *Nature* **488**, 294–303 (2012).

Cruzat D. and C. Jerez-Hanckes. Electrostatic fog water collection. *J. Electrostat.* **96**, 128–133 (2018).

D'Alessandro D. M., B. Smit, and J. R. Long. Carbon dioxide capture: Prospects for new materials. *Angew. Chem. Int. Ed.* **49**, 6058–6082 (2010).

Daniel S., M. K. Chaudhury, and J. C. Chen. Fast drop movements resulting from the phase change on a gradient surface. *Science* **291**, 633–636 (2001).

de Lange M. F., K. J. Verouden, T. J. Vlugt, J. Gascon, and F. Kapteijn. Adsorption-driven heat pumps: The potential of metal–organic frameworks. *Chem. Rev.* **115**, 12205–12250 (2015).

Deepak E. A., T. D. Wilkerson, and L. H. Ruhnke. *Atmospheric Water Vapor,* Elsevier, Saint Louis, USA 2013.

Domen J. K., W. T. Stringfellow, M. K. Camarillo, and S. Gulati. Fog water as an alternative and sustainable water resource. *Clean Technol. Environ. Policy* **16**, 235–249 (2014).

Feng S., Y. Hou, Y. Xue, L. Gao, L. Jiang, and Y. Zheng. Photo-controlled water gathering on bio-inspired fibers. *Soft Matter* **9**, 9294 (2013).

Fessehaye M., S. A. Abdul-Wahab, M. J. Savage, T. Kohler, T. Gherezghiher, and H. Hurni. Fog-water collection for community use. *Renew. Sustain Energy Rev.* **29**, 52–62 (2014).

Furukawa H., K. E. Cordova, M. O'Keeffe, and O. M. Yaghi. The chemistry and applications of metal–organic frameworks. *Science* **341**, 1230444 (2013).

Ghosh R., T. K. Ray, and R. Ganguly. Cooling tower fog harvesting in power plants: A pilot study. *Energy* **89**, 1018–1028 (2015).

Gur I., K. Sawyer, and R. Prasher. Searching for a better thermal battery. *Science* **335**, 1454–1455 (2012).

Hamilton W. and M. Seely. Fog basking by the Namib Desert beetle, Onymacris unguicularis. *Nature* **262**, 284–285 (1976).

Holmes R., J. D. Rivera, and E. Jara. Large fog collectors: New strategies for collection efficiency and structural response to wind pressure. *Atmos. Res.* **151**, 236–249 (2014).

Hou Y., L. Gao, S. Feng, Y. Chen, Y. Xue, L. Jiang, and Y. Zheng. Temperature-triggered directional motion of tiny water droplets on bioinspired fibers in humidity. *Chem. Commun.* **49**, 5253 (2013).

Hou Y., Y. Chen, Y. Xue, L. Wang, Y. Zheng, and L. Jiang. Stronger water hanging ability and higher water collection efficiency of bioinspired fiber with multi-gradient and multi-scale spindle knots. *Soft Matter* **8**, 11236 (2012).

Hu R., N. Wang, L. Hou, Z. Cui, J. Liu, D. Li, Q. Li, H. Zhang, and Y. Zhao. Bioinspired hybrid membrane with wettability and topology anisotropy for high-efficient fog collection. *J. Mater. Chem. A*, **7**, 124 (2019).

Jagadeesh T. and K. S. Reddy. Performance analysis of the natural draft cooling tower in different seasons. *IOSR-J. Mech. Civ. Eng.* **7**, 19–23 (2013).

Jeremias F., D. Fröhlich, C. Janiak, and S. K. Henninger. Water and methanol adsorption on MOFs for cycling heat transformation processes. *New J. Chem.* **38**, 1846 (2014).

Ji K., J. Zhang, J. Chen, G. Meng, Y. Ding, and Z. Dai. Centrifugation-assisted fog-collecting abilities of metal-foam structures with different surface wettabilities. *ACS Appl. Mater. Interfaces* **8**, 10005–10013 (2016).

Ju J., H. Bai, Y. Zheng, T. Zhao, R. Fang, and L. Jiang. A multi-structural and multi-functional integrated fog collection system in cactus. *Nat. Commun.* **3**, 1247 (2012).

Ju J., X. Yao, S. Yang, L. Wang, R. Sun, Y. He, and L. Jiang. Cactus stem inspired cone-arrayed surfaces for efficient fog collection. *Adv. Funct. Mater.* **24**, 6933–6938 (2014).

Kang E., G. S. Jeong, Y. Y. Choi, K. H. Lee, A. Khademhosseini, and S.-H. Lee. Digitally tunable physicochemical coding of material composition and topography in continuous microfibres. *Nat. Mater.* **10**, 877–883 (2011).

Kim H., H. J. Cho, S. Narayanan, S. Yang, H. Furukawa, S. Schiffres, X. Li, Y.-B. Zhang, J. Jiang, O. M. Yaghi, and E. N. Wang. Characterization of adsorption enthalpy of novel water-stable zeolites and metal-organic frameworks. *Sci. Rep.* **6**, 19097 (2016).

Kim H., S. Yang, S. R. Rao, S. Narayanan, E. A. Kapustin, H. Furukawa, A. S. Umans, O. M. Yaghi, and E. N. Wang. Water harvesting from air with metal–organic frameworks powered by natural sunlight. *Science* **356**, 430–434 (2017).

Klemm O., R. S. Schemenauer, A. Lummerich, P. Cereceda, V. Marzol, D. Corell, J. van Heerden, D. Reinhard, T. Gherezghiher, J. Olivier, P. Osses, J. Sarsour, E. Frost, M. J. Estrela, J. A. Valiente, and G. M. Fessehaye. Fog as a fresh-water resource: Overview and perspectives. *Ambio* **41**, 221–234 (2012).

Koch C. Comprehensive notes on the tenebrionid fauna of the Namib Desert. *Ann. Trans. Mus.* **24**, 61–106 (1962).

LaPotin A., H. Kim, S. R. Rao, and E. N. Wang. Adsorption-based atmospheric water harvesting: Impact of material and component properties on system-level performance. *Accounts Chem. Res.* **52**, 1588–1597 (2019).

Lee A., M.-W. Moon, H. Lim, W.-D. Kim, and H.-Y. Kim. Water harvest via dewing. *Langmuir* **28**, 10183–10191 (2012).

Lee J., O. K. Farha, J. Roberts, K. A. Scheidt, S. B. T. Nguyen, and J. T. Hupp. Metal–organic framework materials as catalysts. *Chem. Soc. Rev.* **38**, 1450–1459 (2009).

Li C., Y. Liu, C. Gao, X. Li, Y. Xing, and Y. Zheng. Fog harvesting of a bioinspired nanocone-decorated 3D fiber network. *ACS Appl. Mater. Interfaces* **11**, 4507–4513 (2019).

Lorenceau É. and D. Quéré. Drops on a conical wire. *J. Fluid Mech.* **510**, 29–45 (2004).

Louw G. The role of advective fog in the water economy of certain Namib Desert animals. *Symp. Zool. Soc. Lond.* **31**, 297–314 (1972).

Machin J. The study of evaporation from small surfaces by the direct measurement of water vapour pressure gradients. *J. Exp. Biol.* **53**, 753–762 (1970).

Mekonnen M. M. and A. Y. Hoekstra. Four billion people facing severe water scarcity. *Sci. Adv.* **2**, e1500323 (2016).

Muselli M., D. Beysens, J. Marcillat, I. Milimouk, I. Nilsson, and A. Louch. Dew water collector for potable water in Ajaccio (Corsica Island, France). *Atmos. Res.* **64**, 297–312 (2002).

Nandakumar D. K., Y. Zhang, S. K. Ravi, N. Guo, C. Zhang, and S. C. Tan. Solar energy triggered clean water harvesting from humid air existing above sea surface enabled by a hydrogel with ultrahigh hygroscopicity. *Adv. Mater.* **31**, 1806730 (2019).

Park K.-C., P. Kim, A. Grinthal, N. He, D. Fox, J. C. Weaver, and J. Aizenberg. Condensation on slippery asymmetric bumps. *Nature* **531**, 78–82 (2016).

Park K.-C., S. S. Chhatre, S. Srinivasan, R. E. Cohen, and G. H. McKinley. Optimal design of permeable fiber network structures for fog harvesting. *Langmuir* **29**, 13269–13277 (2013).

Parker A. R. and C. R. Lawrence. Water capture by a desert beetle. *Nature* **414**, 33–34 (2001).

Pinchasik B.-E., M. Kappl, and H.-J. Butt. Small structures, big droplets: The role of nanoscience in fog harvesting. *ACS Nano* **10**, 10627–10630 (2016).

Rayleigh L. On the instability of jets. *Proc. London Math. Soc.* **4**, s1–s10 (1878).

Rivera J. D. and D. Lopez-Garcia. Mechanical characteristics of Raschel mesh and their application to the design of large fog collectors. *Atmos. Res.* **151**, 250–258 (2015).

Roberts C., M. Seely, D. Ward, and D. Mitchell. Body temperatures of Namib Desert tenebrionid beetles: Their relationship in laboratory and field. *J. Campbell, Physiol. Entomol.* **16**, 463–475 (1991).

Sarkar D., A. Mahapatra, A. Som, R. Kumar, A. Nagar, A. Baidya, and T. Pradeep. Patterned nanobrush nature mimics with unprecedented water-harvesting efficiency. *Adv. Mater. Interfaces* **5**(19), 1800667 (2018).

Schemenauer R. S. and P. Cereceda. A proposed standard fog collector for use in high-elevation regions. *J. Appl. Meteorol.* **33**, 1313–1322 (1994).

Schemenauer R. S., P. Cereceda, and P. Osses. Fog water collection manual. *FogQuest: Sustain. Water Solutions* 5–9 (2011).

Schneider S. H. *Encyclopedia of Climate and Weather*, Oxford University Press, 1996.

Seo Y. K., J. W. Yoon, J. S. Lee, Y. K. Hwang, C.-H. Jun, J.-S. Chang, S. Wuttke, P. Bazin, A. Vimont, M. Daturi, S. Bourrelly, P. L Llewellyn, P. Horcajada, C. Serre, and G. Férey. Energy-efficient dehumidification over hierarchically porous metal–organic frameworks as advanced water adsorbents. *Adv. Mater.* **24**, 806–810 (2012).

Shang L., Y. Wang, Y. Yu, J. Wang, Z. Zhao, H. Xua, and Y. Zhao. Bio-inspired stimuli-responsive graphene oxide fibers from microfluidics. *J. Mater. Chem. A*, **5**, 15026 (2017).

Shanyengana E. S., R. D. Sanderson, M. K. Seely, and R. S. Schemenauer. Testing greenhouse shade nets for the collection of fog for use as a water supply. *J Water Supply: Res. Technol.-AQUA* **52**, 237–241 (2003).

Shvartsev S. L. and I. S. Zektser. Fundamental issues of water and water resources. *Water Res.* **28**, 346–349 (2001).

Song C., L. Zhao, W. Zhou, M. Zhang, and Y. Zheng. Bioinspired wet-assembly fibers: From nanofragments to microhumps on string in mist. *J. Mater. Chem. A*, **2**, 9465 (2014).

Team C. W., R. K. Pachauri, and L. Meyer. *Climate Change 2014: Synthesis Report*, IPCC, Geneva, Switzerland 2014.

Tian X., Y. Chen, Y. Zheng, H. Bai, and L. Jiang. Controlling water capture of bioinspired fibers with hump structures. *Adv. Mater.* **23**, 5486–5491 (2011).

Tian Y., P. Zhu, X. Tang, C. Zhou, J. Wang, T. Kong, M. Xu, and L. Wang. Large-scale water collection of bioinspired cavitymicrofibers. *Nat. Commun.* **8**, 1080 (2017).

Veldhuizen H. and J. Ledbetter. Cooling tower fog: Control and abatement. *J. Air Pollut. Control Assoc.* **21**, 21–24 (1971).

Vince G. Out of the mist. *Science* **330**, 750–751 (2010).

Wang C., X. Liu, N. Keser Demir, J. P. Chen, and K. Li. Applications of water stable metal–organic frameworks. *Chem. Soc. Rev.* **45**, 5107–5134 (2016).

Wang S., S. Feng, Y. Hou, and Y. Zheng. Controlling of water collection ability by an elasticity-regulated bioinspired fiber. *Macromol. Rapid Commun.* **36**, 459–464 (2015a).

Wang Y., L. Zhang, J. Wu, M. N. Hedhilib, and P. Wang. A facile strategy for the fabrication of a bioinspired hydrophilic–superhydrophobic patterned surface for highly efficient fog-harvesting. *J. Mater. Chem. A*, **3**, 18963 (2015b).

Watanabe K. Fluid frictions of shark skin and lotus leaf: How does the drag reduction occur? *J. Jap. Soc. Tribol.* **45**, 354–359 (2000).

Xu T., Y. Lin, M. Zhang, W. Shi, and Y. Zheng. High-efficiency fog collector: Water unidirectional transport on heterogeneous rough conical wires. *ACS Nano* **10**, 10681–10688 (2016).

Xue Y., Y. Chen, T. Wang, L. Jiang, and Y. Zheng. Directional size-triggered microdroplet target transport on gradient-step fibers. *J. Mater. Chem. A*, **2**, 7156 (2014).

Yaghi O. M., M. Okeeffe, N. W. Ocking, H. K. Chae, M. Eddaoudi, and J. Kim. Reticular synthesis and the design of new materials. *Nature* **423**, 705–714 (2003).

Yao H., P. Zhang, Y. Huang, H. Cheng, C. Li, and L. Qu. Highly efficient clean water production from contaminated air with a wide humidity range. *Adv. Mater.* **32**, 1905875 (2020).

Zhang L., J. Wu, M. N. Hedhili, X. Yang, and P. Wang. Inkjet printing for direct micropatterning of a superhydrophobic surface: Toward biomimetic fog harvesting surfaces. *J. Mater. Chem. A*, **3**, 2844 (2015).

Zhao F., X. Zhou, Y. Liu, Y. Shi, Y. Dai, and G. Yu. Super moisture-absorbent gels for all-weather atmospheric water harvesting. *Adv. Mater.* **31**, 1806446 (2019).

Zhao L., C. Song, M. Zhang, and Y. Zheng. Bioinspired heterostructured bead-on-string fibers via controlling the wet-assembly of nanoparticles. *Chem. Commun.* **50**, 10651 (2014).

Zheng Y., H. Bai, Z. Huang, X. Tian, F.-Q. Nie, Y. Zhao, J. Zhai, and L. Jiang. Directional water collection on wetted spider silk. *Nature* **463**, 640–643 (2010).

Zhou H., M. Zhang, C. Li, C. Gao, and Y. Zheng. Excellent fog-droplets collector via integrative Janus membrane and conical spine with micro/nanostructures. *Small* **14**, 1801335 (2018).

Zhou H.-C., J. R. Long, and O. M. Yaghi. Introduction to metal-organic frameworks. *Chem. Rev.* **112**, 673–674 (2012).

Zhou X., Y. Guo, F. Zhao, and G. Yu. Hydrogels as an emerging material platform for solar water purification. *Accounts Chem. Res.* **52**, 3244–3253 (2019).

Chapter 4

Anti-Icing/Icephobic Surfaces

In this chapter, anti-icing/icephobic surfaces are introduced, as are bioinspired anti-icing strategies based on the design, theory, and foundations, including micro- and nanostructured surfaces, ice patterns on different wettability surfaces, liquid-infused nanostructured surfaces, and superhydrophobic flexible structured surfaces. Fabrication methods for composite structures and materials are provided with examples, such as superhydrophobic coatings, dual-scale structures, transparent polymer coatings, materials that inhibit heterogeneous ice nucleation, grooves for control of ice formation, and self-lubricating liquid water layers. The characterization of anti-icing/icephobic surfaces is introduced, for example the icing delay times of reference droplets, ice adhesion, self-shedding off, nanostructure-tuned icephobicity, the thermal behavior on icephobic surfaces, and the rate of ice formation. Challenging issues for future applications are introduced, such as magnetically responsive hair arrays, materials with low interfacial toughness, micro- and nanostructured surfaces for anti-icing applications, icephobic coatings and applications, ice adhesion, and material categories.

Bioinspired Materials Surfaces: Nature-Inspired Micro-/Nanostructures
Yongmei Zheng

ISBN 978-981-5129-34-2 (Hardcover), 978-1-003-51388-9 (eBook)
www.jennystanford.com

4.1 Bioinspired Anti-Icing Strategies

Anti-icing/icephobic properties are significant for protecting the surfaces of aircraft from crashes, in the power industry, and so on. Various anti-icing and deicing methods have been developed. In general, bioinspired anti-icing is categorized into three aspects, as shown in Illustration 4.1 (Lv et al., 2014). By learning from biological surfaces with water repellency and low adhesion or by understanding biological surface chemistry, anti-icing biological materials, etc., bioinspired anti-icing strategies can be focused on the design of micro- and nanostructures similar to those on the surfaces of lotus leaves, butterfly wings, strider legs, etc. Such strategies include the use of low-surface-energy materials, anti-icing biological materials (e.g., proteins), and liquid-lubricating surfaces to reduce/prevent interfacial wear, friction, interaction, etc. (Kreder et al., 2016).

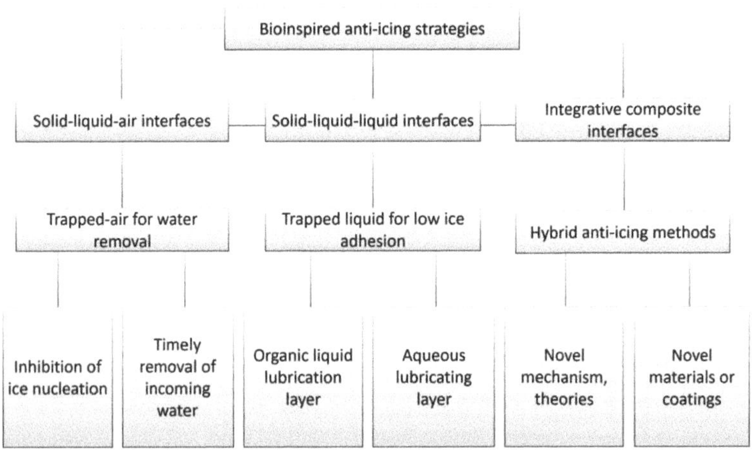

Illustration 4.1 Three aspects of bioinspired anti-icing strategies: solid–liquid–air interfaces (left), solid–liquid–liquid interfaces (middle), and integrative composite interfaces (right).

Passive anti-icing surfaces, or icephobic surfaces, have recently received much attention because of their significant economic, energy, and safety implications for the prevention and easy removal of ice in many facets of society. The complex nature of icephobicity, which requires performance in a broad range of icing scenarios, creates many challenges when designing ice-repellent surfaces.

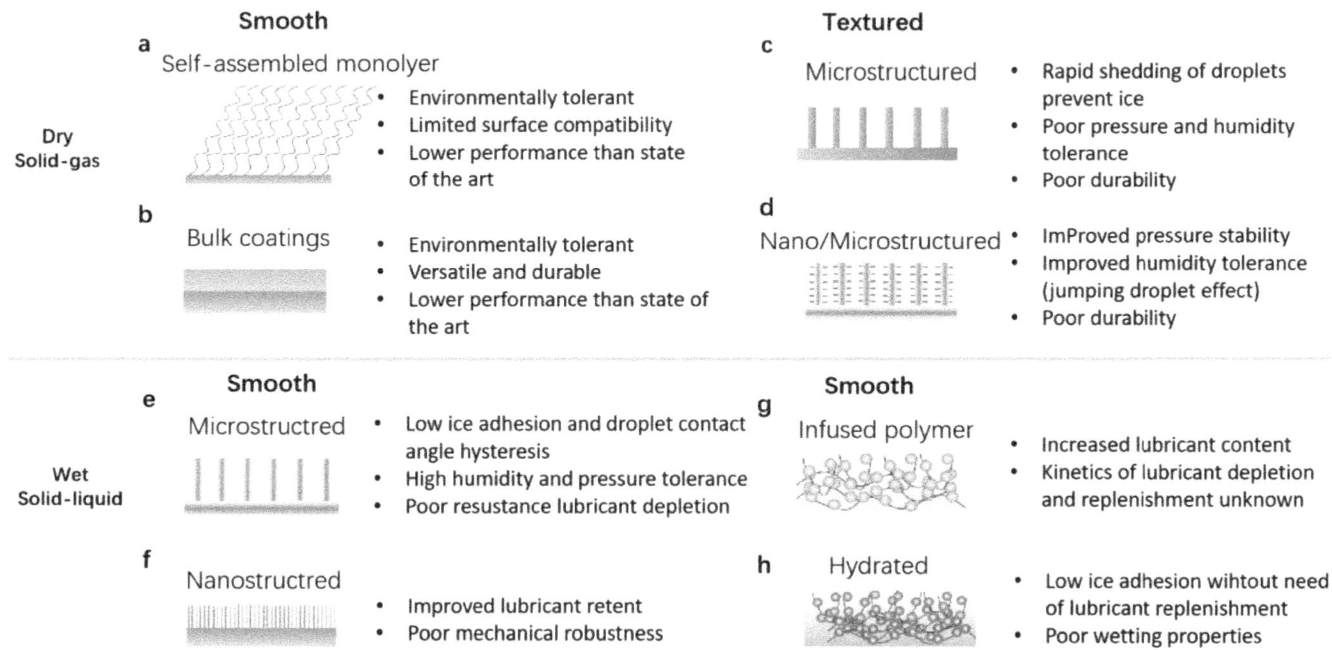

Illustration 4.2 (a–d) Dry mode in the design of materials, including self-assembled monolayers (a), bulk coatings (b), microstructured surfaces (c), and micro-/nanostructured surfaces (d). (e–h) Wet mode in the design of materials, including lubricant microstructures (e), lubricant nanostructures (f), infused polymers (g), and hydrated materials (h) (Kreder et al., 2016).

Although superhydrophobic surfaces incorporating micro- or nanoscale roughness have been shown to prevent ice accumulation under certain conditions, the same roughness can be detrimental in other environments. Surfaces that present a smooth liquid interface can eliminate some of the drawbacks of textured superhydrophobic surfaces, but additional studies are needed to fully realize their potential. As attention begins to shift toward alternative anti-icing strategies, it is important to consider and understand the nature of ice repellency in all environments to identify the limitations of current solutions and to design new materials with robust icephobicity (Kreder et al., 2016).

Various surfaces can be considered to achieve anti-icing/icephobic properties through robust design. To date, some surfaces have been developed with promising applications, as shown in Illustration 4.2. Illustration 4.2a–d shows the dry mode in the design of materials, including self-assembled monolayers, bulk coatings, microstructures with trapped air, and micro-/nanostructures with trapped air. Illustration 4.2e–h shows the wet mode in the design of materials, including lubricant microstructures, lubricant nanostructures, infused polymers, and hydrated materials (Kreder et al., 2016). Based on these studies, the design or development of some novel robust materials is desired to improve the icephobic properties for applications in anti-icing tasks.

4.1.1 Micro- and Nanostructured Surfaces

Surfaces with anti-icing properties have received attention since the 1950s (Raraty and Tabor, 1958). Several methods have been used to fabricate icephobic surfaces based on the superhydrophobicity of surfaces with micro-/nanostructures, such as nanoparticle composite (Guo et al., 2012), laser-treated rough (Liu et al., 2010), and nanocrystalline coverage (Khedir et al., 2011) surfaces. Figure 4.1a shows a micro- and nanostructured surface (i.e., MN-surface) that can be fabricated by mechanical work on a metal surface (to form an array of ratchet structures) and a crystal growth method (to form zinc oxide (ZnO) nanohairs). The ratchet array has a periodicity of ~300 μm and a height of ~80 μm. The ZnO nanohairs are 70–100 nm in diameter and ~3 μm in height. To compare the icephobic properties, other surfaces can be observed at the same time, i.e., a surface including ZnO nanohairs alone (N-surface, Fig. 4.1b), a surface including microratchets alone (M-surface, Fig. 4.1c), and a smooth surface (S-surface, Fig. 4.1d), which are also modified by heptadecafluorodecyltripropoxysilane (FAS-17) to obtain the same low-surface-energy surfaces. Furthermore, the icephobic properties of these surfaces are examined by the in situ observation of reference drops (7 μL) frozen at a temperature of −10°C, as shown in Fig. 4.1e. The MN-surface has a long delay time (DT) for droplet freezing, i.e., DT ~ 7000 s. The longest DT is ~7200 ± 217 s for the MN-surface (see red column, Fig. 4.1f), which is much longer than the others.

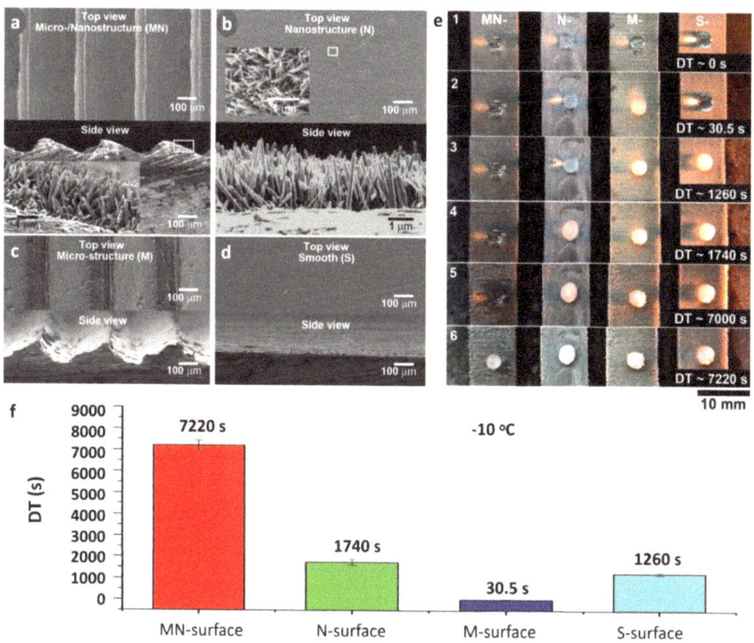

Figure 4.1 (a–d) AFM images of four surfaces: MN-surface (a), N-surface (b), M-surface (c), and smooth surface (d). (e–f) Photos (e) and graph (f) of in situ observations of the four surfaces, where the MN-surface has a long delay time (DT) of 7220 s at −10°C (Guo et al., 2012).

The icephobic property of the MN-surface is attributed to the existence of a high composite contact mode at solid–liquid interfaces, along with a discontinuous three-phase contact line (TCL). Given that the solid–liquid contact density (λ_{sl}) can be defined, the densities follow the order $\lambda_{sl(MN-)} < \lambda_{sl(N-)} < \lambda_{sl(M-)} < \lambda_{sl(S-)}$. A high area fraction of the MN-surface can decrease the influence on adhesion of the electrostatic forces and van der Waals forces compared to the other surfaces (e.g., M-surface, N-surface, and smooth surface).

In terms of the thermodynamics, a system of solid–liquid–air three-phase interfaces is proposed for a droplet on the surface, which gains or losses heat from the air or a cold surface through contact heat conduction and thermal radiation (del Cerro et al., 2010). If Δh is defined as the net heat gain in unit time, then $\Delta h_{(MN-)} > \Delta h_{(N-)} > \Delta h_{(S-)} > \Delta h_{(M-)}$. The larger the Δh, the smaller the t_f, where t_f is the as-decreased temperature of the droplet. This indicates that the MN-surface has a smaller t_f; i.e., the droplet has a smaller change in temperature at low temperatures when the droplet is on the MN-surface. Therefore, the MN-surface generates an excellent icing delay time for icephobicity.

4.1.2 Ice Patterns on Different Wettability Surfaces

As is known, icing on surfaces induces ice patterns due to the interplay between the external macroscopic forces and microscopic interfacial behavior (Aizenberg et al. 1999; Michaelides and Morgenstern, 2007; Algara-Siller et al., 2015; Kiselev et al., 2017). How ice is formed on a surface is important for controlling the anti-icing properties. In the formation of ice on a surface, the wettability is related to the effects of ice patterns. As has been reported, superhydrophobic (i.e., low surface energy) or superhydrophilic (i.e., high surface energy) surfaces can induce distinctly different ice patterns. For example, the ice patterns can be controlled effectively by introducing nanoparticles with inherently low nucleation barriers or distinct wettability (e.g., fluoroalkylsilane modified) on the surface (Liu et al., 2017).

Figure 4.2a illustrates the ice formation resulting from silver iodide (AgI) nanoparticles as ice nucleation active sites under water condensation at a high relative humidity and a temperature below zero. Figure 4.2b–c displays images of the six-leaf clover-like ice formed by the growth process on a hydrophobic surface with a water contact angle of 107.3° (see the inset optical image in Fig. 4.2b), which is defined as the off-surface growth (OSG) mode. By contrast, a hydrophilic surface (with a water contact angle of ~14.5°; see the inset optical image in Fig. 4.2d) leads to sunflower-like ice, and ice crystals continuously spread over the surface (Fig. 4.2d,e) in the growth process, which is defined as the along-surface growth (ASG) mode. The different ice-growth modes generate different icephobic effects. The OSG mode of ice exhibits much lower adhesion to the substrate than the ASG mode of ice. Under wind-blowing action, the ice grown on the superhydrophobic surface (θ = 156.6°) in the OSG mode can be easily blown away (velocity = 5.78 m/s; temperature of −3°C); in contrast, the ice sticks to the hydrophilic surface (θ = 2.9°) with the ASG mode.

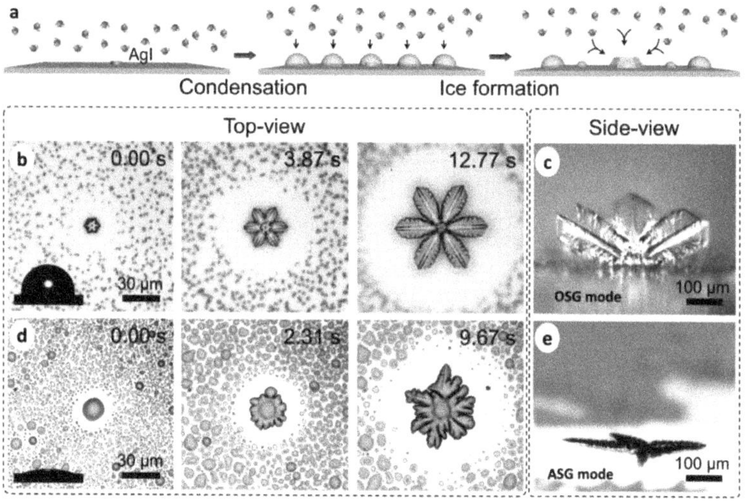

Figure 4.2 (a) Illustration of AgI-induced ice formation under water condensation. (b–c) Optical images of ice formation during off-surface growth (i.e., OSG mode) on hydrophobic surfaces. (d–e) Optical images of ice formation during along-surface growth (i.e., ASG mode) on a hydrophilic surface (Liu et al., 2017).

It can be understood that the superhydrophobic surface repels the water of ice at solid–liquid interfaces, and the hydrophilic surface captures much more of the water of ice. This work reveals the foundation of ice formation on surfaces with different wetting properties, which helps in selecting the types of materials and effectively designing novel anti-icing or icephobic materials for applications.

4.1.3 Liquid-Infused Nanostructured Surfaces

In addition to the design of micro- and nanostructured superhydrophobic surfaces, liquid-infused surfaces display a low adhesion property for ice. In the design of such materials, inspired by pitcher-plant-inspired slippery surfaces (Kim et al., 2012), micro-/nanoporous substrates are infiltrated with a lubricating liquid; thus, a thin ultrasmooth lubricating layer would form on the surface that can repel almost any immiscible materials.

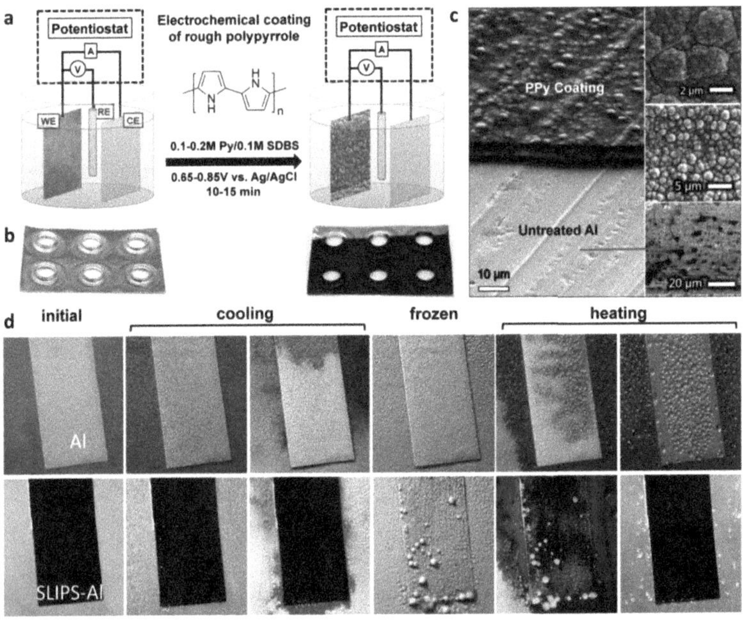

Figure 4.3 (a) Illustration of the fabrication process of a polypyrrole (PPy) coating via oxidative electrochemical deposition. (b) Optical images of the surface of aluminum before (left) and after (right) deposition. (c) SEM image of the PPy coating with a rough and globular morphology. (d) Photos of SLIPS-Al with robust anti-icing compared to Al (Wong et al., 2011).

Figure 4.3a illustrates the fabrication process of a polypyrrole (PPy) coating via oxidative electrochemical deposition. A sheet of aluminum (Al, alloy 1100) is used as the working electrode (WE) in the standard three-electrode configuration for PPy, and sodium dodecylbenzene sulfonate (Na^+DBS^- or SDBS) is used as a surfactant. A constant voltage of 0.85 V vs a silver/silver chloride (Ag/AgCl) reference electrode (RE) is applied for 5–10 min. The surface of aluminum becomes dark blue to black in color (Fig. 4.3b). PPy is predominantly deposited on the surface, forming a uniform PPy film (~3–4 μm in thickness). Figure 4.3c compares the topographies of the resultant PPy/Al surfaces and untreated Al. The PPy coating shows a rough and globular morphology for lubricant infiltration and retention. By fluorinating with (tridecafluoro-1,1,2,2-tetrahydrooctyl)trichlorosilane and infiltrating with a low-viscosity perfluorinated lubricating fluid (Krytox 100, DuPont) (Wong et al., 2011), the as-desired liquid-infused surface can be achieved (named SLIPS-Al).

Figure 4.3d compares the icephobic properties of SLIPS-Al and bare Al surfaces under a very rapid cooling rate of 2°C/min. Droplets above the critical droplet size slide down and leave the SLIPS-Al surface before freezing. However, none of the droplets can grow to the critical size on untreated aluminum, and the droplets freeze before they can slide down.

4.1.4 Superhydrophobic Flexible Structured Surfaces

The use of superhydrophobic surfaces is a general strategy to achieve a desired ice delay time via the design of micro- and nanostructures by learning from biological surfaces (e.g., lotus leaves) with special wettability. Taking this example, flexible structures of superhydrophobic surfaces can be designed to enhance their anti-icing abilities.

Figure 4.4a–c shows an artificial flexible surface with lotus-like hierarchical structures composed of poly(dimethylsiloxane) (PDMS) (Fig. 4.4a) and zinc oxide (ZnO) (Fig. 4.4b), fabricated via soft lithography and an enhanced hydrothermal technique. Such a surface can exhibit a flexible curvature property, and the repellency of a water droplet can be observed at a temperature below zero (i.e., –15°C) at a relative humidity of 90% (Fig. 4.4c). The droplets contact less of the solid due to the micro- and nanostructures with high air-trapping abilities. The frozen droplets are easily removed from the surface, as shown in Fig. 4.4d. The contact mode of solid–liquid interfaces can be specially designed to suspend droplets (Fig. 4.4e). Figure 4.4f,g shows an illustration of a droplet freezing on a flexible micropapillae surface (Fig. 4.10f). The as-frozen ice droplet is suspended by the flexible micropapillae (Fig. 4.10g), where the red region/arrows represent the force direction/strength.

Based on the flexibility of structures, the PDMS-induced elastic energy makes the droplet rebound, along with the repulsive interaction of the ZnO nanohairs between the substrate and the air–water or air–ice interface. The elastic energy (Γ) can be described by $\Gamma \propto \sqrt{K_{PDMS}} / \sqrt{h_0 - \Delta h}$, where K_{PDMS} is the elastic modulus, h is the height between the substrate and the air–water or air–ice interface, Δh is the change in height at point (x), $h_0 = h + \Delta h$, and h_0 is the initial height. A larger Δh leads to a larger Γ. This indicates that the flexible properties of structures favor excellent water repellency even for such a surface in supercooled environments (Wang et al., 2016a).

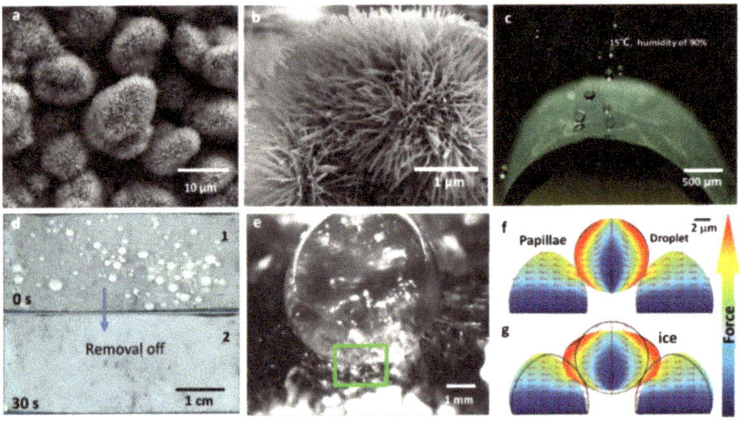

Figure 4.4 (a–b) SEM images of flexible lotus-like PDMS micropapillae covered with ZnO nanohairs. (c) Optical image of water repellency on a flexible surface under 90% humidity at a low temperature of –15°C. (d) Icephobic property. (e) Optical image of an ice droplet suspended by micro- and nanostructures with a robust anti-icing ability. (f–g) Illustration of the icephobic property reduction due to the flexibility of micro- and nanostructures (Wang et al., 2016a).

4.2 Fabrication of Composite Structures and Materials

4.2.1 Superhydrophobic Coatings

A kind of coating with waterborne, nonfluorinated, mechanically robust, and self-healing superhydrophobic properties can be achieved with an anti-icing property in the actual environment (Li et al., 2018).

In this fabrication, suspensions are fabricated via hydrochloric acid (HCl)-catalyzed reactions among hexadecyltrimethoxysilane (HDTMS), tetraethoxysilane (TEOS), and silicon oxide (SiO_2) nanoparticles (i.e., SiO_2@HD-POS). Then, a polyurethane (PU) aqueous solution and the SiO_2@HD-POS suspension are, in turn, spray-coated onto substrates (Fig. 4.5a).

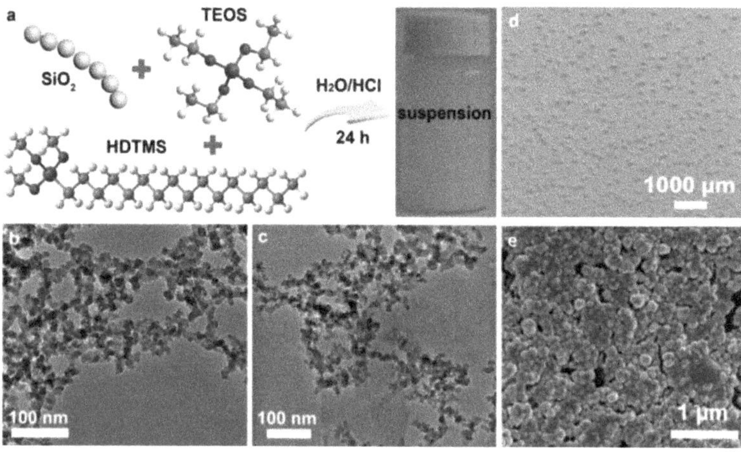

Figure 4.5 (a) Illustration of the fabrication process for the PU/SiO_2@HD-POS suspension. (b–c) Coating with a network (b) of film-like HD-POS to link the SiO_2 particles (c). (d–e) Photo and SEM image of the PU/SiO_2@HD-POS surface (Li et al., 2018).

The as-prepared PU/SiO_2@HD-POS superhydrophobic coatings show excellent superhydrophobicity, very high mechanical stability, and self-healing capability. The SiO_2 nanoparticles are ~15–20 nm in diameter and are in the form of aggregates with a network structure (Fig. 4.5b), and the polycondensation of HDTMS and TEOS forms film-like HD-POS to link the SiO_2 particles together (Fig. 4.5c). A photo and an SEM image of the PU/SiO_2@HD-POS surface are shown in Fig. 4.5d,e, respectively.

The coatings exhibit excellent superhydrophobic and water-repellent properties (Fig. 4.6a,b) for various substrates (polyester fabric, Teflon (PTFE) plate, wood plate, steel plate, and aluminum plate) (Fig. 4.6c–e). There is durability against sandpaper abrasion (Fig. 4.6f,g) and a self-healing property of superhydrophobicity (Fig. 4.6h). In particular, the coatings display anti-icing performance in an outdoor environment (−15°C, relative humidity = 54%). By comparing the bare surface over time, as shown in Fig. 4.6i, ice can form at a long delay time on the as-coated surface, but ice appears in a short time on the bare surface. Such a coating can also achieve a dynamic anti-icing process; the coated surface makes ice roll off, but the bare surface is covered by ice. In addition, the as-coated surface can exhibit wettable durability after the icing/deicing process, and the superhydrophobic properties can be maintained along the robust structures.

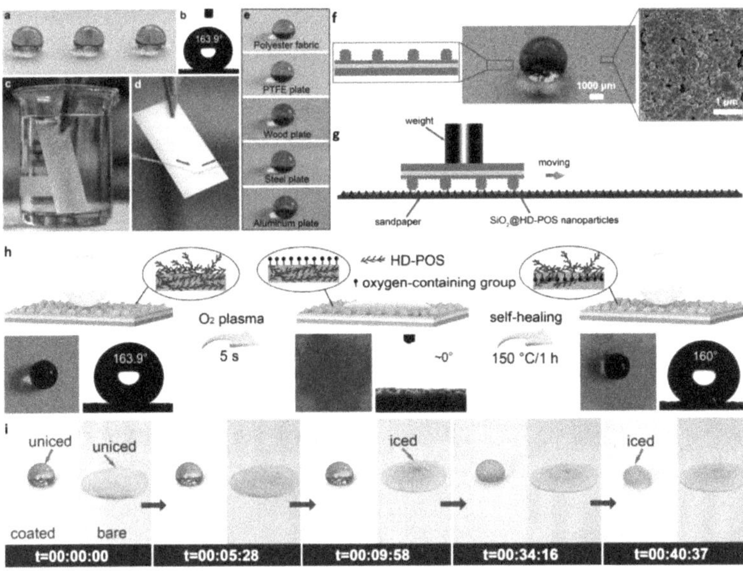

Figure 4.6 (a–e) Photos of coated surfaces with superhydrophobic (a–b) and water-repellent properties (c–d) for various substrates (polyester fabric, PTFE plate, wood plate, steel plate, and aluminum plate) (e). (f–g) Durability against sandpaper abrasion. (h) Self-healing property of the superhydrophobicity. (i) Comparison between coated and bare surfaces. The icephobic properties of the coated surface (left) compared to those of the bare surface (right) (Li et al., 2018).

4.2.2 Dual-Scale Structures

A stable superhydrophobic aluminum alloy surface with dual geometric architectures can be prepared by a combination of the simple processes of chemical etching, dip-coating, and modification with fluorosilicone, as shown in Fig. 4.7a.

In this fabrication, the Al surface with 20 min of acid etching and nanosilica dip-coating has the best superhydrophobicity, showing water contact angles greater than 157° and water sliding angles less than 1°. Such a surface has a roughness (R_{ms}) that ranges from 25.47 at an etching time (ET) of 0 min to 184.54 at 10 min, 267.93 at 15 min, 242.5 at 20 min, and 144.17 at 25 min. The icing test can be performed on the surfaces after etching times of 0 min, 10 min, and 20 min. These surfaces can be defined as surface I (20 min of etching without modification), surface II (0 min of etching with modification), surface III (10 min of etching with modification), and surface IV (20 min of etching without modification), as shown in Fig. 4.7b. For observation times from 20 min to 50 min, the surfaces display different degrees of icing. Surface IV has relatively low icing (Fig. 4.7b).

The ice weight can be estimated as shown in Fig. 4.7c; the values are 72.7 mg, 62.3 mg, 58.1 mg, and 47.7 mg for surfaces I, II, III, IV, respectively. This indicates that surface IV (i.e., 20 min of etching with modification) displays an excellent anti-icing ability (Zhang et al., 2017b).

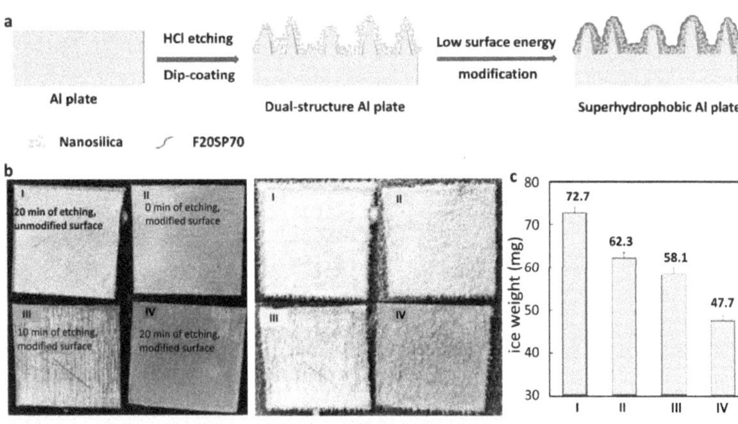

Figure 4.7 (a) Illustration of fabrication of a superhydrophobic aluminum alloy surface with dual geometric architectures via a combination of the simple processes of chemical etching, dip-coating, and modification with fluorosilicone. (b) Icing test versus time on four surfaces: 20 min of etching + unmodified surface (Frame I), 0 min of etching + modified surface (Frame II), 10 min of etching + modified surface (Frame III), and 20 min of etching + modified surface (Frame IV) (Zhang et al., 2017b).

4.2.3 Transparent Polymer Coatings

Highly transparent antifogging/anti-icing coatings have been developed from amphiphilic block copolymers of polyhedral oligomeric silsesquioxane-poly[2-(dimethylamino)-ethylmethacrylate]-block-poly(sulfobetaine methacrylate) (POSS-PDMAEMA-*b*-PSBMA) with a small amount of ethylene glycol dimethacrylate (EGDMA) via UV curing (Li et al., 2017). Figure 4.8a illustrates the synthesis of POSS-PDMAEMA-*b*-PSBMA. POSS-PDMAEMA-*b*-PSBMA block copolymers are synthesized via ATRP of DMAEMA by using POSS-Br as an initiator, followed by ATRP of SBMA by using POSS-PDMAEMA–Br as a macroinitiator. Three POSS-PDMAEMA-*b*-PSBMA block copolymers with different polymerization degrees are prepared, whereas block copolymer PDMAEMA-*b*-PSBMA without POSS is also synthesized as the control by using ethyl-2-bromoisobutanoate as the initiator. Stronger hydrogen-bonded water molecules can form on the zwitterionic SBMA surfaces than on the PEG surface, and a small amount of SBMA in the PDMAEMA-*b*-PSBMA block copolymers still exhibits a dual thermoresponsive performance. Given that PDMAEMA chains play a key role in antifogging, block copolymers with a higher DMAEMA content can be added (Li et al., 2017).

To obtain molar ratios of DMAEMA similar to those of SBMA, three block copolymers, i.e., POSS-PDMAEMA$_{50}$-*b*-PSBMA$_7$, POSS-PDMAEMA$_{70}$-*b*-PSBMA$_{10}$, and POSS-PDMAEMA$_{90}$-*b*-PSBMA$_{13}$, are synthesized. The compositions of the prepared block copolymers are abbreviated as POSS-D$_{50}$-*b*-S$_7$, POSS-D$_{70}$-*b*-S$_{10}$, POSS-D$_{90}$-*b*-S$_{13}$, and D$_{50}$-*b*-S$_7$. The transmittance of the SIPN coatings containing the block copolymers POSS-D$_{50}$-*b*-S$_7$, POSS-D$_{70}$-*b*-S$_{10}$, POSS-D$_{90}$-*b*-S$_{13}$, and D$_{50}$-*b*-S$_7$, denoted as C-POSS-D$_{50}$-*b*-S$_7$, C-POSS-D$_{70}$-*b*-S$_{10}$, C-POSS-D$_{90}$-*b*-S$_{13}$, and C-D$_{50}$-*b*-S$_7$, respectively, is obtained after further airdrying at room temperature. The thickness of the SIPN coatings is ~10 μm. The amphiphilic SIPN coatings exhibit an anti-icing ability with a freezing delay time (T_D) of more than 2 min at −15°C (Fig. 4.8b) owing to the aggregation of the hydrophobic POSS groups and the self-lubricating aqueous layer generated by nonfreezable bound water on the surface. It can also be suggested that the POSS groups play an important role in regulating and controlling the superior hydrophobic/hydrophilic balance of the coating surface to form a self-lubricating aqueous layer with anti-icing performance. Therefore, the C-POSS-D_{50}-*b*-S_7 coating, which has the lowest content of hydrophilic units, exhibits the longest freezing delay time.

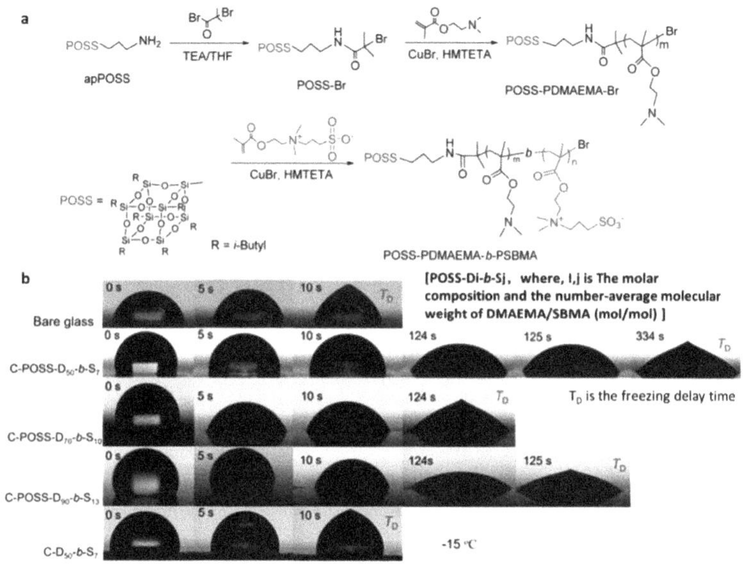

Figure 4.8 (a) Illustration of the synthesis of POSS-PDMAEMA-*b*-PSBMA. (b) Comparison of anti-icing properties via observation of droplet shapes (Li et al., 2017).

Figure 4.9 indicates that the anti-icing mechanism of the amphiphilic SIPN coating including POSS-PDMAEMA-*b*-PSBMA (Fig. 4.9a) arises from the self-lubricating aqueous layer on the surface (see right bottom inset). However, without POSS, i.e., the hydrophilic SIPN coating including PDMAEMA-*b*-PSBMA (Fig. 4.9b), the coating exhibits an antifogging property but not an anti-icing property.

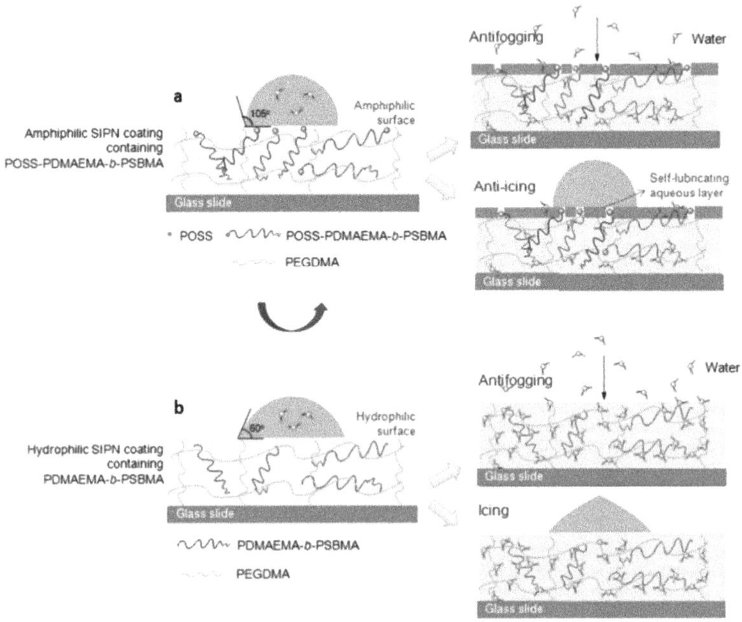

Figure 4.9 (a) Illustration of amphiphilic SIPN coating including POSS-PDMAEMA-*b*-PSBMA for antifogging and anti-icing properties. (b) Illustration of hydrophilic SIPN coating including PDMAEMA-*b*-PSBMA for antifogging but not anti-icing properties (Li et al., 2017).

4.2.4 Inhibition of Heterogeneous Ice Nucleation

Nature has developed its own approaches to control ice formation; for example, antifreeze proteins (AFPs) can protect organisms from freezing damage by controlling ice formation effectively (DeVries and Wohlschlag, 1969; Duman, 2015). Some recent attempts have been undertaken to introduce AFPs onto solid substrates to control heterogeneous ice nucleation (HIN) (Esser-Kahn et al., 2010; Gwak et al., 2015; Liu et al., 2016). To study the HIN behavior on various poly(butylene succinate) (PBS) surfaces ion modified via chain-extension reaction, inspired by AFPs, PBS-based polyampholytes, containing both negative and positive charge groups on a single chain, are employed (He et al., 2017).

Figure 4.10 shows the synthesis process of dihydroxyl-terminated PBS oligomer (PBS-diol) and PBS-based derivatives with various functional groups. PBS-diol can be prepared by esterification and polycondensation of succinic acid with 1,4-butanediol based on the catalyst titanium(IV) butoxide (Fig. 4.10a). These series of PBS-based derivatives are designed and synthesized via a melt chain-extension reaction in the presence of hexamethylene diisocyanate (HDI) as a chain extender at 130°C. By regulating the feed molar ratios of PBS-diol, the carboxyl-based group of 2,2-bis(hydroxymethyl)butyric acid (R_C), the amino-based group of N-methyldiethanolamine (R_N), etc., chain-extended PBS-diol (PBS), carboxylated PBS (PBS-C), aminated PBS (PBS-N), and polyampholytes with random blocky carboxyl and amino groups (PBS-CN) are effectively achieved (Fig. 4.10b). The molar ratio of PBS-diols/R_C for PBS-C and PBS-diols/R_N for PBS-N are both set at 1/1. The molar ratios of PBS-diols/R_C/R_N are set to 1/0.9/0.9, 1/1.5/1.5, and 1/2.8/2.8 for PBS-CN9, PBS-CN15, and PBS-CN28, respectively. Thus, various surfaces can be used to examine ice formation.

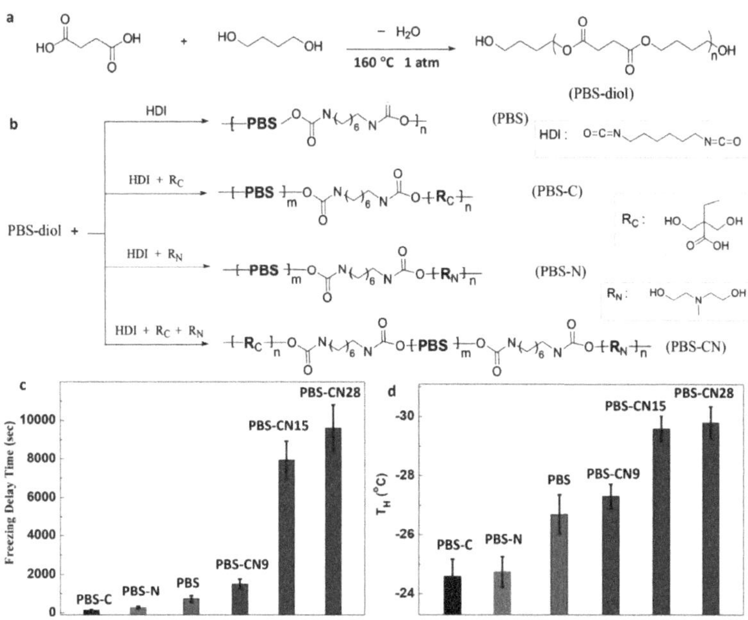

Figure 4.10 (a–b) Illustration of the synthesis of PBS-diol (a) and PBS-based derivatives with various functional groups (b). (c–d) Freezing delay times (c) and different T_H (d) on various surfaces (i.e., PBS-C, PBS-N, PBS, PBS-CN9, PBS-CN15, and PBS-CN28) (He et al., 2017).

Ice nucleation in a water droplet starts at the solid–water interface, generating upward ice growth. The HIN formation in water droplets can be signified by a sudden change in opacity before and after freezing, along with a heterogeneous ice nucleation temperature (T_H). Figure 4.10c,d shows that the different polymers (i.e., PBS-C, PBS-N, PBS, PBS-CN9, PBS-CN15, and PBS-CN28) display different freezing delay times (t_d), among which PBS-CN28 has the longest t_d, i.e., 10 000 s (Fig. 4.10c), and the lowest T_H, i.e., –30°C (Fig. 4.10d).

Figure 4.11 shows the T_H and T_d of the PBS-CN28 coating on different substrates, such as glass, ceramic, silicon, polyethylene, polyimide, polydimethyl siloxane, copper, and aluminum. As shown in Fig. 4.11a, the surfaces coated by PBS-CN28 have generally low T_H (at –30°C); in contrast, the uncoated surfaces have high T_H. The coated surfaces have long t_d, whereas the uncoated surfaces have relatively short t_d (Fig. 4.11b).

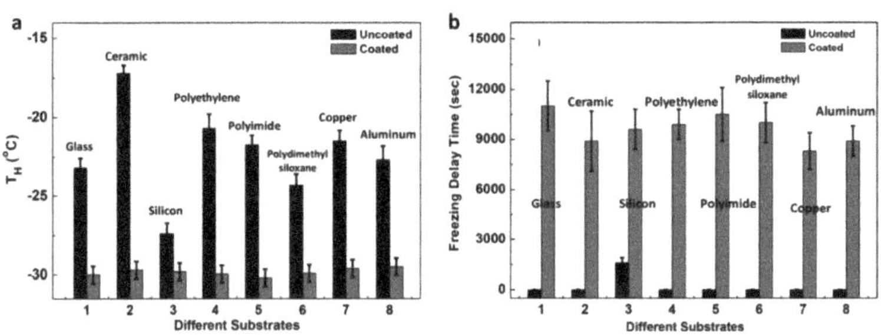

Figure 4.11 Anti-icing properties of PBS-CN28 coatings on different substrates (i.e., glass, ceramic, silicon, polyethylene, polyimide, polydimethyl siloxane, copper, and aluminum) in terms of T_H (a) and freezing delay time (b) (He et al., 2017).

4.2.5 Grooves for Control of Ice Formation

Ice formation is a catastrophic problem that affects our daily lives in a number of ways. The presence of structures on a surface may change the wettability of the surface. Therefore, a structured surface can be designed to control ice formation. Figure 4.12a–c shows three surfaces: Si nanowire (SiNW) array-coated, V-shaped microgroove (VMG), and trapezoid-shaped microgroove patterned

(TMG) surfaces, with water contact angles up to 150°, 140°, and 140°, respectively.

To determine the effect of the structure on ice formation, six kinds of surfaces—plain Si, SiNW, TMG-125, VMG-125, VMG-165, and VMG-250 (here, 125, 165, and 250 indicate groove spacings of 125 μm, 165 μm, and 250 μm)—can be observed by environmental scanning electron microscopy (ESEM). Either the TMG or the VMG surface can spatially control ice formation (Fig. 4.12d–i) with time. The ice formation on the plain Si surface can be randomly distributed, forming normal droplets (Fig. 4.12d). The SiNW surface results in tiny droplets due to the nanowires, and the droplets stay on the top of the nanowires (Fig. 4.12e). The TMG and VMG surfaces generate anisotropic spreading of the droplets along the grooves (Fig. 4.12f–i) (Lo et al., 2017).

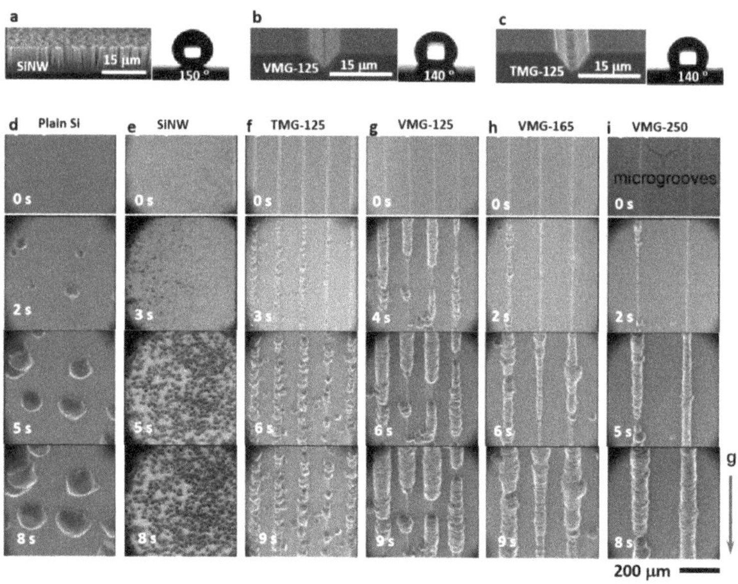

Figure 4.12 (a–c) SEM of three surfaces: SINW (a), VMG-125 (b), and TMG-125 (c), with the hydrophobic properties of CAs of ~150°, 140°, and 140° (right insets), respectively. (d–i) Optical images of in situ observations of ice formation on the structured surfaces of plain Si (d), SiNW (e), TMG-125 (f), VMG-125 (g), VMG-165 (h), and VMG-250 (i) (Lo et al., 2017).

Ice formation can be analyzed by the quantitation of the ice-coverage ratio as a function of time on the studied surfaces. The yellow solid symbols in the figure represent the times of complete ice coverage of the studied surfaces. The ice-coverage ratio (τ, defined as the ratio of the projected area covered by ice to the overall projected area) varies with time during the frosting process. The ice-coverage times on the TMG-125, VMG-125, VMG-165, and VMG-250 surfaces are ~31, ~29, ~40, and ~68 s, respectively. The ice-coverage time on the VMG-250 surface is ~2.7 and ~34 times those on the plain Si and SiNW surfaces, respectively (Fig. 4.13).

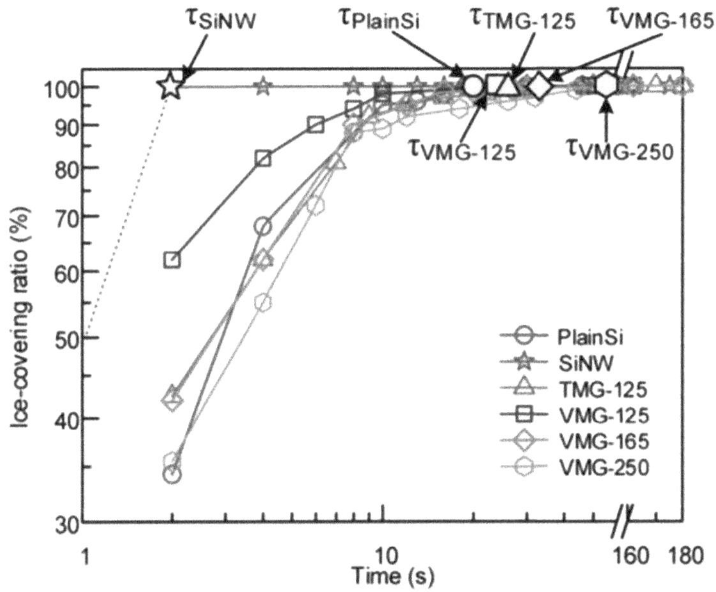

Figure 4.13 Relationship of ice-coverage ratios versus time on different surfaces (plain Si, SiNW, TMG-125, VMG-125, VMG-165, and VMG-250) (Lo et al., 2017).

4.2.6 Self-Lubricating Liquid Water Layer

A robust prototypical anti-icing coating with a self-lubricating liquid water layer (SLWL) can be fabricated by grafting crosslinked hygroscopic polymers inside the micropores of silicon wafer surfaces. Figure 4.14a shows the preparation process of a micropore array silicon wafer surface impregnated with crosslinked poly(acrylic acid) (PAA). 3-(Methacryloxypropyl)-trimethoxysilane (MPS) is attached to the hydroxylated silicon wafers treated with a piranha solution (1:3 w/w H_2O_2/H_2SO_4). Then, crosslinked PAA is grafted inside the silicon wafer by free radical polymerization (Diao et al., 2011) using water-soluble polyethylene glycol diacrylate (PEGDA) as the crosslinker and bis(2,4,6-trimethylbenzoyl)phenylphosphine oxide (BAPO) as the photoinitiator under UV irradiation.

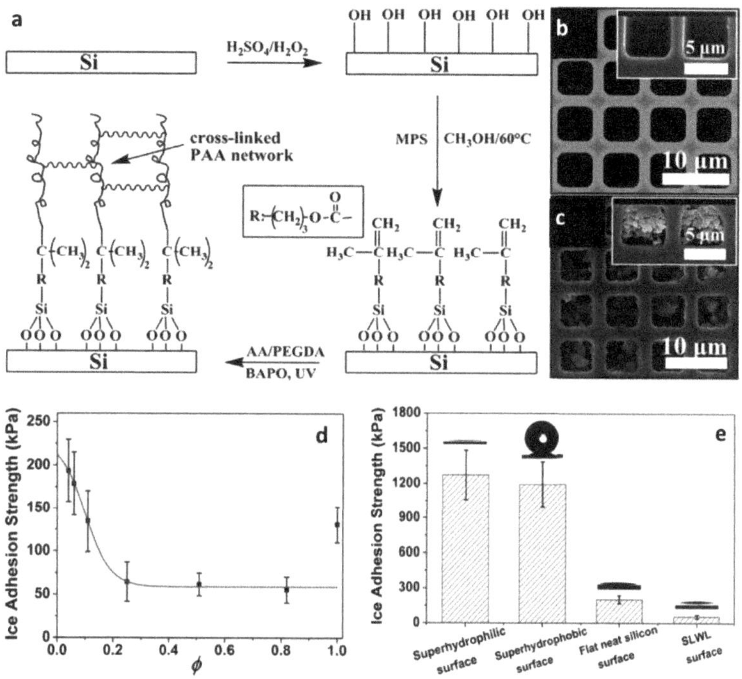

Figure 4.14 (a) Illustration of the fabrication of a micropore array silicon wafer surface impregnated with crosslinked poly(acrylic acid) (PAA). (b–c) SEM images of the micropore array silicon wafer surface before (b) and after grafting with the crosslinked hygroscopic polymer (c). (d–e) Ice adhesion strength versus area fraction (ϕ) (d) and different wettability surfaces (e) (Chen et al., 2013).

Figure 4.14b,c shows SEM images of the micropore array silicon wafer surface before and after grafting with the crosslinked hygroscopic polymer. The crosslinked hygroscopic polymer can be obviously observed inside the micropores after grafting of PAA (insets of Fig. 4.14b,c). The ice adhesion property can be examined by ice shedding off under the action of wind. This method has been widely used to give a reproducible value of the ice adhesion on various surfaces, in which tests are done under freezing rain conditions (in a wind tunnel).

Figure 4.14d shows the variation in the ice adhesion strength with the area fraction (ϕ), which is defined as $\phi = a^2/(a + b)^2$, where a is the side length of the pore and b is the space between two neighboring pores. Figure 4.14d shows that ice adhesion on the SLWL surface decreases sharply as ϕ increases from 0 to 0.2 and is then maintained at a constant value (60 ± 16 kPa) when ϕ is higher than 0.2 (Fig. 4.14d). Figure 4.14e shows a comparison of ice adhesion between surfaces such as superhydrophilic, superhydrophobic, flat neat silicon, and SLWL surfaces. The SLWL surface has the lowest ice adhesion, far less than 150 kPa (Fig. 4.14e).

4.3 Characterization of Anti-Icing/Icephobic Surfaces

Generally, the properties of anti-icing/icephobic surfaces can be characterized by the icing delay times of reference droplets, ice adhesion for ice droplet shedding off via wind action, sliding-off angle, etc.

4.3.1 Long Icing Delay Times of Reference Droplets

Superhydrophobic surfaces can induce a long delay time for droplet freezing, i.e., a droplet on a surface can take a long time to freeze into ice. Different wettability surfaces induce different delay times of droplet freezing. The icing delay time can be defined as the time over which the droplet state changes from transparent to opaque. The icing delay times can be controlled by the design of surfaces in terms of their chemical properties and structures.

A composite micro-/nanostructure surface can be designed by using poly(vinylidene difluoride) (PVDF) polymer in combination with zinc oxide (ZnO) materials to realize a ZnO-on-PVDF micro-/nanostructure (ZP-MN) (Wen et al., 2014). In this fabrication, a heat-pressing pattern-template method is adopted to obtain a microratchet PVDF polymer surface as the substrate. Then, ZnO nanohairs are planted on the microratchet PVDF polymer substrate. Figure 4.15a,b shows an SEM image of a microratchet PVDF surface covered with nanoscale ZnO pillars (i.e., ZP-MN). The microratchets have a periodicity of 300 μm, and the nanoscale pillars have a diameter of 100–200 nm. To characterize the icephobic property of surfaces, four surfaces with smooth (S), microratchet (M), nanopillar (N), and ZP-MN structures are observed in situ. Reference droplets can be placed on the four surfaces at the same time under the same low temperature of −10°C, as shown in Fig. 4.15c,d. The icing delay time can be tested by observing the droplet state change from transparent to opaque. It is observed that the first change in droplet state appears on the M-surface at 300 s, followed by the S-surface at 720 s. The N-surface starts to induce a change in the droplet state at 2280 s. ZP-MN has the longest icing delay time of 6360 s, at which the droplet state starts to be opaque. Complete freezing of the droplet occurs at 7360 s. This demonstrates that the micro- and nanostructured surface achieves an excellent icephobic property.

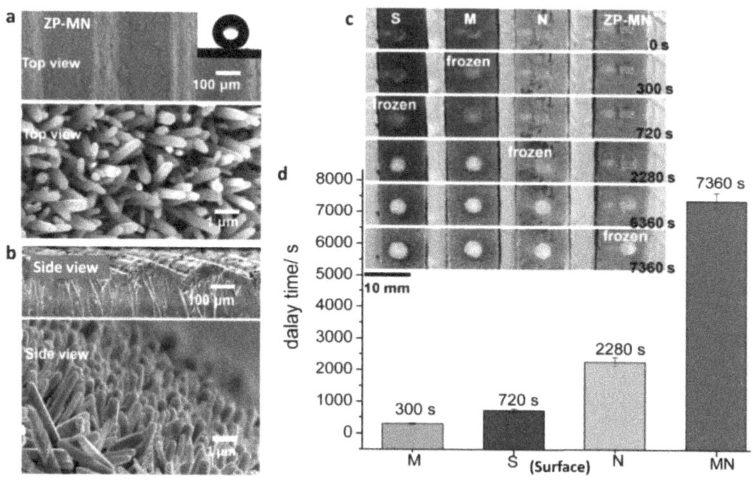

Figure 4.15 (a–b) SEM images of a micro- and nanostructured surface composed of PVDF microratchets and ZnO nanopillars (ZP-MN). (c–d) Different icing delay times (300 s, 720 s, 2280 s, and 7360 s) for four surfaces (M-surface, S-surface, N-surface, and MN-surface) (Wen et al., 2014).

Anti-icing abilities are achieved on surfaces of micropillar arrays with nanohairs fabricated by soft replication and crystal growth methods (Shi et al., 2015), i.e., on surfaces of different micropillar arrays with similar nanohairs and of different nanohairs with similar micropillar arrays. The silicon master pattern is obtained through the Bosch-etching technique. The micropillar arrays are realized through two replication steps: An elastomeric polymer polydimethylsiloxane (PDMS) mold of negative well arrays is obtained by soft lithography (Howarter and Youngblood, 2008; Tricoli et al., 2009), replicated from the Si master, and then, an epoxy replica of positive micropillars is procured via a second replication. Figure 4.16a–g shows SEM images of micropillar arrays with nanohairs on seven surfaces. Figure 4.16a, b, c, and d shows images of micro-/nanostructures with widths (W) of ≈3.71, ≈4.88, ≈5.29, and ≈6.47 μm, respectively, but nearly the same length (L) of ≈11.62 μm on average and nearly the same diameter of nanohairs (d_{nano} ≈ 70 nm). Figure 4.16e, f, and g shows images of nanohairs with different diameters of ≈100, ≈490, and ≈260 nm, respectively, but nearly the same micro-/nanostructures of W ≈ 3.93 μm and L ≈ 12.0 μm on average.

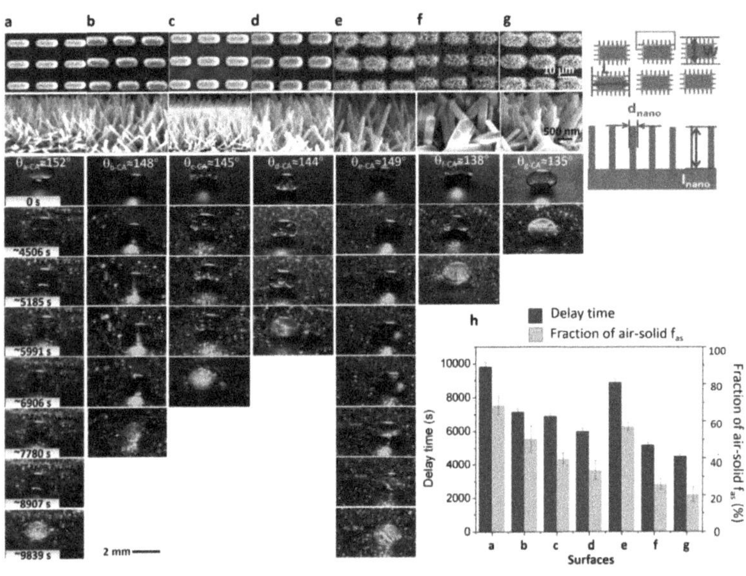

Figure 4.16 (a–g) SEM images of different-sized micro-/nanostructured surfaces, and optical images of icing delay properties via observation of opaque droplets. (h) Graph of delay time versus air–solid fraction on different surfaces (Shi et al., 2015).

The icing delay times can be examined when a droplet (volume of 7 μL) is placed on these surfaces at a low temperature of −10°C using a cooling stage. As the results indicate, the a-surface has the longest delay time (DT) of ≈9839 s, and the e-surface has the second longest DT of 8907 s. The other surfaces have relatively short DTs, for example, b-surface (DT ≈ 7780 s), c-surface (DT ≈ 6906 s), d-surface (DT ≈ 5991 s), f-surface (DT ≈ 5185 s), and g-surface (DT ≈ 4506 s). As shown in Fig. 4.16h, these surfaces with different DTs have different air-to-solid ratios. The liquid–solid fraction (f_{ls}) changes with low temperature. According to Cassie's equation, $f_{ls} = (\cos \theta_{CA} + 1)/(\cos \theta_0 + 1)$, where $f_{ls} = 1 - f_{as}$, and f_{as} is the air–solid fraction; θ_{CA} and θ_0 are the apparent contact angles on a rough surface and a smooth surface (given $\theta_0 = 150°$). For the a-surface (Fig. 4.16a), it can be estimated that f_{ls} is 15.08%; thus, $f_{as} \approx 84.92\%$. In contrast, the other surfaces have low air–solid fractions, as shown in Fig. 4.16h. This study indicates that the optimal micro- and nanostructures on the as-designed surface would induce a robust icephobic property along with a long icing delay time.

4.3.2 Ice Adhesion

Ice buildup and its subsequent pinning on exposed surfaces of infrastructure, such as on ship hulls, aircraft, wind turbines, and power lines, have become severe safety challenges (Parent and Ilinca, 2011; Fakorede et al., 2016). Consequently, extraordinary efforts have been dedicated to preventing or alleviating ice adhesion to substance surfaces (Kreder et al., 2016; Jin et al., 2017; Dash et al., 2018). In general, the adhesion of ice on smooth hydrophobic coatings is lower than that on hydrophilic coatings (Coady et al., 2018). Therefore, hydrophobic surfaces have been greatly investigated to decrease ice adhesion.

As an example, a novel icephobic coating with matrix enhancement and abrasion resistance integrates branched poly(dimethylsiloxane) (PDMS), vinyl-functionalized polyhedral oligomeric silsesquioxane (POSS) partially modified with short PDMS chains, and a specific lubricant (Gao et al., 2019). The functionalized POSS is tailored and intended for compatibility and miscibility with the bulk PDMS matrix as well as fulfillment of co-crosslinking, as shown in Fig. 4.17a. Branched PDMS can be prepared by mixing the multiple silicon hydrogen-functional linear PDMS copolymer PDMS-PMHS with the monovinyl-terminated linear PDMS polymer PDMS-V through the hydrosilylation reaction (Fig. 4.17a). PDMS-functionalized POSS (V_3-POSS-$PDMS_5$) can be synthesized as a co-crosslinker via hydrosilylation of OVPOSS and PDMS-H (Tao et al., 2017; Gao et al., 2019), as depicted in Fig. 4.17b. Figure 4.17c illustrates the fabrication of the PDMS slippage coating through the hydrosilylation reaction.

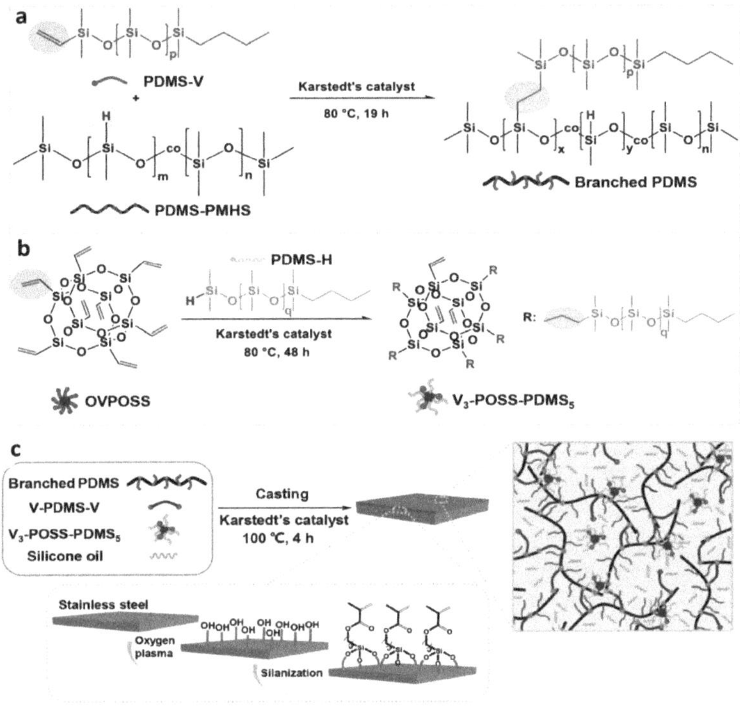

Figure 4.17 Illustration of the preparation of branched PDMS (a), functionalized POSS V_3-POSS-$PDMS_5$ (b), and a PDMS slippage coating through the hydrosilylation reaction (c) (Gao et al., 2019).

The ultralow ice shear strength of slippery wavy coatings can increase after 15 icing/deicing cycles, demonstrating a relatively short-lived icephobicity. The ice shear strength is measured on prepared icephobic coatings at −15°C (Li et al., 2014). The thickness of the coatings is usually measured before ice shear strength tests in view of the close relationship between the thickness and the ice shear strength (Chaudhury and Kim, 2007; Kim et al., 2007; Kim and Chaudhury, 2009). The icephobic coatings are crosslinked by the original vinyl-terminated PDMS and functionalized POSS as the co-crosslinker, in which silicone oil is entrapped around the side chains of the branched PDMS in a homodisperse form.

When water droplets set in the cuvette turn into solid ice, a shear force is applied horizontally in an attempt to remove the ice from the as-prepared coatings (Fig. 4.18a). Silicone oil migrates toward the coating surface gradually owing to its intrinsic osmotic pressure, forming a gradient distribution along the depth of the coatings, represented by arrows at the bottom of the as-prepared coating in Fig. 4.18b. This consequently leads to the formation of a layer of oil lubricant, illustrated in yellow, on the surface, with a relatively weak interaction at the interface. At sufficiently high tensile stress, stick–slip motion of ice occurs under indirect contact between the ice and the surface, resulting in slow slip-off until complete detachment from the coatings.

To describe the as-designed different materials, herein, the materials are defined as follows: $^{b}PDMS^{\circ}$ refers to the branched PDMS-based coating containing silicone oil, whereas for $^{b}PDMS^{\circ}{}_{x}$, the number x represents the integral mass percentage of the functionalized POSS in the as-prepared icephobic coating. Interfacial slippage is indicative of adhesive failure as expected, which can be manifested by the insets shown in Fig. 4.18b with no residual icicles on the surface of the $^{b}PDMS^{\circ}4$ coating (with a crosslink density of ~74 mol/m^3 and a gel fraction of ~55.3%) after the first and fiftieth icing/deicing cycles. The relatively mild abrasion measurement (Fig. 4.18c) is executed to mimic severe wear conditions and assess the abrasion resistance durability of the icephobic materials, as illustrated in Fig. 4.18d. It is remarkable that the as-prepared PDMS slippage elastomers retain their abrasion resistance and icephobicity, with up to 175 abrasion cycles conducted with no prominent change in the ice shear strength values, all of which are below 25 kPa. To be more precise, the abrasion resistance of the as-prepared PDMS slippage coatings increases gradually with increasing POSS co-crosslinker content, with up to 175 cycles being reached for the sample $^{b}PDMS^{\circ}15$ (with a crosslink density of ~150 mol/m^3 and a gel fraction of ~57.5%).

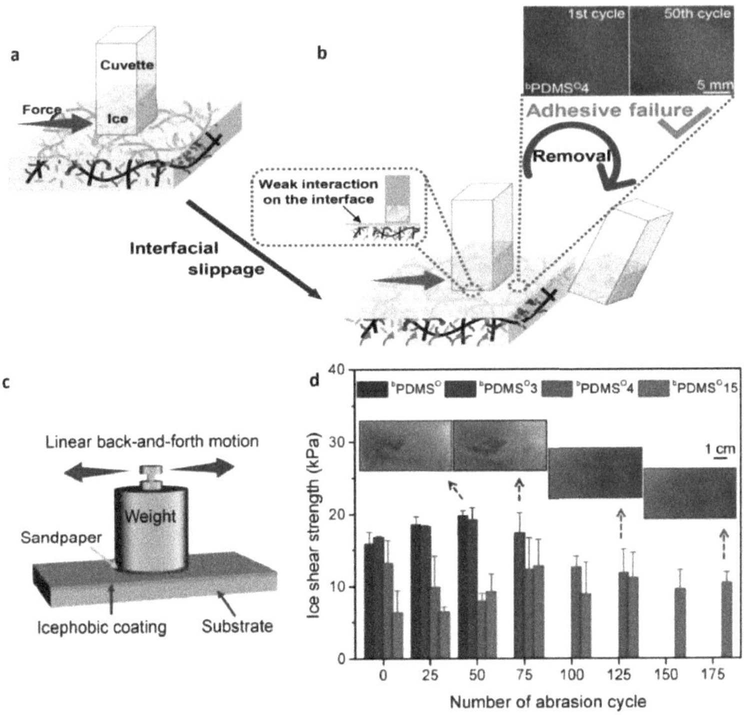

Figure 4.18 (a) Illustration of ice removal from PDMS slippage coatings under stick–slip motion. (b) Silicone oil migrates toward the coating surface gradually (see arrows at the bottom), along with the formation of a thin layer of oil lubricant with a relatively weak interaction. The insets indicate the sample of ^{b}PDMS°4. (c–d) Schematic of the linear abrasion test (c), and changes in the ice shear strength of the PDMS slippage coatings (d) measured at −15°C after abrasion cycles. The insets show optical images of the samples (^{b}PDMS°, ^{b}PDMS°3, ^{b}PDMS°4, and ^{b}PDMS°15) after abrasion damage (Gao et al., 2019).

4.3.3 Self-Shedding Off

As is known, a passive anti-icing property prevents freezing of incoming water without any power input, while an active deicing property provides an alternative function of ice removal when the environmental conditions exceed anti-icing limits.

As an example, a perfluorododecylated graphene nanoribbon (FDO-GNR) coating (Wang et al., 2016b) can be synthesized by splitting multiwalled carbon nanotubes (MWCNTs) using an Na/K alloy (Genorio et al., 2012) and functionalizing the edges with perfluorododecyl groups (as shown in Fig. 4.19a). Potassium intercalates between the walls of the MWCNTs and longitudinally splits the structure to form carboanion edges (Genorio et al., 2012; Lu et al., 2013), which easily react with 1-iodoperfluorododecane to become chemically bonded to the GNR edges. The hydrophobicity of the GNRs is largely improved by the modification of perfluoroalkyl groups compared to hexadecylated GNRs (Fig. 4.19b), and then, liquid (e.g., oil) is infused into the structures of the FDO-GNRs to achieve the lubricant property (i.e., slippage property). Furthermore, as shown in Fig. 4.19c, the lubricant FDO-GNR surface displays the self-shedding-off property for ice droplets on the surface (③-④), in contrast to the surface without the lubricant (①-②).

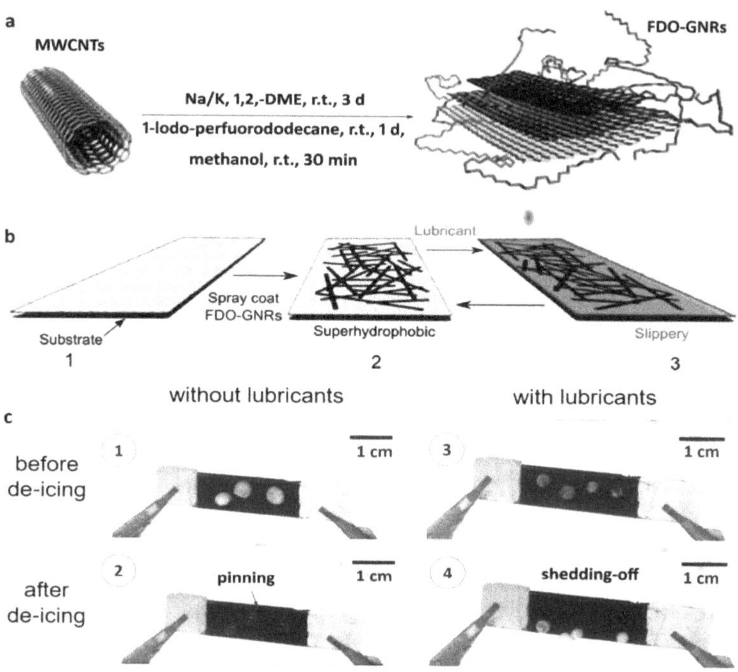

Figure 4.19 (a) Illustration of the synthesis of FDO-GNRs and preparation of a superhydrophobic slippery surface. (c) Excellent icephobic property for surfaces with lubricants (③-④), and non-icephobic property for surfaces without lubricants (①-②) (Wang et al., 2016b).

4.3.4 Nanostructure-Tuned Icephobicity

To obtain robust ice-repellent and lightweight materials, icephobic gummed tape is fabricated by pressing PVDF microspheres onto common gummed tape and then fabricating different ZnO nanostructures on the PVDF microspheres using an improved crystal growth technique by controlling the reaction time (Wang et al., 2014).

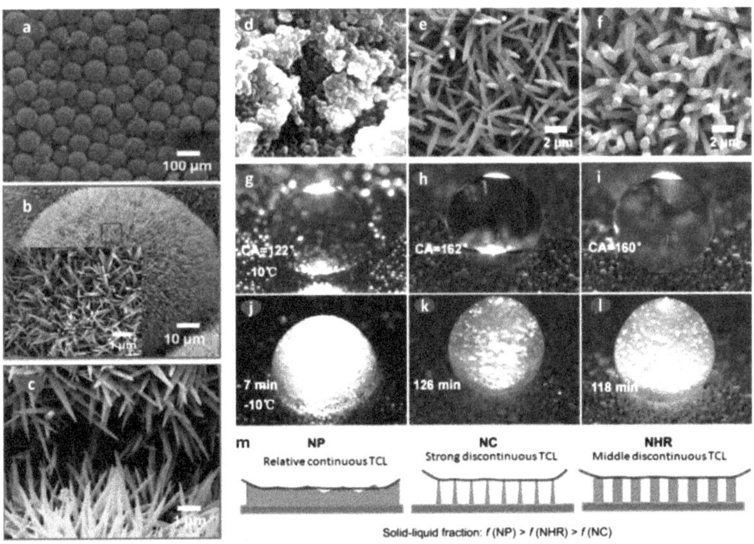

Figure 4.20 (a–c) SEM image of icephobic gummed tape covered with PVDF microspheres containing ZnO nanoscale cones. (d–l) Comparison between surfaces with nanoparticles (NPs) (d), nanocones (NCs) (e), and nanohexagonal rods (NHR) (f) and their superhydrophobic properties (g–i) and icephobic properties (j–l). (m) Illustration of the interface effect with different TCLs. The solid–liquid fractions follow f(NP) > f(NHR) > f(NC) (Wang et al., 2014).

As shown in Fig. 4.20a, microspheres with diameters of ~80–120 μm cover the whole surface (Fig. 4.20a). Every microsphere is covered by nanoscale cones (NCs) (Fig. 4.20b). The spacings of the microspheres are ~2 μm (Fig. 4.20c), and the NC length is ~3 μm. Such a surface possesses a large air-trapping ability due to the micro- and nanostructures (Boreyko and Chen, 2009; Hu et al., 2011). The icephobic property can be examined by comparing the icing delay times of three nanostructured surfaces: those with nanoparticles (NPs), nanocones (NCs), and nanohexagonal rods (NHRs), as shown in Fig. 4.20d–f. A droplet can be placed on these surfaces at –10°C, with different contact angles of 122°, 162°, and 160° (Fig. 4.20g–i). The icing delay times are different, i.e., 7 min, 126 min, and 118 min (Fig. 4.20j–l). This can be attributed to the difference in the solid–liquid–air TCL, i.e., a relative continuous TCL for NPs, a strongly discontinuous TCL for NCs, and a relatively discontinuous TCL for NHRs (Fig. 4.20m). Thus, the solid–liquid fractions (*f*) are different, i.e., *f*(NP) > *f*(NHR) > *f*(NC). The ice-shedding-off property can be observed on surfaces containing NCs (Fig. 4.21a), but pinning occurs on surfaces containing NPs (Fig. 4.21b).

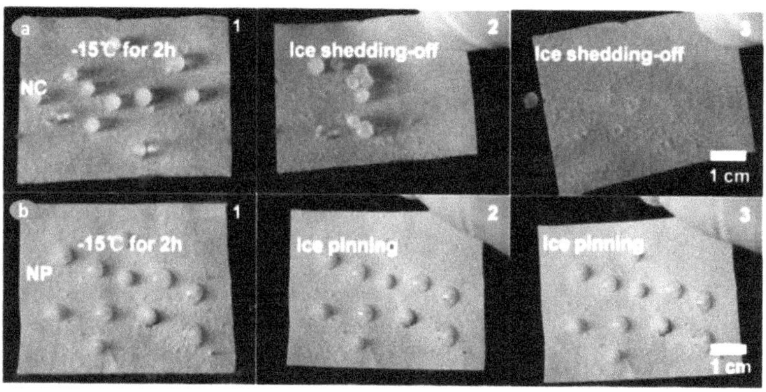

Figure 4.21 Excellent icephobic property of the nanocone (NC) surface (a), but not of the nanoparticle (NP) surface (b) (Wang et al., 2014).

4.3.5 Thermal Behavior on Icephobic Surfaces

The thermal behavior on icephobic surfaces can be observed. As an example, an excellent durable micro- and nanostructured surface film is fabricated by combining the airless spray and crystal growth hydrothermal methods (Fig. 4.22a), which is composed of soda lime borosilicate glass hollow (SGH) microspheres and ZnO nanorods (denoted as ZnO-on-SGH) (Zhang et al., 2017a). The average diameter of the SGH microspheres is ~40 μm (Fig. 4.22b,c). The ZnO nanoroads have diameters of 50–100 nm and heights of ~2 μm on average.

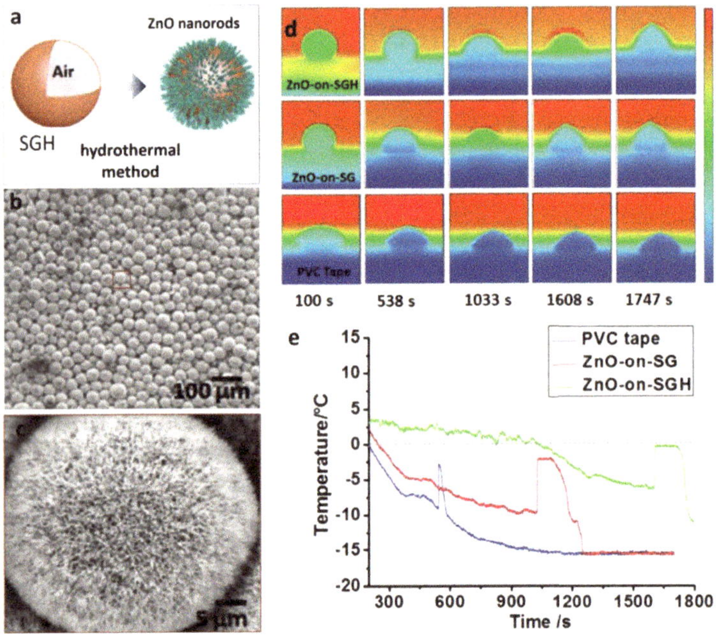

Figure 4.22 (a) Illustration of the fabrication of ZnO-on-SGH. (b–c) SEM images of a structured surface with uniformly distributed microspheres (b) covered by nanorods (c). (d) Thermal infrared images of droplet states on three surfaces (i.e., ZnO-on-SGH, ZnO-on-SG, and PVC tape). The color bar indicates the different temperatures. (e) Temperature curve for the icephobic property versus time (Zhang et al., 2017a).

To further study the impact of the micro- and nanostructured surface and the SGH spheres on the low-temperature behavior, a frozen macroscopic droplet experiment is designed for the analysis of the dynamic heat transfer process. Three kinds of films—ZnO-on-SGH film, superhydrophobic ZnO-on-SG film (SG denotes soda lime borosilicate glass microspheres), and normal PVC tape—are placed on a cooling stage at −18°C. After 10 min, one droplet with a volume of 10 μL is placed on the surface using a microsyringe. In situ observation is performed by using a thermal infrared imaging camera. As shown in Fig. 4.22d, the color changes indicate the surface and droplet temperature change, and the color bar indicates that blue color corresponds to less than −10°C. At the contact interface location, ZnO-on-SGH displays stability over time, whereas the other surfaces exhibit droplet changes of a large degree. The recorded curves in Fig. 4.22e indicate that the droplet on the ZnO-on-SGH film remains at a temperature above the ice point until ~1050 s. At 1608 s, it begins to freeze and has a longer freezing time than the droplets on the ZnO-on-SG film.

4.3.6 Rate of Ice Formation

The rate of ice formation can be improved; for example, inspired by the poison dart frog's skin that secretes functional liquid after being stimulated, a liquid-infused double-layer surface (LDS) is designed by integrating a superhydrophobic layer with a liquid-infused layer (Guo et al., 2019). As shown in Fig. 4.23a, the top layer is fabricated by electrospinning PET (polyester) fibrous film, and superhydrophobicity can be obtained by spray coating. The liquid-infused layer consists of an MA-infused part and PET/CHI (chitosan) fibrous film, and the MA-infused part forms naturally when the MA underneath infiltrates into the bare fibrous PET. Figure 4.23b shows SEM images of the top layer on the LDS with fibril-crossed porous structures, where the average diameter of the fibers is 900 nm. After spray coating with superhydrophobic SiO_2 nanoparticles, water droplets achieve a contact angle of more than 150° on the surface (Fig. 4.23c) at 3645 s when the surface is in high relative humidity (RH) environments (e.g., 60%) and at low temperature ($T \approx -10°C$). In particular, the LDS achieves water droplet (10 μL) bounce off (Fig. 4.23d). The frost formation can be compared between the LDS and a normal surface, as shown in Fig. 4.23e,f. When observed at 75 min, the LDS generates less frost (Fig. 4.23e), whereas the normal surface has much more frost (Fig. 4.23f).

In detail, the antifrosting sites of the LDS are examined by an infrared thermal camera. Figure 4.24a shows time-lapse infrared images of LDS sites at a low temperature ($T = -15°C$, RH = 60%). Apparently, the LDS forms a less scattered distribution than the normal surface until ~3240 s, which indicates that the antifrosting rate can be reflected by the temperature change at sites on the surfaces. The maps corresponding to the infrared images display three-dimensional distributions of the temperature, where the changes in color and vertical coordinates can indicate frost formation. The frosting rate (W) is defined to reflect the proportion of the entire surface area covered by frost, along with the surface average temperature (Γ). When observed at ~3240 s, W_{LDS} = 3.5% on the LDS at Γ_{LDS} = -11.6°C.

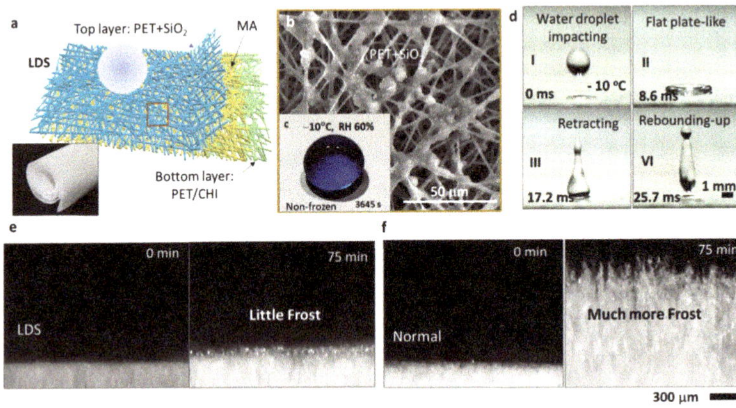

Figure 4.23 (a) Illustration of fabrication of a liquid-infused double-layer surface (LDS). The inset shows a photo of the LDS on a large scale. (b) SEM image of the LDS with upper PET + SiO_2. (d) Photo of the water repellency. (e) Excellent anti-icing/frosting properties for LDS compared to those of a normal surface (Guo et al., 2019).

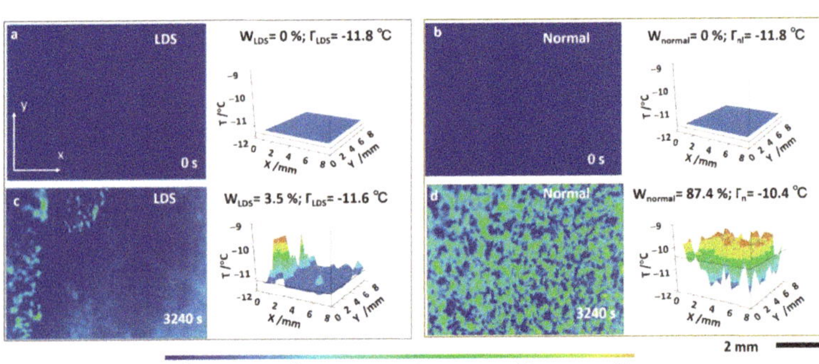

Figure 4.24 (a–b) Analysis of ice formation versus time by infrared thermal mapping. The LDS surface has a strong antifrosting ability with a low rate of ice formation (a), whereas the normal surface has a high rate of ice formation (Guo et al., 2019).

In contrast, for the normal surface, as shown in Fig. 4.24b, there is a much more scattered distribution at ~3240 s, along with

W_{normal} = 87.4% and Γ_{normal} = −10.4°C, which indicates a weak antifrosting ability on the normal surface. Apparently, the LDS has considerably low values of W and Γ in the process over the observed time, i.e., the LDS displays a much stronger antifrosting property than the normal surface.

4.4 Challenging Issues for Future Applications

4.4.1 Magnetically Responsive Hair Arrays

The formation of ice on the surfaces of structures, ranging from roads, buildings, and power lines to cars, trains, and aircraft, brings about extensive social, technical, and economic problems (Mishchenko et al., 2010; Stone, 2012). Although Joule heating, chemical release, and mechanical removal techniques have been commonly utilized for the prevention of ice accumulation and removal (Laforte et al., 1998), a dynamic interface would decrease the ice adhesion, enhancing the anti-icing ability of materials.

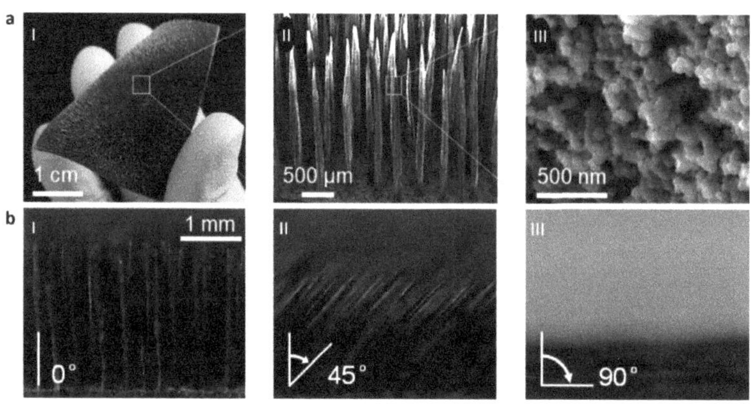

Figure 4.25 (a) Photos of a material with a magnetically responsive hair array (Frames I–II). Every hair is composed of nanoparticles (Frame III). (b) Hair array with different tilt angles of 0° (Frame I), 45° (Frame II), and 90° (Frame III) under the control of a permanent magnet (Lee et al., 2018).

For example, an anti-icing material can actively modulate the droplet bouncing dynamics on a magnetically responsive hierarchical hair array (Lee et al., 2018). The magnetically responsive hair array exhibits an immediate and reversible structural bending motion in response to an external magnetic field. Figure 4.25a shows the detailed structure of the fabricated magnetically responsive hair array (MRA). As reported, an array of hairs with a high aspect ratio (AR) is uniformly constructed on a flexible elastomeric substrate through a moldless self-assembly process (Kim et al., 2016; Huang et al., 2017). As the fabrication process relies on magnetic-field-induced self-alignment of carbonyl iron (CI) particles, a high AR MRA can be formed over a large area. The individual hairs have a top diameter of ~20 μm, a base diameter of ~135 μm, and a height of ~2.3 mm on average. The surfaces of the hairs take on a nanoscale roughness due to the carbon nanoparticles (CNPs) coated over the array to provide superhydrophobic wetting properties. Due to the CI particles embedded in the array, the MRA exhibits dynamic structural changes and actuating motion with a tilt angle (e.g., 45° or 90°) when the magnetic field is modulated by using a permanent magnet (Fig. 4.25b).

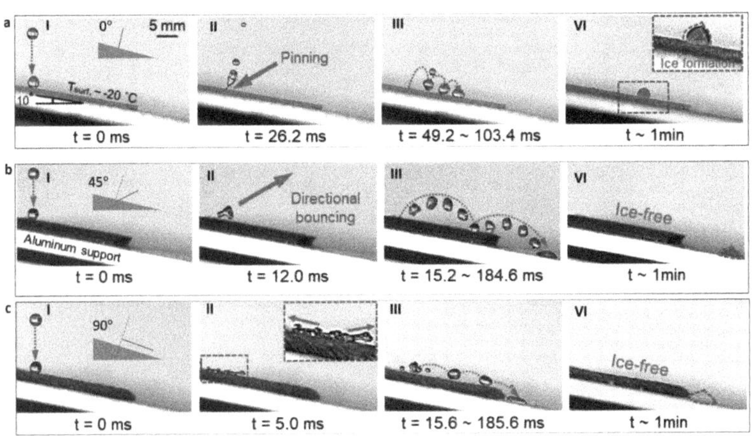

Figure 4.26 Icephobic properties via in situ observation of different surfaces with tilt angles of 0°, with pinning of ice (a), 45°, ice free (b), and 90°, ice free (c) (Lee et al., 2018).

The array also exhibits superhydrophobicity, regardless of its tilt angle, due to the tapered geometry of the hairs and the multiscale surface roughness of the array. Due to its dynamic structure and water-repellent characteristics, the array can induce distinct multiple modes of droplet bouncing behavior by adjusting its structural bending state in a reversible fashion. As for the icephobic properties of these surfaces at a low temperature (–10°C) (Fig. 4.26a–c), the tilt-angle arrays can induce ice droplet shedding off via rebounding (surface slope of 10° to the base). Three different types of bouncing behavior, namely, quasi-pancake bouncing, directional bouncing, and macrotexture-induced droplet fragmentation, can be obtained for surfaces with vertical (0°), tilted (45°), and fully bent (90°) hair arrays, respectively. The ice droplets can be removed from the tilt-angle surfaces, for example, from the tilted (45°) and fully bent (90°) hair arrays (Fig. 4.26b,c).

4.4.2 Low-Interfacial-Toughness Materials

The accretion of ice on surfaces has a severe detrimental impact on a range of commercial and residential activities (Makkonen, 2012). There have been efforts to create coatings that protect against the buildup of ice. Typically, the efficacy of these coatings has been evaluated by measuring the force (F) required to debond a specified area (A) of ice and by defining an ice adhesion strength $f_{ice} = F/A$ as a characteristic property of the system (Meuler et al., 2010). The term icephobic is generally used to describe surfaces for which $f_{ice} < 100$ kPa (Hejazi et al., 2013; Sojoudi et al., 2016), in contrast to structural materials such as aluminum and steel, for which $f_{ice} > 1000$ kPa (Mulherin and Haehnel, 2003; Menini and Farzaneh, 2011). Using f_{ice} to characterize an interface inevitably requires that the force necessary to remove ice scales with the iced area. Many engineering structures susceptible to icing, such as airplane wings, wind turbine blades (Golovin et al., 2019), and boat hulls (Burton et al., 2011), have surface areas that can approach thousands of square meters. Consequently, even with extremely icephobic coatings, structures with large surface areas would require prohibitively high forces to detach entire sheets of ice from the surface.

Table 4.1 Values for interfacial properties measured between ice and 20 different surfaces

No.	Surface	$\theta_{rec}/\theta_{adv}$ (Water)	f_{ice} (kPa) at L = 20 cm	$F_{ice(cr)}$ (N/cm) at $\Gamma = 0.1$ (J/m^2)	Γ(J/m^2)	E (GPa)
1.	Silicone A	120°/87°	107 ± 12	≥198	≥3.8	$(3.5 \pm 1.2) \times 10^{-1}$
2.	Plasticized silicone A	112°/102°	55 ± 11	≥128	≥1.6	$(2.7 \pm 0.33) \times 10^{-2}$
3.	Silicone B	113°/96°	29 ± 2	≥297	≥8.7	$(5.3 \pm 0.99) \times 10^{-4}$
4.	Plasticized silicone B	115°/86°	12 ± 2	≥126	≥1.6	$(1.2 \pm 0.14) \times 10^{-4}$
5.	PVC	92°/63°	114 ± 26	229 ± 52	5.1 ± 2.3	3.6 ± 0.11
6.	PC	92°/69°	103 ± 11	207 ± 22	4.2 ± 0.9	3.3 ± 0.01
7.	Nylon	53°/15°	89 ± 11	179 ± 22	3.1 ± 0.76	3.5 ± 1.2
8.	ABS	99°/59°	86 ± 30	174 ± 23	3.0 ± 0.79	3.4 ± 0.15
9.	PMMA	78°/52°	121 ± 11	170 ± 22	2.8 ± 0.27	2.9 ± 0.07
10.	PETG	89°/62°	84 ± 20	169 ± 40	2.8 ± 1.3	3.3 ± 0.18
11.	CPVC	99°/55°	67 ± 7	133 ± 39	1.7 ± 0.37	3.7 ± 0.07
12.	Polypropylene	104°/86°	64 ± 5	127 ± 10	1.6 ± 0.26	3.0 ± 0.53
13.	Garolite	85°/35°	70 ± 29	124 ± 16	1.5 ± 0.39	4.2 ± 0.07
14.	PS	103°/64°	76 ± 25	120 ± 17	1.4 ± 0.40	3.0 ± 0.07
15.	LDPE	108°/83°	52 ± 6	105 ± 12	1.1 ± 0.24	1.0 ± 0.10

No.	Surface	$\theta_{rec}/\theta_{adv}$ (Water)	f_{ice} (kPa) at L = 20 cm	$F_{ice(cr)}$ (N/cm) at $\Gamma = 0.1$ (J/m^2)	Γ(J/m^2)	E (GPa)
16.	UHMWPE	103°/76°	49 ± 5	97 ± 10	0.9 ± 0.19	1.9 ± 0.04
17.	PTFE	123°/91°	39 ± 4	77 ± 8	0.6 ± 0.12	0.2 ± 0.03
18.	LIT PS	104°/82°	30 ± 3	66 ± 6	0.43 ± 0.08	$(5.9 \pm 1.0) \times 10^{-4}$
19.	LIT PVC	97°/77°	23 ± 4	52 ± 7	0.27 ± 0.07	$(1.2 \pm 0.04) \times 10^{-1}$
20.	LIT PDMS	113°/96°	18 ± 2	35 ± 4	0.12 ± 0.03	$(1.2 \pm 0.14) \times 10^{-4}$
21.	Pure van der Waals	—		32 (calc)	0.1	—

Ref. Golovin et al., 2019.

In contrast, for materials with low interfacial toughness (LIT; interfacial toughness G < 1 J/m^2), the force required to remove adhered ice from large areas (few cm^2 or greater) is both low and independent of the interfacial area (Golovin et al., 2019). The force required to remove an ice layer reaches an asymptotic value. Table 4.1 shows the comparison results for substrates made from common plastics such as polyethylene, polypropylene, and polystyrene (substrate thickness t = 1.6 mm) without any additional modification. A comparison is made between 1-mm-thick coatings of icephobic silicones on aluminum, 1.6-mm-thick plastic substrates, and 1- to 2-μm-thick LIT coatings on aluminum. $F_{(ice\text{-}cr)}$ for the pure van der Waals interaction is calculated using $\Gamma = 0.1\ J/m^2$ for a 6-mm-thick ice sheet. E_{ice} is the modulus of ice of thickness h. $2E_{ice}\,\Gamma/\sigma^2_Y < h < L$, where σ_Y is the yield strength of ice.

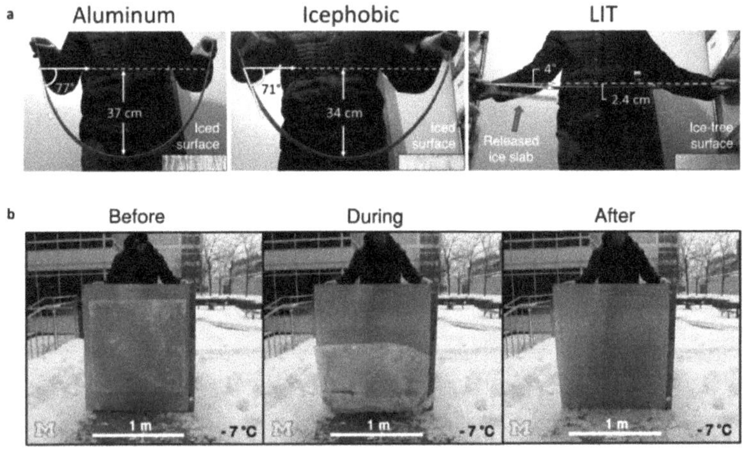

Figure 4.27 (a) Ice-free properties observed for an aluminum surface (left), an icephobic surface (middle), and LIT coatings (right) (Golovin et al., 2019). (b) Durability of an LIT coating on a surface, observed before (left), during (middle), and after (right) fracture of a large sheet of ice.

Figure 4.27 shows the large-scale testing of LIT materials. Figure 4.27a presents a comparison of uncoated, icephobic, and LIT aluminum beams adhered to a sheet of ice (100 cm × 2.5 cm × 0.8 cm) undergoing off-center load flex tests inside a walk-in freezer held at −20°C. Ice fractures from the LIT-coated surface with a remarkably low apparent ice adhesion strength of 0.39 kPa, whereas ice remains adhered to the uncoated aluminum and icephobic specimens even at severe deflections. Figure 4.27b shows the aluminum sheet coated with LIT PDMS before, during, and after fracture of a large sheet of ice (0.95 m × 0.95 m × 0.01 m). The weight of the ice sheet alone is sufficient to cause fracture, displaying an exceedingly low apparent ice adhesion strength of 0.09 kPa (Golovin et al., 2019).

4.4.3 Micro- and Nanostructured Surfaces for Anti-Icing Applications

Aluminum is the most abundant metal on earth and is widely used in mechanical apparatuses, the aerospace industry, and overhead power lines. The accumulation of ice on transmission lines can cause a drastic decrease in the mechanical and electrical performance, leading to serious problems such as the collapse of towers, flashovers of insulator strings, galloping of conductors, and consequent power outages (Jiang et al., 2007). Electrical power outages due to icing have been widely reported in many countries, such as the United States, England, Canada, Norway, and China (Cherney, 1980; Fikke et al., 1993).

The general mechanism of the formation of micro-/nano-structures is depicted in Fig. 4.28. After sanding, the oxide layer of the aluminum surface is removed. When etching in $CuCl_2$ solution, the substitution reaction occurs: $2Al + 3Cu^{2+} \rightarrow 2Al^{3+} + 3Cu$ and $2Al + 6H^+ \rightarrow 2Al^{3+} + 3H_2\uparrow$. The presence of H_2 bubbles prevents Cu deposition because of the absence of Cu^{2+}, leading to a micrometer-scale surface roughness. Large numbers of grain boundaries and dislocations exist in common polycrystalline materials. These areas are prone to dissolve due to their relatively high surface energy, which leads to the generation of deep ravines. The reaction between Al and H_2O at the initial stage can be as follows: $Al + H_2O \rightarrow Al_2O_3 \cdot xH_2O + H_2\uparrow$.

Moreover, the generated $Al_2O_3{\cdot}xH_2O$ can further react with H_2O to form AlO(OH). It can be clearly concluded that the micro-nano-binary structures can trap a large amount of air, which can prevent the penetration of water into the grooves and lead to superhydrophobicity on the surfaces (Fig. 4.28a, the inset is an SEM image of the surface). Figure 4.28b shows the artificial climate laboratory, and Fig. 4.28c presents an inner schematic of the artificial climate laboratory used to simulate glaze ice with a control of the temperature and humidity. Figure 4.28d shows in situ observations of the anti-icing properties of a superhydrophobic aluminum surface and a bare aluminum surface (Zuo et al., 2015).

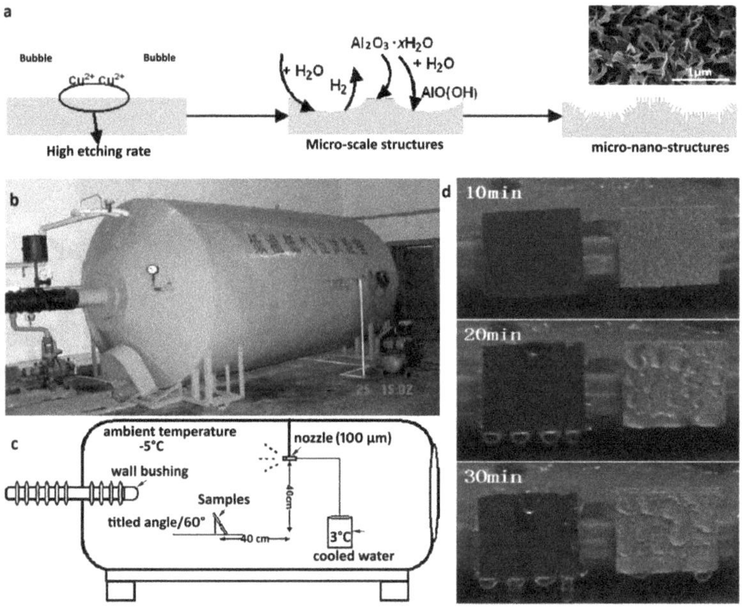

Figure 4.28 (a) Illustration of fabrication of a micro-/nanostructured surface. The inset shows an SEM image of the nanostructure. (b–c) Photo and illustration of an artificial climate laboratory for tests on low-temperature materials. (d) Anti-icing properties of the observed superhydrophobic aluminum surface (left) and bare aluminum surface (right) in the artificial climate laboratory (Zuo et al., 2015).

4.4.4 Icephobic Coatings and Applications

A simple and low-cost technique for the preparation of silicon-oil-infused PDMS coatings with different silicon oil contents can be achieved for icephobic applications. In this design, the highly hydrophobic property of PDMS is maintained by avoiding any polar groups such as C=O and OH in the structure. Thus, the polymer main chain is attached with vinyl and Si–H groups to obtain a crosslinking capability while ensuring a nonpolar chemical structure. The prepared coating surface shows a low surface energy and a very low ice adhesion strength of 50 kPa, only approximately 3% of the value on a bare aluminum (Al) surface. In the silicon-oil-infused PDMS coatings, the low surface energy of the silicon oil and PDMS and the high mobility of silicon oil play important roles in the icephobic property. Both these factors endow the surface with a large water contact angle and hence a small contact area, leading to the formation of a loose ice layer. In addition, the oil-infused polymer structure significantly reduces the contact area of the ice with the solid substrate since the ice mostly contacts the mobile oil. This leads to a very weak interaction between the substrate and ice, consequently significantly reducing the ice adhesion strength on the surface (Zhu et al., 2013).

Figure 4.29 shows the anti-icing properties of both PDMS-coated and bare Al surfaces. Due to the superhydrophobic PDMS coating, a water droplet on the PDMS-coated surface at −5°C can still sufficiently constrict to a spherical shape, then start to freeze, and finally become an ice ball, with a small contact area. As a result, this leads to the formation of a loose ice layer on the PDMS-coated surface. However, on the Al surface, the droplets spread out and freeze, forming an ice layer, and a compact ice layer attaches to the bare Al surface. As shown in Fig. 4.29, loose ice forms on the coated surface (Fig. 4.29a), whereas compact ice forms on the uncoated surface (Fig. 4.29b).

Figure 4.29 (a) Enhanced icephobic property of a coating on the surface of an insulator. (b) Non-icephobic property of the uncoated surface of the insulator (Zhu et al., 2013).

4.4.5 Ice Adhesion and Material Categories

Ice adhesion on the surface of materials is a problem in reality. Ice eventually forms on even the best icephobic surfaces under extreme conditions. Enabling the easy removal of ice from a surface becomes a considerably challenging issue for icephobic materials. Fundamentally, the strong interaction of ice with most solids can be attributed to van der Waals forces (Wilen et al., 1995) and electrostatic interactions (Ryzhkin and Petrenko, 1997). The electrostatic interaction is proposed as the dominant mechanism due to the interaction of electrical charge at the ice surface and induced charge on the solid substrate (Hays, 1991). Surfaces with more hydroxyl groups can also exhibit increased ice adhesion through hydrogen bonding (Petrenko and Peng, 2003). Covalent chemical bonding directly associated with the ice surface can also be considered, but it is limited to very short distances (0.1–0.2 nm) and is only a factor for solids with specific chemical and crystal arrangements (Ryzhkin and Petrenko, 1997).

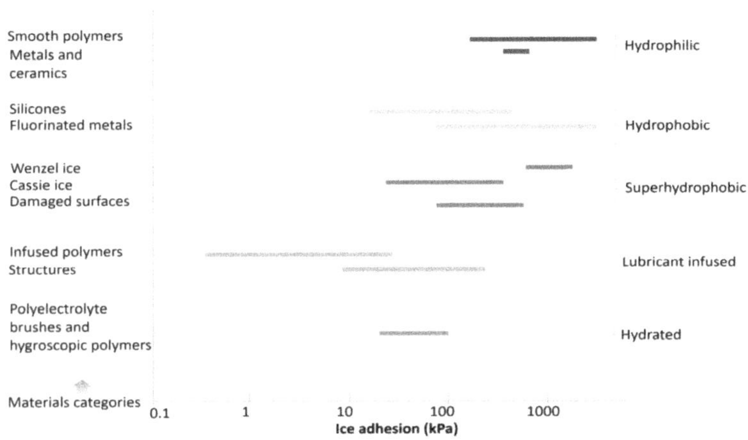

Figure 4.30 Ice adhesion for different material categories.

A comprehensive study on a large number of smooth surface coatings identified that the practical work of adhesion for water, i.e., $W_a = \gamma_{lv}(1 + \cos\theta_r)$, has the strongest correlation with ice adhesion (Meuler et al., 2010), where γ_{lv} is the surface tension of the water–vapor interface and θ_r is the receding contact angle. The various surfaces can exhibit ice adhesion in difference ranges, as shown in Fig. 4.30, in terms of hydrophilic, hydrophobic, superhydrophobic, lubricant-infused, and hydrated materials, which can be used as a reference for the design of materials with anti-icing/icephobic properties.

In sum, self-removal of ice accumulation poses significant challenges for building infrastructure, marine applications, the aerospace industry, refrigeration, power transmission, telecommunications, and other industries.

Thinking Questions

1. Can you elaborate on the significance of anti-icing and icephobic properties?
2. What are the key points of materials and surfaces with the anti-icing or icephobic ability?
3. Can you give examples of methods and techniques for the fabrication of materials and surfaces with anti-icing or icephobic properties?
4. Do you know of some bioinspired materials and surfaces that display anti-icing and icephobic properties?
5. What are some robust materials and surfaces for the application of anti-icing and icephobic properties?

References

Aizenberg J., A. J. Black, and G. M. Whitesides. Control of crystal nucleation by patterned self-assembled monolayers. *Nature* **398**, 495–498 (1999).

Algara-Siller G., O. Lehtinen, F. C. Wang, R. R. Nair, U. Kaiser, H. A. Wu, A. K. Geim, and I. V. Grigorieva. Square ice in graphene nanocapillaries. *Nature* **519**, 443–445 (2015).

Boreyko J. B. and C.-H. Chen. Restoring superhydrophobicity of lotus leaves with vibration-induced dewetting. *Phys. Rev. Lett.* **103**, 174502–174506 (2009).

Burton T., N. Jenkins, D. Sharpe, and E. Bossanyi. *Wind Energy Handbook*, edn 2, Wiley, 2011.

Chaudhury M. K. and K. H. Kim. Shear-induced adhesive failure of a rigid slab in contact with a thin confined film. *Eur. Phys. J. E* **23**, 175–183 (2007).

Chen J., R. Dou, D. Cui, Q. Zhang, Y. Zhang, F. Xu, X. Zhou, J. Wang, Y. Song, and L. Jiang. Robust prototypical anti-icing coatings with a self-lubricating liquid water layer between ice and substrate. *ACS Appl. Mater. Interfaces* **5**, 4026–4030 (2013).

Cherney E. Flashover performance of artificially contaminated and iced longrod transmission line insulators. *IEEE Trans. Power App. Syst.* **99**, 46–52 (1980).

Coady M. J., M. Wood, G. Q. Wallace, K. E. Nielsen, A. M. Kietzig, F. L. Labarthet, and P. J. Ragogna. Icephobic behavior of UV-cured polymer networks incorporated into slippery lubricant-infused porous surfaces: Improving slips durability. *ACS Appl. Mater. Interfaces* **10**, 2890–2896 (2018).

Dash S., J. de Ruiter, and K. K. Varanasi. Photothermal trap utilizing solar illumination for ice mitigation. *Sci. Adv.* **4**, eaat0127 (2018).

del Cerro D. A., G. R. B. E. Romer, and A. J. H. in't Veld. Erosion resistant anti-ice surfaces generated by ultra short laser pulses. *Phys. Proc.* **5**, 231 (2010).

DeVries A. L. and D. E. Wohlschlag. Freezing resistance in some Antarctic fishes. *Science* **163**, 1073–1075 (1969).

Diao Y., T. Harada, A. S. Myerson, T. Alan Hatton, and B. L. Trout. The role of nanopore shape in surface-induced crystallization. *Nat. Mater.* **10**, 867–871 (2011).

Duman J. G. Animal ice-binding (antifreeze) proteins and glycolipids: An overview with emphasis on physiological function. *J. Exp. Biol.* **218**, 1846–1855 (2015).

Esser-Kahn A. P., V. Trang, and M. B. Francis. Incorporation of antifreeze proteins into polymer coatings using site-selective bioconjugation. *J. Am. Chem. Soc.* **132**, 13264–13269 (2010).

Fakorede O., Z. Feger, H. Ibrahim, A. Ilinca, J. Perron, and C. Masson. Ice protection systems for wind turbines in cold climate: Characteristics, comparisons and analysis. *Renew. Sustain. Energy Rev.* **65**, 662–675 (2016).

Fikke S. M., J. E. Hanssen, and L. Rolfseng. Long range transported pollutants and conductivity of atmospheric ice on insulators. *IEEE Trans. Power Deliv.* **8**, 1311–1321 (1993).

Gao S., B. Liu, J. Peng, K. Zhu, Y. Zhao, X. Li, and X. Yuan. Icephobic durability of branched PDMS slippage coatings co-crosslinked by functionalized POSS. *ACS Appl. Mater. Interfaces* **11**, 4654–4666 (2019).

Genorio B., W. Lu, A. M. Dimiev, Y. Zhu, A.-R. O. Raji, B. Novosel, L. B. Alemany, and J. M. Tour. In situ intercalation replacement and selective functionalization of graphene nanoribbon stacks. *ACS Nano* **6**, 4231–4240 (2012).

Golovin K., A. Dhyani, M. D. Thouless, and A. Tuteja. Low-interfacial toughness materials for effective large-scale deicing. *Science* **364**, 371–375 (2019).

Guo P., Y. Zheng, M. Wen, C. Song, Y. Lin, and L. Jiang. Icephobic/anti-icing properties of micro/nanostructured surfaces. *Adv. Mater.* **24**, 2642–2648 (2012).

Guo Z., C. Gao, J. Li, Y. Liu, and Y. Zheng. Anti-icing properties of bioinspired liquid-infused double-layer surface with internal wetting transport ability. *Adv. Mater. Interfaces* **6**, 1900244 (2019).

Gwak Y., J.-I. Park, M. Kim, H. S. Kim, M. J. Kwon, S. J. Oh, Y.-P. Kim, and E. Jin. Creating anti-icing surfaces via the direct immobilization of antifreeze proteins on aluminum. *Sci. Rep.* **5**, 12019 (2015).

Hays D. A. *Fundamentals of Adhesion* (Lee L.-H., ed.), Springer, 1991, pp. 249–278.

He Z., L. Zheng, Z. Liu, S. Jin, C. Li, and J. Wang. Inhibition of heterogeneous ice nucleation by bioinspired coatings of polyampholytes. *ACS Appl. Mater. Interfaces* **9**, 30092–30099 (2017).

Hejazi V., K. Sobolev, and M. Nosonovsky. From superhydrophobicity to icephobicity: Forces and interaction analysis. *Sci. Rep.* **3**, 2194 (2013).

Howarter J. A. and J. P. Youngblood. Self-cleaning and next generation anti-fog surfaces and coatings. *Macromol. Rapid Commun.* **29**, 455 (2008).

Hu H. M., J. A. Wastn, B. W. Cribb, and G. S. Wason. Fouling of nanostructured insect cuticle: Adhesion of natural and artificial contaminants. *Bioadhesion Biofilm Res.* **27**, 1125–1137 (2011).

Huang Y., B. B. Stogin, N. Sun, J. Wang, S. Yang, and T. S. Wong. A switchable cross-species liquid repellent surface. *Adv. Mater.* **29**, 1604641 (2017).

Jiang X., S. Wang, Z. Zhang, S. Xie, and Y. Wang. Study on AC flashover performance and discharge process of polluted and iced IEC standard suspension insulator string. *IEEE Trans. Power Deliv.* **22**, 472–480 (2007).

Jin Y., Z. He, Q. Guo, and J. Wang. Control of ice propagation by using polyelectrolyte multilayer coatings. *Angew. Chem., Int. Ed.* **56**, 11436–11439 (2017).

Khedir K. R., G. K. Kannarpady, H. Ishihara, J. Woo, C. Ryerson, and A. S. Biris. Design and fabrication of Teflon-coated tungsten nanorods for tunable hydrophobicity. *Langmuir* **27**, 466127 (2011).

Kim J. H., S. M. Kang, B. J. Lee, H. Ko, W. G. Bae, K. Y. Suh, M. K. Kwak, and H. E. Jeong. Remote manipulation of droplets on a flexible magnetically responsive film. *Sci. Rep.* **5**, 17843 (2016).

Kim J., B. J. Chisholm, and J. Bahr. Adhesion study of silicone coatings: The interaction of thickness, modulus and shear rate on adhesion force. *Biofouling* **23**, 113–120 (2007).

Kim K. H. and M. K. Chaudhury. Shear-induced fracture at the interface of PDMS and a rigid slab modified with polyelectrolyte layers. *J. Adhes.* **85**, 792–811 (2009).

Kim P., T.-S. Wong, J. Alvarenga, M. J. Kreder, W. E. Adorno-Martinez, and J. Aizenberg. Liquid-infused nanostructured surfaces with extreme anti-ice and anti-frost performance. *ACS Nano* **6**, 6569–6577 (2012).

Kiselev A., F. Bachmann, P. Pedevilla, S. J. Cox, A. Michaelides, D. Gerthsen, and T. Leisner. Active sites in heterogeneous ice nucleation-the example of K-rich feldspars. *Science* **355**, 367–371 (2017).

Kreder M. J., J. Alvarenga, P. Kim, and J. Aizenberg. Design of anti-icing surfaces: Smooth, textured or slippery? *Nat. Rev. Mater.* **1**, 15003 (2016).

Laforte J. L., M. A. Allaire, and J. Laflamme. State-of-the-art on power line de-icing. *Atmos. Res.* **46**, 143–158 (1998).

Lee S.-H., M. Seong, M. K. Kwak, H. Ko, M. Kang, H. W. Park, S. M. Kang, and H. E. Jeong. Tunable multimodal drop bouncing dynamics and anti-icing performance of a magnetically responsive hair array. *ACS Nano* **12**, 10693–10702 (2018).

Li C., X. Li, C. Tao, L. Ren, Y. Zhao, S. Bai, and X. Yuan. Amphiphilic antifogging/anti-icing coatings containing POSS-PDMAEMA-*b*-PSBMA. *ACS Appl. Mater. Interfaces* **9**, 22959–22969 (2017).

Li X., Y. Zhao, H. Li, and X. Yuan. Preparation and icephobic properties of polymethyltrifluoropropylsiloxane-polyacrylate block copolymers. *Appl. Surf. Sci.* **316**, 222–231 (2014).

Li Y., B. Li, X. Zhao, N. Tian, and J. Zhang. Totally waterborne, nonfluorinated, mechanically robust, and self-healing superhydrophobic coatings for actual anti-icing. *ACS Appl. Mater. Interfaces* **10**, 39391–39399 (2018).

Liu B. Q., M. He, J. X. Wang, H. L. Li, X. L. Jin, J. J. Wang, and Y. L. Song. Superhydrophobic film retards frost formation. *Soft Matter.* **6**, 2396 (2010).

Liu J., C. Zhu, K. Liu, Y. Jiang, Y, Song, J. S. Franciscoc, X. Zeng, and J. Wang. Distinct ice patterns on solid surfaces with various wettabilities. *PNAS*, **114**, 11285–11290 (2017).

Liu K., C. Wang, J. Ma, G. Shi, X. Yao, H. Fang, Y. Song, and J. Wang. Janus effect of antifreeze proteins on ice nucleation. *Proc. Natl. Acad. Sci. USA* **113**, 14739–14744 (2016).

Lo C.-W., V. Sahoo, and M.-C. Lu. Control of ice formation. *ACS Nano* **11**, 2665–2674 (2017).

Lu W., G. Ruan, G. Genorio, Y. Zhu, B. Novosel, G. Peng, and J. M. Tour. Functionalized graphene nanoribbons via anionic polymerization initiated by alkali metal-intercalated carbon nanotubes. *ACS Nano* **7**, 2669–2675 (2013).

Lv J., Y. Song, L. Jiang, and J. Wang. Bio-inspired strategies for anti-icing. *ACS Nano* **8**, 3152–3169 (2014).

Makkonen L. Ice adhesion: Theory, measurements and countermeasures. *J. Adhes. Sci. Technol.* **26**, 413–445 (2012).

Menini R. and M. Farzaneh. Advanced icephobic coatings. *J. Adhes. Sci. Technol.* **25**, 971–992 (2011).

Meuler A. J., J. D. Smith, K. K. Varanasi, J. M. Mabry, G. H. McKinley, and R. E. Cohen. Relationships between water wettability and ice adhesion. *ACS Appl. Mater. Interfaces* **2**, 3100–3110 (2010).

Michaelides A. and K. Morgenstern. Ice nanoclusters at hydrophobic metal surfaces. *Nat. Mater.* **6**, 597–601 (2007).

Mishchenko L., B. Hatton, V. Bahadur, J. A. Taylor, T. Krupenkin, and J. Aizenberg. Design of ice-free nanostructured surfaces based on repulsion of impacting water droplets. *ACS Nano* **4**, 7699–7707 (2010).

Mulherin N. D. and R. B. Haehnel. *Ice Engineering: Progress in Evaluating Surface Coatings for Icing Control at Corps Hydraulic Structures* (Engineer Research and Development Center, Cold Regions Research and Engineering Lab, Hanover, NH, 2003).

Parent O. and A. Ilinca. Anti-icing and de-icing techniques for wind turbines: Critical review. *Cold Reg. Sci. Technol.* **65**, 88–96 (2011).

Petrenko V. F. and S. Peng. Reduction of ice adhesion to metal by using self-assembling monolayers (SAMs). *Can. J. Phys.* **81**, 387–393 (2003).

Raraty L. E. and D. Tabor. The adhesion and strength properties of ice. *Proc. R. Soc. A: Math., Phys. Engineer. Sci.* **245**, 184 (1958).

Ryzhkin I. A. and V. F. Petrenko. Physical mechanisms responsible for ice adhesion. *J. Phys. Chem.* **5647**, 6267–6270 (1997).

Shi W., L. Wang, Z. Guo, and Y. Zheng. Excellent anti-icing abilities of optimal micropillar arrays with nanohairs. *Adv. Mater. Interfaces* **2**, 1500352 (2015).

Sojoudi H., M. Wang, N. D. Boscher, G. H. McKinley, and K. K. Gleason. Durable and scalable icephobic surfaces: Similarities and distinctions from superhydrophobic surfaces. *Soft Matter* **12**, 1938–1963 (2016).

Stone H. A. Ice-phobic surfaces that are wet. *ACS Nano* **6**, 6536–6540 (2012).

Tao C., X. Li, B. Liu, K. Zhang, Y. Zhao, K. Zhu, and X. Yuan. Highly icephobic properties on slippery surfaces formed from polysiloxane and fluorinated POSS. *Prog. Org. Coat.* **103**, 48–59 (2017).

Tricoli A., M. Righettoni, and S. E. Pratsinis. Anti-fogging nanofibrous SiO_2 and nanostructured SiO_2-TiO_2 films made by rapid flame deposition and in situ annealing. *Langmuir* **25**, 12578 (2009).

Wang L., M. Wen, M. Zhang, L. Jiang, and Y. Zheng. Ice-phobic gummed tape with nano-cones on microspheres. *J. Mater. Chem. A* **2**, 3312 (2014).

Wang L., Q. Gong, S. Zhan, L. Jiang, and Y. Zheng. Robust anti-icing performance of a flexible superhydrophobic surface. *Adv. Mater.* **28**, 7729–7735 (2016a).

Wang T., Y. Zheng, A. O. Raji, Y. Li, W. K. A. Sikkema, and J. M. Tour. Passive anti-icing and active deicing films. *ACS Appl. Mater. Interfaces* **8**, 14169–14173 (2016b).

Wen M., L. Wang, M. Zhang, L. Jiang, and Y. Zheng. Antifogging and icing-delay properties of composite micro- and nanostructured surfaces. *ACS Appl. Mater. Interfaces* **6**, 3963–3968 (2014).

Wilen L. A., J. S. Wettlaufer, M. Elbaum, and M. Schick. Dispersion-force effects in interfacial premelting of ice. *Phys. Rev. B* **52**, 12426–12433 (1995).

Wong T.-S., S. H. Kang, S. K. Y. Tang, E. J. Smythe, B. D. Hatton, A. Grinthal, and J. Aizenberg. Bioinspired self-repairing slippery surfaces with pressure-stable omniphobicity. *Nature* **477**, 443–447 (2011).

Zhang M., L. Wang, S. Feng, and Y. Zheng. A strategy of antifogging: Air-trapped hollow microsphere nanocomposites. *Chem. Mater.* **29**, 2899–2905 (2017a).

Zhang Q., B. Jin, B. Wang, Y. Fu, X. Zhan, and F. Chen. Fabrication of a highly stable superhydrophobic surface with dual-scale structure and its antifrosting properties. *Ind. Eng. Chem. Res.* **56**, 2754–2763 (2017b).

Zhu L., J. Xue, Y. Wang, Q. Chen, J. Ding, and Q. Wang. Ice-phobic coatings based on silicon-oil-infused polydimethylsiloxane. *ACS Appl. Mater. Interfaces* **5**, 4053–4062 (2013).

Zuo Z., R. Liao, C. Guo, Y. Yuan, X. Zhao, A. Zhuang, and Y. Zhang. Fabrication and anti-icing property of coral-like superhydrophobic aluminum surface. *Appl. Surf. Sci.* **331**, 132–139 (2015).

Chapter 5

Surfaces for Droplet Transport

In this chapter, the surfaces for droplet transport are introduced, including the theory of driving a droplet on a surface, the static gradient of the surface, the dynamic gradient of the surface, and the role of multigradient coordination. Gradients can be fabricated on a surface by various methods, such as dip-coating, fluid coating, microfluidics, self-assembly, electrospinning, electrochemistry, soft lithography, and 3D printing. Examples of controlled droplet transport are provided, such as unidirectional spreading on asymmetric nanostructured surfaces, unidirectional spreading of droplets on a fibrous surface, *Nepenthes-peristome*-inspired unidirectional liquid transport, droplet transport on an integrative Janus membrane and conical spine, oriented-structure-induced liquid transport, unidirectional drop cargo transport, time-dependent liquid transport, and liquid transport in hierarchical microchannels. In addition, droplet transport is controlled by external actions, such as magnet-induced dynamic arrays for microdroplet transport, droplet transport on a photoinduced dynamic gradient, magnetic tubular microactuators, oil droplet self-transportation, electrochemistry-controlled droplet transport, and surface tension transport.

Bioinspired Materials Surfaces: Nature-Inspired Micro-/Nanostructures
Yongmei Zheng

ISBN 978-981-5129-34-2 (Hardcover), 978-1-003-51388-9 (eBook)
www.jennystanford.com

5.1 Theory of Driving a Droplet on a Surface

It is interesting that a surface can drive a droplet to move in a certain direction. In nature, such surfaces exist and are capable of transporting liquid droplets as biological functions, such as butterfly wings for directional water repellency (Zheng et al., 2007; Liu et al., 2014), rice leaves for directional self-cleaning (Feng et al., 2002; Zheng et al., 2008), feathers of ducks and geese for droplet movement (Kennedy, 1970), seabird pecking for ratchet motion of liquid (Prakash et al., 2008), the pitcher of *Sarracenia purpurea* for object sliding (Chen et al., 2018), the peristome surface of *Nepenthes alata* for liquid transport (Chen et al., 2016), wetted capture spider silk for directional water collection (Zheng et al., 2010), and the spine of cactus for fog harvesting (Ju et al., 2012; Liu et al., 2015). The surfaces of these biological materials capable of directionally transporting liquid droplets have aroused considerable interest because of their potential applications in microfluidics, textile dyeing, filtration, smog removal, etc.

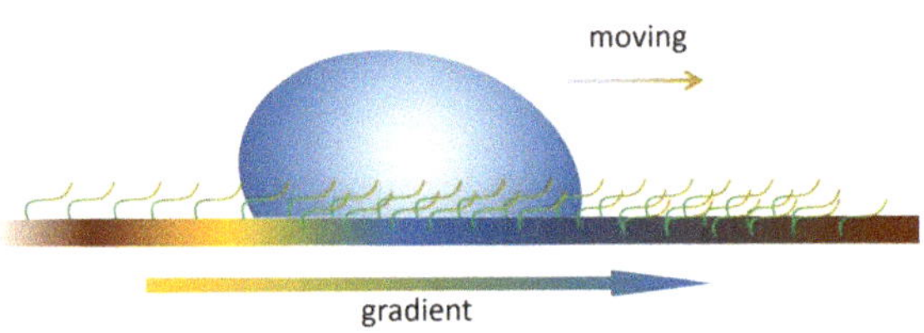

Illustration 5.1 Liquid droplet transport on a gradient surface.

To achieve droplet transport in a certain direction, surfaces of biological materials feature unique micro- and nanostructures, for example having gradients in oriented and anisotropic overlapping structures (Zheng et al., 2007), shapes of ratchets, cones, or spindle-knots (Ju et al., 2012), or slippery and liquid-infused surfaces in multistructures (Chen et al., 2016). This indicates that biological surfaces carry out a unique mechanism for droplet transport. Based on the theory and mechanism of biological functions, many ideas have been proposed, and novel methods have been developed to prepare bioinspired surfaces to achieve the functions of directional liquid transport (Wu et al., 2008; Malvadkar et al., 2010). Additionally, droplet transport can be manipulated by external actions, such as light (Gao et al., 2018) and magnetism (Wang et al., 2015a), when surfaces and structures are designed by using responsive materials beyond biological functions for droplet transport with high efficiency. Here, the theory of driving a droplet can be introduced from three types of gradients: static gradients, dynamics gradient, and multigradients.

5.1.1 Surface Static Gradient

The term "static gradient" indicates that the gradient is stably formed on the surface after fabrication and that the as-formed gradient does not change with time during droplet movement. Generally, a droplet on a surface is subjected to counterbalanced external forces. Once the external forces cannot be balanced, the droplet will move along a direction. In general, the driving forces for the movement of droplets can be generated by gradients, which are classified into two types: One is a wettability gradient (WG) induced by directional wettability, for instance from hydrophobic to more hydrophobic (superhydrophobic); from hydrophilic to more hydrophilic (superhydrophilic); and from hydrophobic to hydrophilic (Yang et al., 2008). Another is a structural gradient induced by an asymmetrical geometric shape (e.g., oriented, conical) or increasing (or decreasing) roughness in some direction.

Nonuniform surface wettability can be generated in many ways, such as unevenly pre-modifying a homogeneous surface (Chaudhury and Whitesides, 1992), taking advantage of chemicals contained in moving droplets to promote covalently bonding (Bain et al., 1994; Sumino et al., 2005; Yao et al., 2012) or physically adsorbing (Lee and Laibinis, 2000) on the underlying substrate and causing asymmetric wettability in the front and rear of the droplet, asymmetrically photoirradiating surfaces covered with photoisomerizable monolayers (Lchimura et al., 2000), heating or cooling surfaces locally (Yarin et al., 2002), and creating surface roughness variation (Fang et al., 2008) (such as the different arrangement of fibrils on the spindle-knot and joint of spider silk and width-changing grooves on the cactus spine).

In these cases, the driving force F_c is proportional to the gradient of the surface wettability. As the liquid contact angle is usually used to assess the wettability of a solid surface, the driving force can thus be described as:

$$F_c \sim \pi r_0 \gamma(\cos\theta_B - \cos\theta_A) \quad (5.1)$$

where r_0 is the radius of the droplet, γ is the surface tension, and θ_A and θ_B are the contact angles at the less wettable and more wettable sides of the droplet, respectively (Fig. 5.1a).

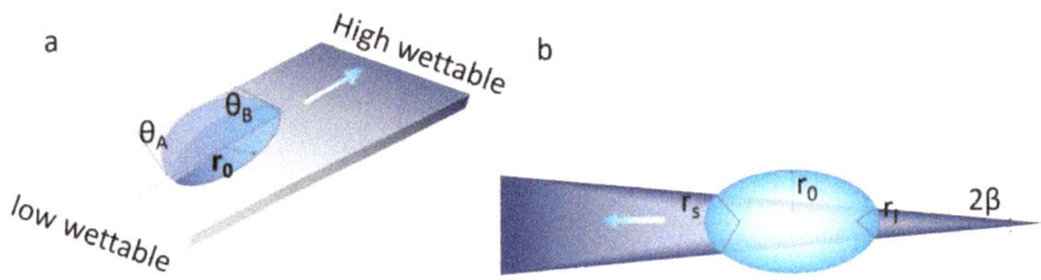

Figure 5.1 Illustration of static gradient surfaces: (a) surface energy gradient and (b) curvature gradient.

With regard to droplet movement on the asymmetric substrate, such as cone fibers (Lorenceau and Quéré, 2004), cone tubes (Bico and Quéré, 2002), and other space-confined regions (Zhang and Han, 2007), the driving force (F_L) arises from the gradient in the Laplace pressure inside the droplet. Taking a water droplet on a cone-structured object as an example, the driving force F_L can be expressed as:

$$F_L \sim -\int_{r_s}^{r_l} \frac{2\gamma}{(r + r_0)^2} \sin\beta dz \quad (5.2)$$

where r is the local radius of the cone-structured object, r_s and r_l are the local radii of the object at the two opposite sides of the droplet, and r_0 is the radius of the droplet. In addition, β is the half apex angle of the cone, and dz is the minute incremental radius along the cone (Fig. 5.1b). In addition to these two main situations above, a chemically homogeneous flat surface with a stiffness gradient can also drive liquid droplets toward a more compliant region due to the different deformation degrees of the substrate and the resulting change in apparent contact angles (Style et al., 2013).

In addition, by means of an electric field, an ionic liquid droplet can be pumped along the conductive nanowire smoothly from the cathode to the anode (Huang et al., 2013). All of these driving forces can be solely or cooperatively harnessed by a 1D material for directional liquid transport. On most occasions, due to surface inhomogeneity, the motions of droplets with low viscosity are subjected to resistance force F_R. This force is caused by the contact angle hysteresis (CAH) (difference between the advancing contact angle θ_{Adv} and receding contact angle θ_{Rec} of a droplet) and can be calculated as:

$$F_R \sim \pi R_0 \, \gamma(\cos\theta_{Rec} - \cos\theta_{Adv}) \tag{5.3}$$

As a result, a smaller hysteresis in contact angles favors a droplet's directional movement.

5.1.2 Surface Dynamic Gradient

When the gradient can be generated by external actions such as light or magnetism, the surface wettability can be changed through phenomena such as a surface thermal gradient (Gao et al., 2018) or surface tension with dynamic tilted angles (Wang et al., 2015a).

A time-varying driving force can also be generated by dynamic gradients. An example can be introduced as follows (Fig. 5.2). The material surface is composed of photothermal Fe_3O_4 nanoparticles in the gel matrix. A photothermal-responsive surface via light actions can generate a thermal gradient ∇T on the surface along with the direction of light action, and the surface can form a dynamic gradient with the light action movement (Fig. 5.2a,b); thus, a droplet on such a surface can be driven in a time-dependent manner by the dynamic gradient (Fig. 5.2c) (Gao et al., 2018).

Due to the heat transfer between the surface and the droplet (Fig. 5.2d), ∇T apparently forms on the side of the droplet. Spatial variation in the surface tension caused by ∇T results in added tangential stresses, that is, the Marangoni force (F_M), which can be described as follows (Bjelobrk et al., 2016; Li et al., 2017; Yang et al., 2018):

$$F_M = \pi R^2 \frac{d\gamma_{la}}{dT} \frac{dT}{dx} \tag{5.4}$$

where the direction of the force is toward the low-temperature side of the droplet (i.e., the X-direction). The $d\gamma_{la}/dT$ is the variation in the liquid–air interfacial tension with temperature, and dT/dx is a thermal gradient on the surface along the X-direction, where X indicates the direction of droplet motion.

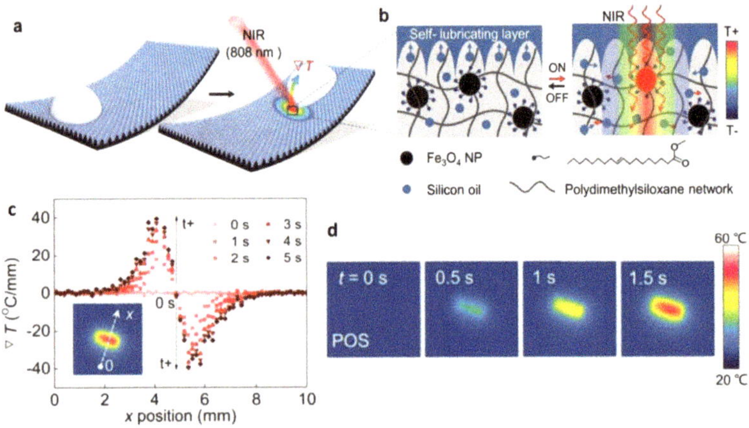

Figure 5.2 Design of a dynamic gradient surface. (a) Illustration of droplet motion on a photoresponsive organogel surface with a dynamic gradient (∇T) driven by light (NIR). (b) The material surface can be composed of photothermal Fe_3O_4 nanoparticles in the gel matrix. (c) Gradient formation of the surface acting on the two sides of the droplet by light action. (d) Observation of the flow inside the droplet during the surface by light irradiation (Gao et al., 2018).

5.1.3 Role of Multigradient Cooperation

Multigradient surfaces are found on wetted spider silk or cactus spines, ryegrass seta, strider legs with bristles, etc., including roughness and curvature in a gradient or small cone oriented on a cone surface; as a result, these surfaces can efficiently control droplet transport in multiple directions.

The cooperative forces ($\overrightarrow{F_{Co}}$) can result from the vector sum of multigradient forces in various directions, i.e.,

$$\overrightarrow{F_{Co}} = \overrightarrow{F_C} + \overrightarrow{F_L} + \overrightarrow{F_R} + \overrightarrow{F_M} + \cdots \tag{5.5}$$

where $\overrightarrow{F_C}$ is the force of the surface tension gradient resulting from surface chemistry varying with position along one direction, e.g., along a more wettable direction from a hydrophobic location to a hydrophilic location. The liquid droplet moves toward the wettable direction. However, smaller droplets need a larger degree in the gradient direction to be propelled for movement due to capillary adhesion. The multigradient forces can enhance the driving force to propel the movement of tiny droplets if the multigradient forces act together along the same directions. $\overrightarrow{F_L}$ is the force of the Laplace pressure due to the geometric curvature acting on the droplet (e.g., a droplet on the surface of a cone), and the droplet is curved at its two sides, which induces an unbalanced force. $\overrightarrow{F_R}$ is the force of CAH opposite to the direction of droplet movement. $\overrightarrow{F_M}$ is the dynamic force resulting from external action, such as light and magnetism. These gradients can be selected or incorporated onto a surface via physics and chemistry methods.

We consider an example of a size-gradient spindle-knot on a fiber with droplet transport. As shown in Fig. 5.3, such transport to two main aspects forms the driving force (as illustrated in Fig. 5.3a,b): the force from the Laplace pressure and capillary force.

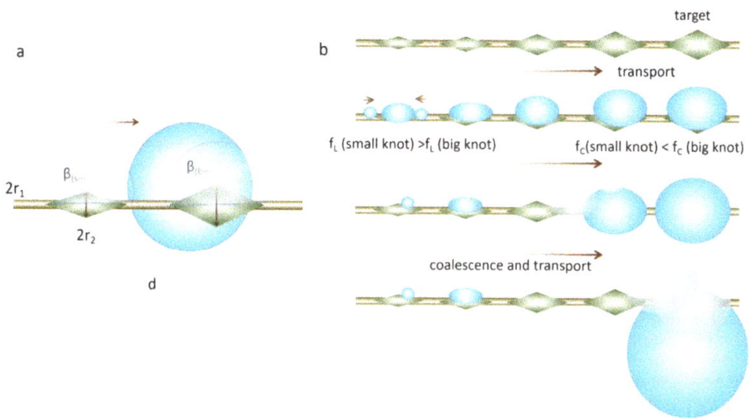

Figure 5.3 Cooperative forces by multigradients on a size-gradient spindle-knotted fiber. (a) Two spindle-knots with different semi-apex angles of $\beta_{(small)} < \beta_{(large)}$; d is the outer distance of the small spindle-knot and large spindle-knot, r_1 is the radius of the fiber, and r_2 is the maximum radius of the spindle-knot. The arrow indicates the direction of droplet coalescence. (b) Droplet transport via coalescence of droplets along size-gradient spindle-knots on a fiber. The small knot has a larger Laplace pressure force f_L (small knot) than the large knot f_L (large knot) and a smaller capillary force f_C (large knot) than the large knot f_C (small knot).

The force (f_L) from the difference in Laplace pressure is formed along the gradient of the spindle-knots, which can be described as follows:

$$f_L = \int_{r_1}^{r_2} \frac{2\gamma}{(r+R)^2} \sin\beta dz \tag{5.6}$$

where γ is the surface tension of the liquid, r is the local radius, R is the drop radius, β is the half apex angle of the spindle-knot (where β(large knot) > β(small knot)), and z is the integration variable along the heights of the gradient spindle-knots. In addition, r_1 is the local radius of the high-curvature site at the minimum diameter of the fiber, and r_2 is the local radius of the low-curvature site at the spindle-knot. We have $r_1 < r_2$; thus, the Laplace pressure on the high-curvature site (r_1) is larger than that on the low-curvature site (r_2). Because r_1(small knot) = r_1(large knot), r_2(small knot) < r_2(large knot), then f_L(small knot) > f_L(large knot) and the nonequilibrium Laplace pressure will propel the water drop to move from the small knot to the large knot.

The capillary adhesion force (f_C) is enhanced with the increasing geometric features of the spindle-knots along the fiber as follows:

$$f_C = 2\int_{x_1}^{x_1+d} \gamma\phi(x)dx \tag{5.7}$$

where x is the axial coordinate variable along the fiber; x_1 is the initial site for the integration, located at the outer end of the left spindle-knot; d is the distance between the outer ends of the two adjacent knots; and ϕ(x) is the local crossing angle between the liquid surface and the vertical reference plane along the fiber. As d is proportional to the knot size, the larger spindle-knot has a larger d, so f_C(small knot) < f_C(large knot). It is inferred that the geometry gradient of the spindle-knots induces the gradient of the capillary adhesion force, thus eliminating the shift toward the larger sized spindle-knot. Based on both the above factors, the total force (F) can be described as $F \sim F_L + F_C$. As a result, droplets directionally coalesce from the joint to the spindle-knot, and continuous coalescence induces droplet transport from the small spindle-knot to the large spindle-knot.

5.2 Fabrication of Gradients on a Surface

5.2.1 Dip-Coating Method

The dip-coating method can be controlled to achieve the gradient geometry in structures, such as gradient humps on fibers (Chen et al.,

2013). Figure 5.4a shows optical images of the fabrication process of bioinspired fibers with gradient humps via the tilt-angle dip-coating route (Chen et al., 2013). A uniform nylon fiber (~18 mm) at a tilt angle θ was immersed in a polymer solution, polyvinylidene fluoride (PVDF, Mw 5 530 000, Aldrich), dissolved in N,N-dimethylformamide (DMF), and then drawn out at a tilted angle at a given velocity. Thus, the slope thickness of the PVDF film covers the fiber surface (at 0 ms). Following the effects of Rayleigh instability, the film with different gradient thickness breaks into different polymer droplets along the direction of the gradient thickness (at ~8 ms). The result forms gradient polymer droplets (at ~92 ms). It is estimated that the heights of polymer droplets range from ~186 to ~51 mm at ~92 ms and that their widths range from ~289 to ~160 mm. After the solvent evaporates, PVDF gradient humps are formed along the fiber.

Figure 5.4b–e shows the SEM images of fibers (Fig. 5.4b) composed of four humps of different sizes from smaller to larger, where the heights of the humps are ~59, ~74, ~88, and ~118 μm and the widths are ~162, ~206, ~235, and ~268 μm, respectively, along with half apex angles of ~25.0°, ~28.0°, ~28.5°, and ~31.0° for geometric features (from smaller ones to larger ones). The gradient of roughness appears on the surface of the humps from the center site (Fig. 5.4d) to the side site (Fig. 5.4e) with different nanostructures.

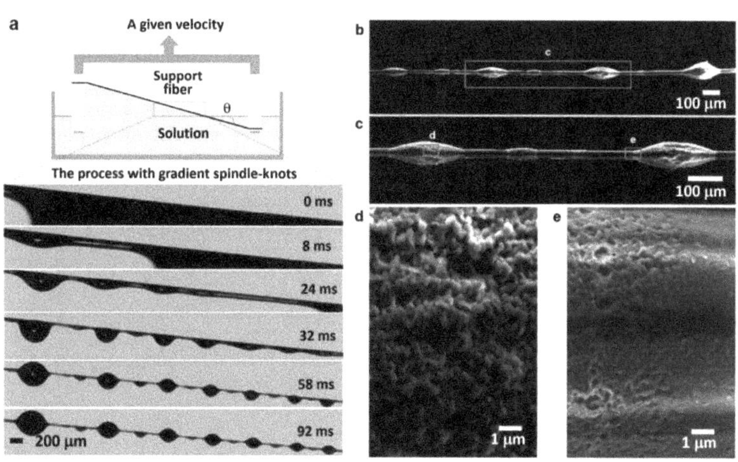

Figure 5.4 (a) Fabrication of a tilt-angle dip-coating and (b–d) SEM images of a bioinspired gradient-humped fiber (Chen et al., 2013).

The dip-coating method can be used to fabricate a gradient-wettability surface, termed the lifting-dissolution method (Zhang et al., 2015). For instance, a hydrophilic component (chitosan, CHI) gradient can be achieved on an aligned PET (poly ester)/CHI fibrous surface. A CHI-composition gradient is formed along the aligned fibers by using the lifting-dissolution method (Fig. 5.5a). In this design, an aligned PET (hydrophobic)/CHI (hydrophilic) fibrous surface can be fabricated by electrospinning coupled with a high-speed collection drum (Fig. 5.5b). Thus, the anisotropy of the PET/CHI surface can be achieved by tailoring the PET/CHI composition. When PET/CHI is 10/2, a droplet deforms to a stick-like shape (strong anisotropy with a length–width ratio of ~6.6), which demonstrates that anisotropic wettability is formed effectively. Furthermore, to achieve the gradient wettability of the surface, the lifting-dissolution method is used. In this process, the CHI dissolves over time in the acid solution (HAc, CH_3COOH). In contrast, the bottom part takes more time to release CHI to be less hydrophilic. As a result, a WG is formed in the parallel direction (Fig. 5.5c). When water droplets (3 μL) are dripped on different regions along the direction of gradient wettability, the droplets exhibit decreasing contact angles from 116° to 22°, and the sites with smaller contact angles are more hydrophilic.

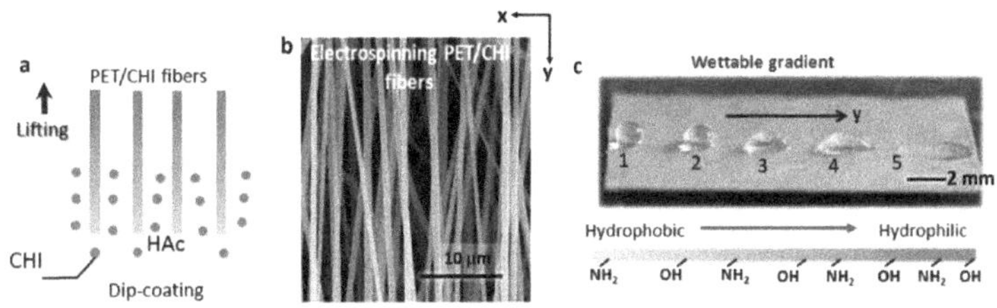

Figure 5.5 (a) Illustration of dip-coating for gradient wettability on PET/CHI fiber array anisotropic structured surfaces. (b) SEM image of the surface with electrospun PEI/CHI fibers. (c) The surface with gradient wettability due to gradient chemical contents (Zhang et al., 2015).

5.2.2 Fluid-Coating Method

The fluid-coating technique can be developed to fabricate novel structures on fibers (Bai et al., 2011). As illustrated in Fig. 5.6a, a fiber is withdrawn from a polymer solution in a reservoir tube by a motor with controllable fluid-coating velocity.

Figure 5.6b shows the process of fluid coating on a fiber (75 μm in diameter) by a polymer solution composed of polymethyl methacrylate (PMMA) dissolving in DMF with a weight percentage of 11%, along with a drawn-out velocity of ~3.21 mm/s. At the location of uniform solution film thickness (e.g., ~53.8 μm), the Rayleigh instability causes the film to be broken up into a polymer droplet (Fig. 5.6b, Frame IV). Subsequently, droplets are continuously formed (Fig. 5.6b) after ~0.57 s. The compete polymer droplet pattern forms within 14 s, where the spindle-knots have a periodicity (D) of ~921.7 μm and joint radii (R) ≈ 91.3 μm, i.e., $D \approx 10$ R for the array. A critical velocity (V_{Critical}) of fluid coating is defined as the initial formation of spindle-knots versus the solution concentration. According to the relationship of Ca = ηV/γ, a smaller velocity can coat a thick film with a more viscous solution. As a result, V_{Critical} decreases from ~12.81 to ~0.29 mm/s as the solution concentration increases from 7% to 15% for the fabrication of excellent spindle-knot fibers (Bai et al., 2011). Figure 5.6c–e shows SEM images of the spindle-knot and joint, along with the gradient porous structures between the spindle-knot and joint.

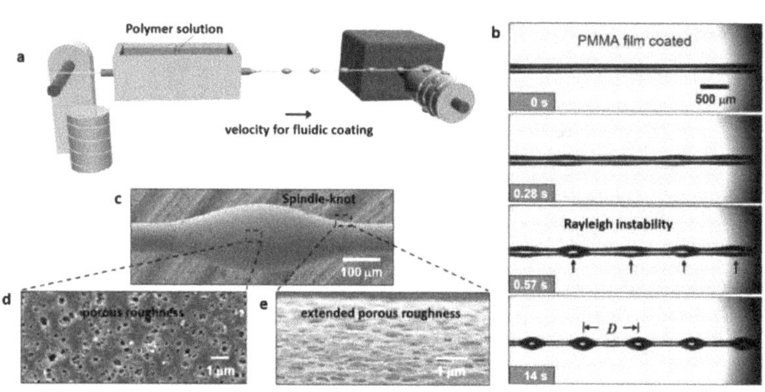

Figure 5.6 (a) Illustration of the fluid-coating process and (b) formation of spindle-knots on fibers via Rayleigh instability. (c–e) SEM images of gradient structures on the spindle-knot and joint (Bai et al., 2011).

The velocity-change coating method can be introduced to fluid coating to fabricate unique gradient-step (GS) spindle-knots on fibers (Xue et al., 2014). As shown in Fig. 5.7a, the optical images indicate in situ GS formation on the nylon fiber (110 μm in diameter) by coating a solution of PMMA in DMF (10 wt.%) by changing the drawing velocity from the initial velocity ($V_0 \sim 0$ mm/s) to the terminal velocity ($V_t \sim 21.4$ mm/s) with a constant acceleration $a = 8.3$ mm/s^2. After a short time (1.2 s), the cone solution film appears to be convex–concave, subsequently generating periodic polymer droplets with a GS height from ~208.8 μm to ~595 μm at 2.5 s. Generally, the local gradient morphology of the spindle-knots is a highly porous structure at the center of the spindle-knot, with a gradient porous structure at the middle and a relatively smooth structure at the side.

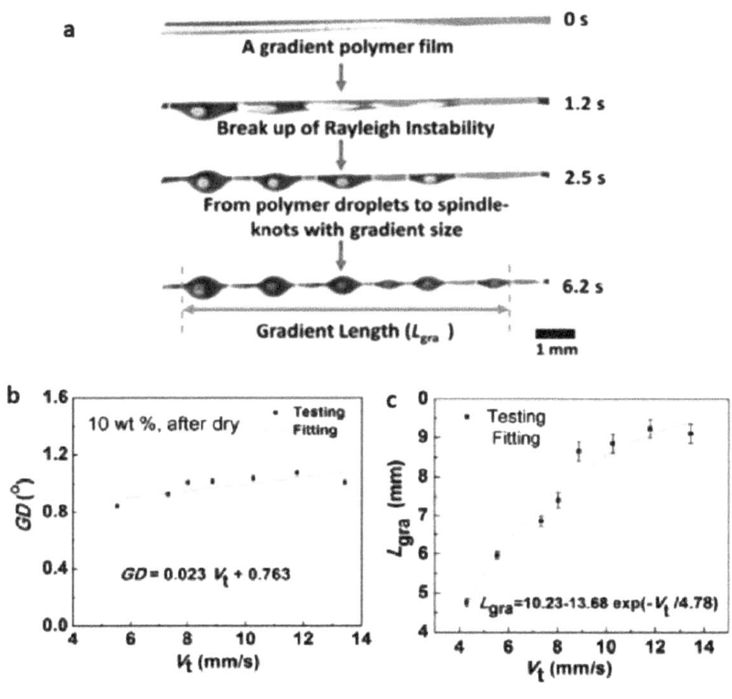

Figure 5.7 (a) Optical images of in situ observations of the fabrication process (Xue et al., 2014). The gradient length can be defined with L_{gra}. (b–c) Relationship of the gradient degree (GD) and gradient length (L_{gra}) versus the terminal drawing velocity (V_t).

Figure 5.7b,c shows the relationship between the gradient degree (GD) and drawing velocity (V_t) for a solution concentration of 10 wt.% after drying. The linear relationship of GD = 0.023 V_t +

0.763 (Fig. 5.6b) and the nonlinear relationship of $L_{gra} = 10.23 - 13.68\exp(-V_t/4.78)$ (mm) were obtained by fitting (Fig. 5.6c), where L_{gra} indicates the gradient length along the fiber and V_t indicates the terminal drawing velocity. Such unique gradient spindle-knots can control droplet transport in directions.

5.2.3 Microfluidic Method

Microfluidic systems were used early to develop microdevices for medical applications (Whitesides, 2006). Recently, based on fluid features controlled in channels of microfluidics, microfluidic methods have been developed to fabricate novel materials (Kang et al., 2011). Taking the example of the as-designed microfluidic systems, by investigating the spinning apparatus of a spider and the mixing of fluids from glands in its abdomen via an open–close valve to restart spinning after internal rupture (Fig. 5.8a), a microfluidic fabricating system was designed for the fabrication of bioinspired silks. Figure 5.8b illustrates the as-designed microfluidic chip and integration of small pneumatic valves into the hybrid channels to control the flow, where a rotating spool twists to achieve triple-helix fibers.

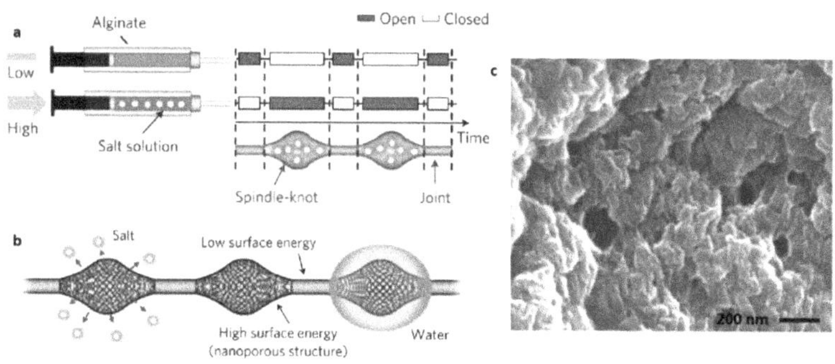

Figure 5.8 (a) Illustration of the design of a microfluidic system for bioinspired microfiber production via open and closed modes versus time (Kang et al., 2011). (b) Illustration of gradient coding of fibers with high surface energy at the spindle-knot and low surface energy at the joint for aggregation of water droplets. (c) SEM image of rough structures on the spindle-knot.

In this design, two alternating fluids stream via an open–close valve to generate continuous fibers with periodic nanoporous spindle-knots, where an alginate solution containing salt (0.5% w/w) is controllably delivered into one of the channels at a high flow rate and where the length of the spindle-knot can be controlled by altering the open-valve time. Salt dissolution forms nanoporous structures on spindle-knots, leading to a higher surface energy to attract water more than the joints do. The driving force of the Laplace pressure gradient resulting from the conical spindle-knot causes the water droplets to self-collect at the spindle-knots on the fiber.

5.2.4 Self-Assembly

Self-assembly can be developed to fabricate novel structures by controlling the wetting time and water condensation, defined as wet assembly (Song et al., 2014; Zhao et al., 2014). Based on the method of wet assembly, a kind of humped fiber can be fabricated by using composite materials composed of PMMA and titanium tetrachloride ($TiCl_4$)-hydrolyzed nanoparticles or others by integrating PMMA electrospinning in a nanoparticle-included fog with a humidity of 50–90% and a wetting time of 2–16 s. This wet-assembly process introduces a driving force that results from both cooperation of the surface energy gradient and Laplace pressure for coalescence and transport of nanoparticle-included droplets in certain directions. This driving force (df) is expressed as $df = \int_{x_{joint}}^{x_{hump}} \gamma(\cos\theta_A - \cos\theta_R)dx$, where γ is the surface tension of water, x is the integrating variable along the axis of the fiber from joint to hump, and θ_A, and θ_R are the advancing and receding angles of droplets with higher and lower nanoparticle concentrations ($\theta_A < \theta_R$), respectively. In addition, the Laplace pressure (ΔP) acts on the droplet, i.e., $\Delta P = \int_{r_1}^{r_2} \frac{\gamma}{(r + R_0)^2} \sin\alpha dz$, where r is the local radius of the fiber, R_0 is the droplet radius ($R_0 = (3\ V/4\pi)^{1/3}$, V is the droplet volume), α is the half apex angle of the hump, and z is the integrating variable along the diameter of the hump. The Laplace pressure on the joint with its high curvature (with a local radius r_1) is larger than that on the hump with its low curvature (with a local radius r_2; $r_1 < r_2$). Thus, humps form, corresponding to larger droplets on the fiber after drying. Figure 5.9a shows optical images of the in situ observation of the wet-assembly process from water condensation to nanofragment assembly with time. Water condensation occurs at 56 s, and hump structures form at 362 s on the fibers after they dry. Figure 5.9b illustrates the process of wet assembly.

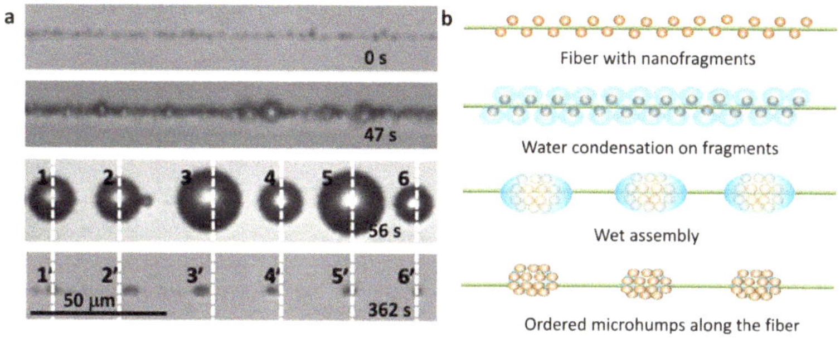

Figure 5.9 Optical image and illustration of the wet-assembly process (Song et al., 2014). (a) Optical images of the formation of hump structure on the fiber under water condensation versus time. (b) Illustration of hump formation on the fiber via wet assembly.

5.2.5 Electrospinning Method

Electrospinning can be used to fabricate unique structures, such as bead-on-string (Tian et al., 2011), similar to the capture silk of ecribellate spiders with a uniform coating of glycoprotein glue. Figure 5.10a shows the experimental setup that is designed by two blunt metallic needles in the axis. A conductive inner needle was connected to the cathode of a high-voltage generator, and a collecting substrate of aluminum foil was connected to the anode. This strategy applies a sprayable outer fluid to imprint a series of heterogeneous beads on a fiber from the spinnable inner fluid. The spinnable inner fluid is highly viscous, including 35% (w/v) polystyrene (PS, Mw = 350 000) solution in DMF. The sprayable outer fluid has a low viscosity, including 20% (w/v) poly(ethylene glycol) (PEG, M = 20 000) solution in a mixed solvent of DMF and methylene chloride (MC) (v/v = 1:1). The distance of the needles and collecting substrate was ~22 cm, and the applied voltage was 19 kV. The flow rate was tuned from 3 to 6 mL/h. Figure 5.10b is an SEM image of the bead-on-string fiber. Figure 5.10c shows a fluorescent image of the fiber with emitting beads (composed of PEG) to demonstrate the heterogeneity, where the fluorescein isothiocyanate (FITC) could be grafted onto the PEG via a reaction between the isothiocyanate and the PEG hydroxyl groups. Figure 5.10d shows the bead-on-string profiles following water collection on the beads inside an environmental scanning electron microscope (ESEM).

The achievement of bead-on-string structures can be considered a result of the surface energy difference between the core and shell components; i.e., poly(ethylene glycol) with high surface energy and polystyrene with low surface energy were the components of the outer and inner fluids, respectively. The higher surface energy of the shell polymer favors the spontaneous Rayleigh breakup of the outer fluid; thus, the beads imprint on the fiber. The formation of a bead can raise both forces on the fiber, i.e., the Laplace pressure from the curvature of the bead and string and the surface energy gradient between the bead and string.

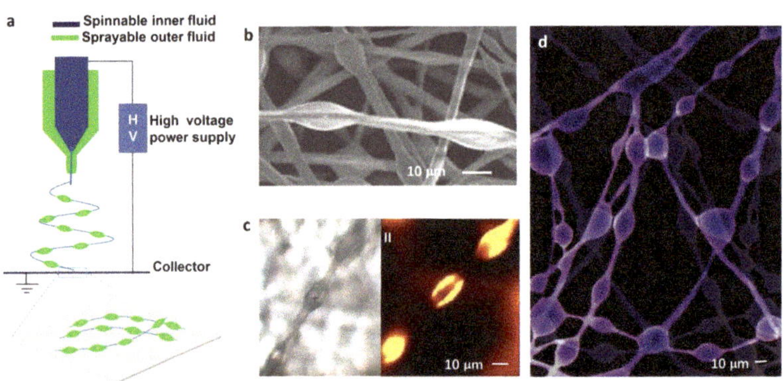

Figure 5.10 Fabrication of bead-on-string fibers (Tian et al., 2011). (a) Illustration of the combined electrospinning/electrospraying approach and (b) SEM image of the bead-on-string fibers. (c) Optical images of heterogeneous beads (see light parts, in Frame II). (d) ESEM image of bead-on-string fibers with water collection on the beads.

5.2.6 Electrochemistry

Electrochemistry techniques can be developed for the fabrication of nanostructures in gradients. For instance, the cathodic electrodeposition method is applied to realize a WG (Hou et al., 2016), as shown in Fig. 5.11.

Taking an example, a copper plate can be an anode to produce a current (I) with a gradient ($I_3 > I_2 > I_1$). A drain is used to adjust the deposition time (t) with a gradient ($t_3 > t_2 > t_1$) (Fig. 5.11a). By combining both gradients via the control of the voltage (V) and the flow velocity of the electrolyte (f), the surface generates gradients of geometrical structures (i.e., density of structures) and chemical compositions (including different types of Cu, Cu^+, Cu^{2+}), which induce WG in different directions (Fig. 5.11b). By controlling the current and time in gradients, surfaces of three types, i.e., type I (superhydrophobic → hydrophobic), type II (superhydrophobic → hydrophilic), and type III (hydrophobic → hydrophilic), can be achieved.

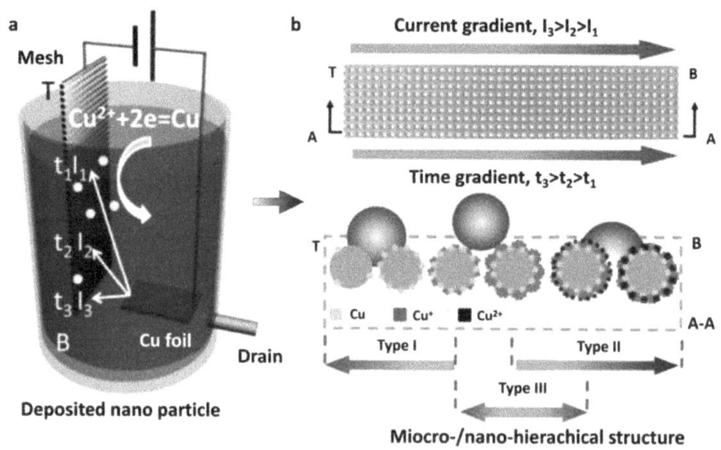

Figure 5.11 Illustration of the formation of various wettability gradient surfaces (Hou et al., 2016). (a) System of electrochemistry to generate a gradient on a surface. (b) Illustration of a gradient micro-/nanohierarchical surface via control of current and time in different directions.

Surface properties with adhesion can be realized by electrochemistry methods, such as an improved anodic oxidation method with current gradient control to adjust the extent of the process (Feng et al., 2014). Figure 5.12a illustrates the design of an electrochemistry technique to realize a wettable adhesive surface via the control of resistance between different areas (i.e., A_1, A_2, A_3) on a graphite plate; thus, the cathode resistance is different (i.e., $R_1 > R_2 > R_3$; Fig. 5.12b). According to Ohm's law, $I = U/R$ (given a voltage U), a current gradient forms in the oxidation process (i.e., $I_1 < I_2 < I_3$).

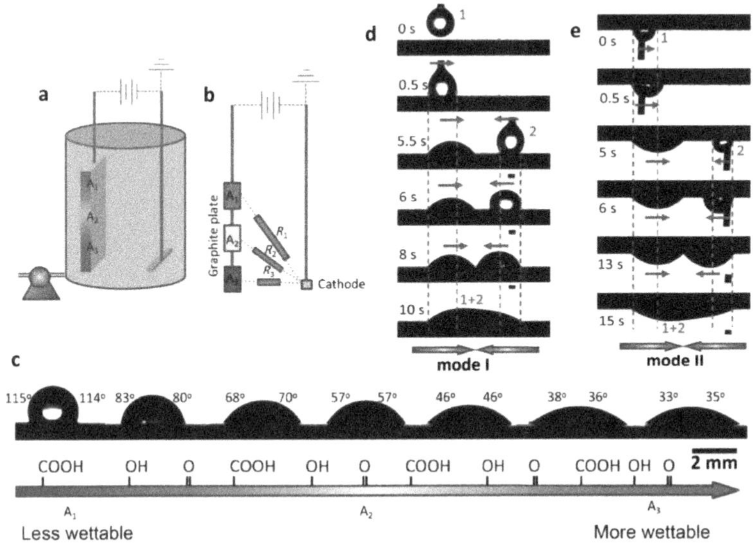

Figure 5.12 Fabrication of gradient surface by the electrochemistry technique (Feng et al., 2014). (a–b) Illustration of the formation of wettability gradients on a graphite plate and (c) optical images of droplet states with different water contact angles along the gradient surface due to gradient chemical components. (d–e) Stability of droplet gathering on gradient surfaces under two modes: horizontal or upside down.

Figure 5.12c shows the result of the surface with gradient wettability, i.e., contact angle of water from 115° (hydrophobic one) to 34° (hydrophilic one). Based on the method, a WG (i.e., hydrophobic–hydrophilic–hydrophobic mode) surface can be achieved effectively (Fig. 5.12d,e). Due to the gradient in contrast directions, the two water droplets move together along the directions of the WG (Fig. 5.12d). The adhesive property can be observed on the surface when the surface is turned upside down; the moving behaviors are the same (Fig. 5.12e). This indicates that the droplet behaviors can be controlled by high adhesion with a WG.

5.2.7 Soft Lithography

Biological structures are usually complex in configuration, presenting a challenge to biomimetic research. Recently, soft lithography has been used to fabricate biological structures by using various polymer materials, such as polystyrene (PS), poly(vinyl alcohol) (PVA), polydimethylsiloxane (PDMS), poly(etherimide) (PEI), and polyvinylidene fluoride (PVDF), which have been developed. In studies on grass leaves with ratchet structures, the soft lithography technique can be used to fabricate biomimetic surfaces (Guo et al., 2012). Figure 5.13 illustrates the fabrication of artificial surfaces with biomimetic structures from natural leaves (e.g., fresh ryegrass leaves) as templates. In detail, for instance, such a complex structure can be fabricated by using PVDF via modified soft lithography. First, the PDMS prepolymer is used to copy negative ratchet arrays of fresh ryegrass leaves after treatment at a temperature less than 60°C for 3 h. Then, a PVDF polymer is used to emboss the positive ratchet arrays from the negative ratchet arrays of the PDMS surface after the pressing process at a temperature of ~270°C, along with a heat-pressing force of $\sim 2 \times 10^6$ Pa. Finally, the artificial PVDF surface is quite similar to the surface of fresh ryegrass leaves.

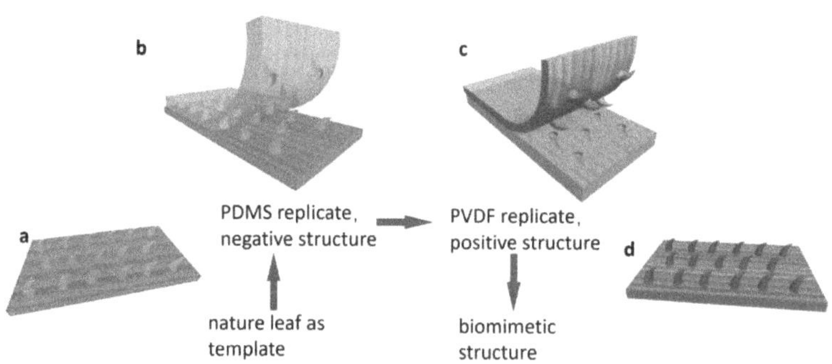

Figure 5.13 Illustration of the fabrication process from natural leaf surfaces to artificial surfaces (a→b→c→d) via the use of polymers (e.g., PDMS, PVDF) (Guo et al., 2012).

Tilt-angle pillar arrays can be fabricated via soft lithography with multimolding steps coupled with a controlled-temperature heat-pressing process (Lin et al., 2018a), as shown in Fig. 5.14a. In the fabrication processes, shape memory polymer (SMP) (T_g = 75°C) composed of bisphenol A diglycidyl ether (BADGE), poly(propylene glycol) bis(2-aminopropyl ether) (Jeffamine D230), and decylamine (DA) (molar ratio = 4:1:2) is utilized to fabricate micropillars with tilt angles, which then act as the molds. At the phase transition temperature, i.e., glass transition temperature (T_g), the elastic modulus changes by two to three orders of magnitude. The vertical SMP micropillar arrays are first obtained by the replica-molding technique from the template of Si micropillar arrays and then heated to 85°C (> $T_g \approx 75$°C) on a hot plate for 5 min, followed by covering with clean PDMS and shearing along one direction using a cylindrical iron bar until being cooled down completely. The pillars are tilted completely along the shearing direction. By controlling the reheating temperature to less than T_g (e.g., 60–70°C) and the reheating time (e.g., 0–300 s), the pillars are recovered to an equilibrium position with tilt angles ranging from 0° to 60°.

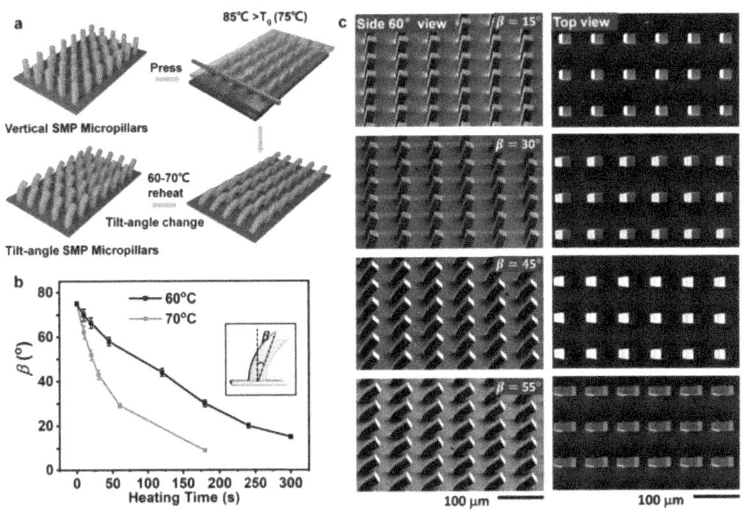

Figure 5.14 Fabrication of tilt-angle pillar arrays by soft lithography (Lin et al., 2018a). (a) Illustration of the fabrication of tilt-angle pillar arrays using SMP properties in soft lithography. (b) The relationship of the tilt angle (β) and heating time for the tilt-angle pillars. (c) SEM images of tilt-angle pillar arrays with different β: 15°, 30°, 45°, 55°, side view (left) and top view (right).

Figure 5.14b shows the tilt angles (β ranging from 0° to 70°) of micropillar arrays versus the reheating temperature (60 and 70°C) and reheating time (0–300 s). For the same recovery time (e.g., 100 s), the tilt angle decreases when the reheating temperature increases. For a certain reheating temperature (e.g., 70°C), the tilt angle decreases with increasing heating time. Figure 5.14c shows SEM images of tilt-angle pillar arrays (pillar diameter of 17 μm, height of 46 μm, and spacing of 53 μm) with different tilt angles (β = 15°, 30°, 45°, and 55°). The method can also be used to fabricate other arrays, e.g., cylindrical and conical arrays.

5.2.8 3D Printing

Three-dimensional (3D) microstructures have aroused great interest and been applied in integrated electronics, tissue engineering, high-efficiency collectors and sensors, photonic materials, etc. Various 3D fabrication techniques have been proposed in decades, including self-assembly of in-plane units, layer-by-layer printing, and direct writing (Wang et al., 2015b).

Figure 5.15a illustrates the fabrication of 3D microstructures on the liquid–solid composite surface guided by a cubic magnet. Magnetic ink (M-ink) droplets are printed onto the surface and deformed from hemispherical to pillared microstructures with droplet edge retracing. The 3D microstructures can be oriented by altering the tilt angle by deflecting the magnet. Figure 5.15b shows the liquid–solid composite surface constructed by swelling PDMS with silicone oil. The dynamic dewettability of the surface can be accurately controlled by the swelling ratio (defined as the ratio of absorbed silicone oil and PDMS), which essentially alternates the amount of liquid component. In particular, the receding angle (θ_R) of the surface presents fine adjustability of the swelling ratio (Fig. 5.15c). As a result, the M-ink droplets possess controllable retracing behavior. The formation of the 3D microstructure is illustrated in Fig. 5.15d. Pulled by a magnetic force F_m and distorted by a resistance force F_r, the droplet is deformed into a conic shape, along with a fast reduced instantaneous contact angle (CA). When the CA is less than the receding angle, the droplet edge starts to retrace. With droplet heightening and diameter reduction, the instantaneous CA increases gradually to reach the receding angle value, and the raised droplet will not be drawn away due to the substrate resistance. Eventually, a 3D high-aspect-ratio microstructure is formed, and the base angle of the 3D structure on the substrate is equal to its M-ink receding angle. A typical deformation process for one M-ink droplet on the composite surface with a receding angle of ca. 95° was captured by a CCD camera (Fig. 5.15e).

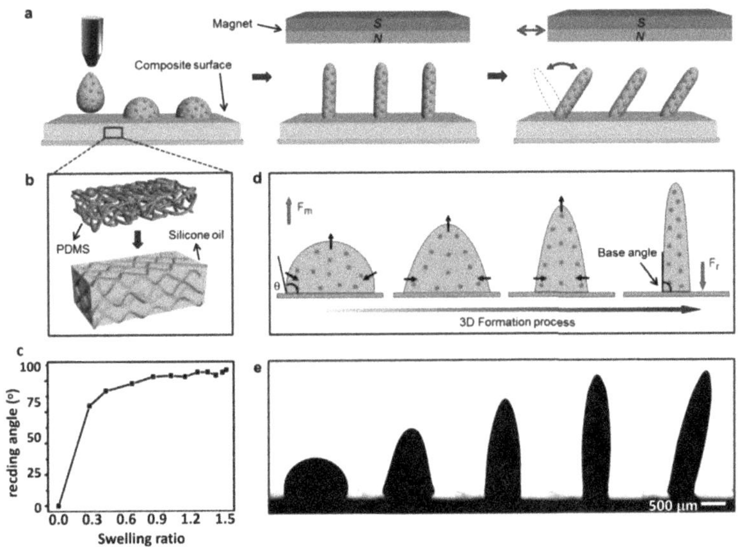

Figure 5.15 Fabrication of 3D microstructures by printing (Wang et al., 2015b). (a) Illustration of the fabrication of 3D microstructures on a liquid–solid composite surface guided by a cubic magnet. (b) The substrate composed of PDMS and silicone oil. (c) The relationship of the receding angle versus the swelling ratio. (d) The process of 3D structure formation by controlling forces (F_m and F_r). (e) Optical images of 3D microstructures.

This strategy can be carried into the fabrication of asymmetric microstructures (e.g., tilt pillars), which are realized by laterally moving or rotating the magnet (Fig. 5.16a). The tilted pillar is affected by both F_m and F_r. As $F_r > F_m$, the pillar droplet is tilted by a magnetic anisotropic torque without slipping on the surface. As a result, tilted 3D microstructures with tilting angles can be obtained (Fig. 5.16b–e). Moreover, synchronized-tilted pillar arrays with certain angles can be facilely printed (Fig. 5.16f). An anisotropic tilted-angle pillar row possessing tilting angles from 0° to 90° is achieved by rotating the magnet (Fig. 5.16g).

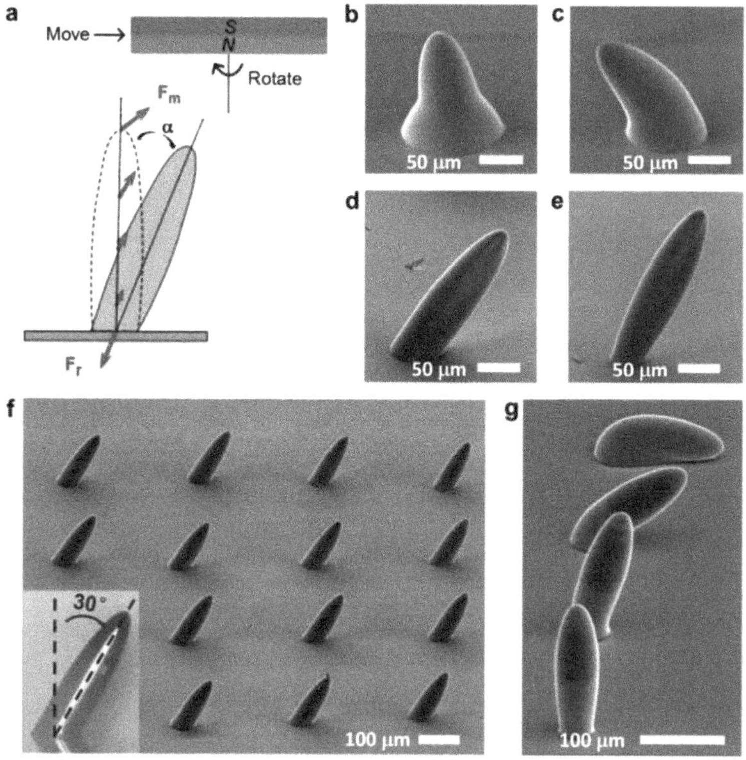

Figure 5.16 Fabrication of anisotropic tilted-angle pillars by rotating the magnet (Wang et al., 2018). (a) Illustration of fabrication for tilt pillars. (b–g) Versatile printing of asymmetric 3D pillars with tilt angles.

5.3 Control of Droplet Transport

5.3.1 Unidirectional Spreading on Asymmetric Nanostructured Surfaces

Asymmetric nanostructures can achieve liquid transport in different directions. It is demonstrated that the design of asymmetric nanostructured surfaces can be harnessed to realize unidirectional liquid spreading (Chu et al., 2010), where the liquid propagates in a single preferred direction and is pinned in all others (Fig. 5.17b), rather than the effects of a uniform array (Fig. 5.17a), because the asymmetric nanostructures can generate energy barriers to limit the liquid transport in structure-directed directions. Figure 5.17c shows the surfaces of asymmetric nanopillar arrays with deflection angles (φ) of 7° (the left) and 12° (the right) along the positive X (+X) direction.

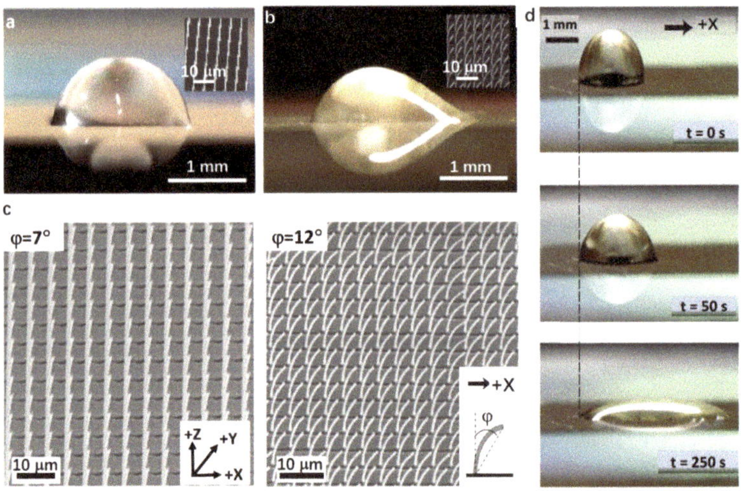

Figure 5.17 Asymmetric nanostructured surfaces for unidirectional liquid spreading (Chu et al., 2010). (a–b) Liquid spreading on a normal structured surface (a) and unidirectional liquid spreading on an asymmetric nanostructured surface (b). (c) SEM images of surfaces of pillars with tilt angles (φ) of 7° and 12°. The tilt-angle direction can be defined as the +X direction. (d) Optical side-view images of droplet spreading in different directions on the asymmetric (i.e., tilt-angle) pillar surface.

Figure 5.17d shows time-lapse images of droplets on a polymer surface (water CA of 59°) with tilted φ-12° nanopillars. A mixing droplet containing deionized water with Triton X-100 (a surface tension of 39.7 dyn/cm) is placed on the surfaces. Unidirectional spreading appears along the nanopillar geometry in the +X direction.

A simple two-dimensional model (along the x- and z-axes) explains the liquid behavior on asymmetric tilt-angle pillar arrays (Fig. 5.18). The asymmetric pillars can be idealized as slanted rectangles (the inset of Fig. 5.18). According to Young's equation, the local CA of the liquid, i.e., the intrinsic CA, θ_{eq}, must be equal to or smaller than a critical angle, θ_{cr} (the inset of Fig. 5.18). The film propagates at the height of the pillars; therefore, $\theta_{cr} = \tan^{-1}(H_{eff}/l_{eff})$, where H_{eff} and l_{eff} are the effective height of the pillars and their spacing, respectively. The critical angle in the +X direction is given by

$$\theta_{cr,+X} = \tan^{-1}\frac{\left(\frac{H}{l}\right)\cos\varphi}{1-\left(\frac{H}{l}\right)\sin\varphi} \tag{5.8}$$

where H and l are the height and spacing of the pillars, respectively. In the –X direction, the critical angle is given by:

$$\theta_{cr,-X} = \tan^{-1}\frac{\left(\frac{H}{l}\right)\cos\varphi}{1+\left(\frac{H}{l}\right)\sin\varphi} \tag{5.9}$$

If $\theta_{cr} > \theta_{eq}$, the liquid film transports within the nanopillars; otherwise, the liquid film is pinned. The unidirectional spreading regions can be indicated with a blue color region.

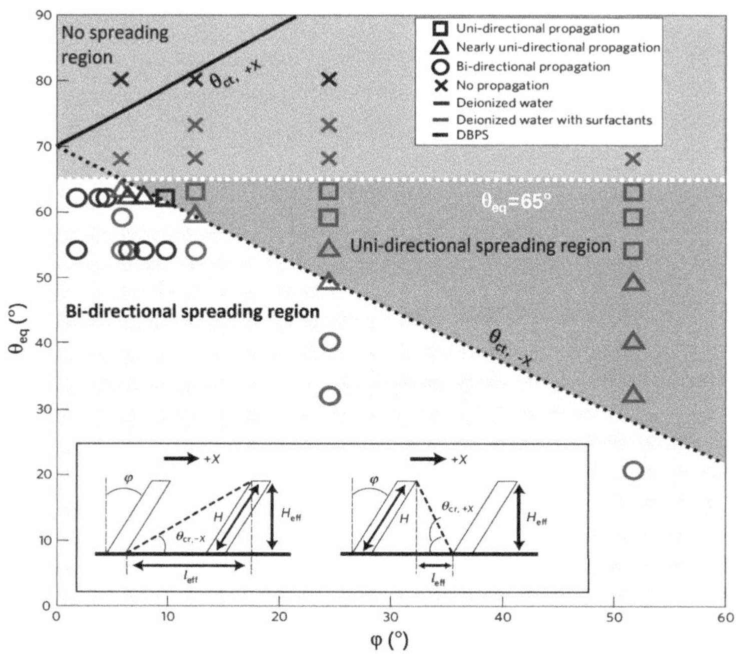

Figure 5.18 Illustration of the properties of liquid spreading based on experiments and theories (Chu et al., 2010), including the bidirectional spreading region (yellow), unidirectional spreading region (blue), and non-spreading region (gray).

5.3.2 Unidirectional Spreading of Droplets on a Fibrous Surface

Biological surfaces display smart wettability to control the behavior of liquid, such as wetted spider silk for water transport (Zheng et al., 2010) and butterfly wings for directional water repellency (Zheng et al., 2007; Liu et al., 2014). The anisotropic micro- and nanostructures (e.g., grooves, stripes) along with WG can effectively regulate the behaviors of droplets on a surface (Zhang et al., 2015).

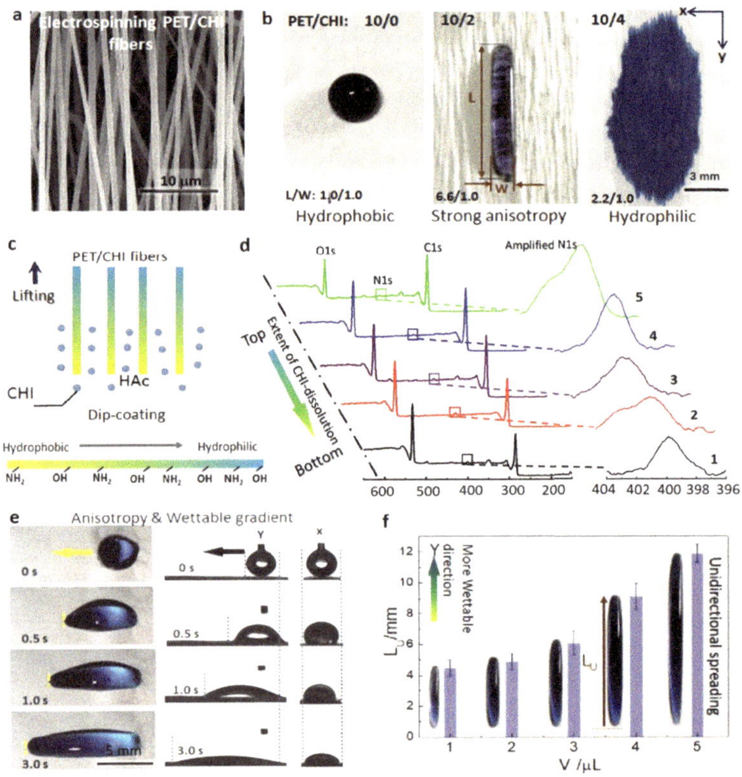

Figure 5.19 Gradient anisotrpic wettable surface with directional spreading of liquid (Zhang et al., 2015). (a) SEM image of the electrospun PET/CHI fiber array. (b) Different wettabilities for different ratios of PET/CHI: 10/0 for hydrophobic, 10/2 for strongly anisotropic, and 10/4 for hydrophilic properties. (c–d) Gradient fabrication and formation of gradient wettability (c) and XPS analysis (d). (e) Optical images for the control of wetting anisotropy and wettability gradients on the aligned fibrous surface and the liquid-spreading behaviors. (f) Relationship of the unidirectional spreading length (Lu) versus the volume of the liquid.

Based on the gradient effect, a smart liquid-spreading surface can be designed by modulating the gradient of the hydrophilic component (chitosan, CHI) on the aligned PET (poly ester)/CHI fibrous surface (Zhang et al., 2015). First, an electrospinning technique is used to fabricate an aligned PET (hydrophobic)/ CHI (hydrophilic) fibrous surface (Fig. 5.19a). After tailoring the PET/CHI composition (Fig. 5.19b), liquid spreading in a stick-like shape can be achieved with a shape ratio of 1:6 (width:length). A WG is formed on the above anisotropic surfaces after the CHI-dissolution process in the acid solution (HAc, CH_3COOH) (Fig. 5.19c). Figure 5.19d illustrates the WG along the Y direction (from the hydrophobic side to the hydrophilic side) on a fibrous surface due to increasing OH group density. The chemical components (at sites 1, 2, 3, 4, and 5 of Fig. 5.19d) show the gradient peaks by XPS analysis. Figure 5.19e,f demonstrates the unidirectional droplet-spreading behavior on the WG surface.

It can be concluded that the WG force (F_W) drives the droplet toward the more wettable region of the surface, described as follows (Brochad, 1989; Daniel and Chaudhury, 2002):

$$F_W = \pi R^2 \gamma_{lv} k \sin\theta \quad (5.10)$$

where R is the base radius of the droplet in the X direction just before spread, γ_{lv} is the water tension, θ is the water CA along the WG, k is the degree of WG equal to $-d\theta/dy$, and dy is the integrating variable along the WG. For a given R, a higher k or θ leads to a larger F_W.

The hysteresis force (F_H) is always opposite to the moving direction and can be described as:

$$F_H = 2R\gamma_{lv}\left(\cos\theta_{r0} - \cos\theta_{a0}\right) \quad (5.11)$$

where θ_{r0} and θ_{a0} are the receding and advancing CAs, respectively. When $F_W > F_H$, the droplet can spread along the WG direction, while the droplet is pinned when $F_W < F_H$ (Li and Amirfazli, 2007; Bliznyuk et al., 2011).

5.3.3 Nepenthes-Peristome-Inspired Unidirectional Liquid Transport

The slippery peristome of the pitcher plant *Nepenthes* has attracted much attention due to its unique function for preying on insects. Recent findings on the peristome surface of *Nepenthes alata* demonstrate fast and continuous unidirectional liquid transport, which is enabled by the combination of a pinning effect at the sharp edges and a capillary rise in the wedge, derived from the multiscale structure, which provides inspiration for designing and fabricating functional surfaces for unidirectional liquid transport (Zhang et al., 2017a,b). Inspired by these results, functional surfaces with similar peristome or peristome-mimetic structures can be designed, for example, by directly using the natural peristome as a template and two steps of replication using PDMS (Li et al., 2016a), or by two-step UV photolithography for biomimetic surfaces with an inclined pit array along the microgrooves (Chen et al., 2017b).

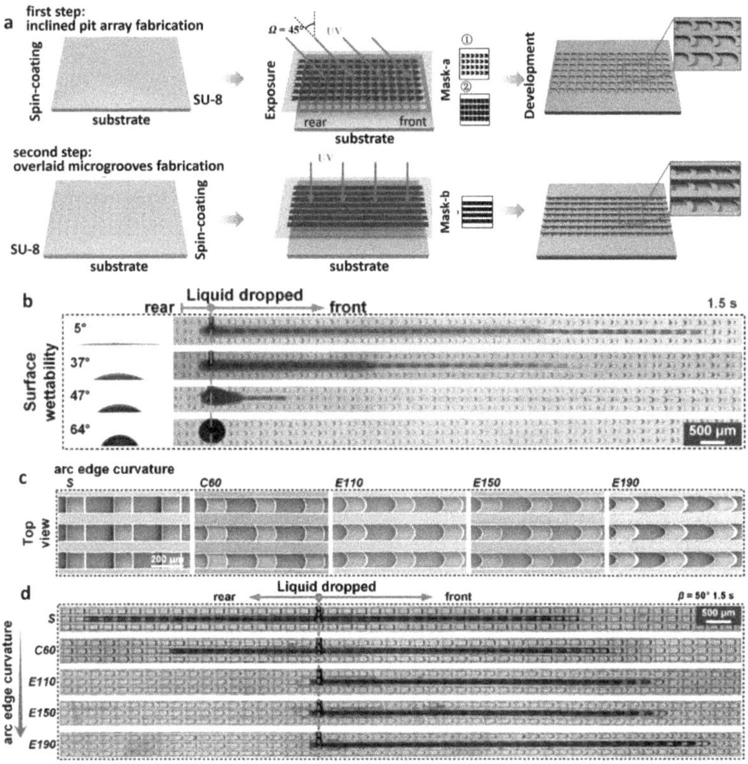

Figure 5.20 (a) Illustration of fabrication on the *Nepenthes*-peristome-inspired structures (Chen et al., 2017a). (b) Water-spreading abilities controlled by surface energies (CA: 5°, 37°, 47°, 64° from top to bottom), (c) and the arc edge curvatures (i.e., S, C60, E110, E150, E190 from top to bottom) for liquid spreading in different styles. The E190 structure makes the smart liquid spread in a specific direction (Chen et al., 2017b).

Figure 5.20a illustrates the fabrication process of *Nepenthes*-peristome-inspired structures (Chen et al., 2017a). Figure 5.20b shows the water spreading for unidirectional transport versus wettability, where the spreading ability decreases with decreasing surface energy (e.g., characterized by CAs from 5°, 37°, 47° to 64°). There is a threshold CA (~64°) for unidirectional liquid transport. Furthermore, by changing the parameters of arc edge curvature or wedge angle (β), the liquid spreading can be made more distinct (Chen et al., 2017b). As shown in Fig. 5.20c,d, for β = 50°, smart unidirectional liquid transport occurs at the ellipse arc edge (E) at 1.5 s, for example, the surfaces of E190, E150, E110 (the bottom regions, Fig. 5.20d), but nearly no unidirectional liquid transport occurs at the straight edge (S) or circle arc edge (C) (the top regions, Fig. 5.20d).

5.3.4 Droplet Transport on an Integrative Janus Membrane and Conical Spine

It is feasible to mimic biological structures that can achieve water-collecting ability. A key is that bioinspired structures can be endowed with physical and chemical multigradients to propel droplet transport in the process of water collection (Zheng et al., 2010; Ju et al., 2012). The greater desired property of high droplet transport efficiency is obtained by designing a surface shape gradient (Zheng et al., 2017), magnetic stimuli-responsive wettability asymmetry (Tian et al., 2016), and temperature-responsive lubricating interfaces (Wang et al., 2018).

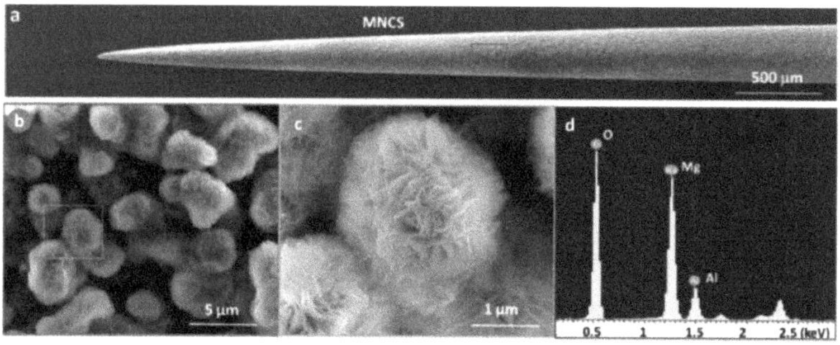

Figure 5.21 Characterization of the surface on the MNCS (Zhou et al., 2018). (a–c) SEM images of the MNCS with micro- and nanostructures. (d) Energy dispersive spectroscopy (EDS) analysis.

Inspired by the integrated fog-collection system involving cactus spine and trichomes, a micro-/nanostructured conical spine along with a Janus membrane integrative system (MNCS+JM) is fabricated by using aluminum (Al) wire via an integrative electrochemical corrosion and hydrothermal method (Zhou et al., 2018). Figure 5.21a–c shows SEM images of the MNCS with conical spines covered with nanoscale lamella-stacked microscale papillae. An energy dispersive spectroscopy (EDS) analysis reveals the presence of oxygen (O)-, magnesium (Mg)-, and Al-dominated components (Fig. 5.21d).

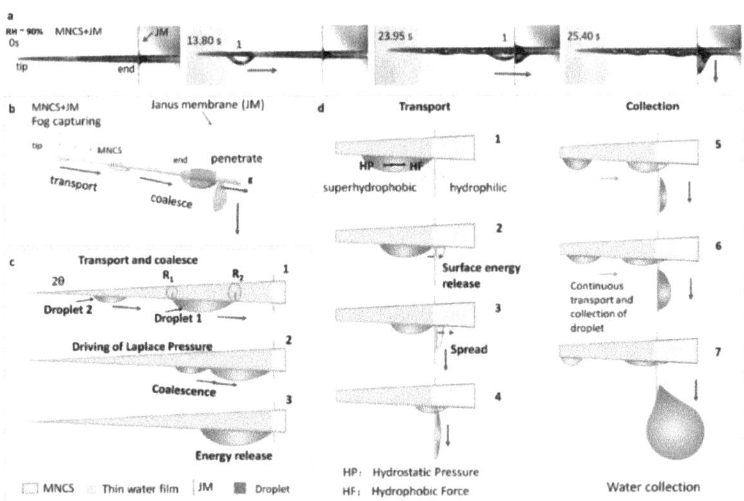

Figure 5.22 Optical images and mechanism of droplet transport (Zhou et al., 2018). (a) Optical images for in situ observation of droplet transport along an MNCS and further transport through a JM. (b–d) Illustration of the mechanism for droplet transport (b) as the cooperative result of Laplace pressure in different directions. (c) Coalescence of a droplet and the wettability gradient between the superhydrophobic and hydrophilic regions of the JM (d).

The moisture-harvesting ability can be examined on the MNCS+JM in a chamber with a fog flow velocity (v) of 1 m/s and a relative humidity (RH) of 90% (Fig. 5.22). It is estimated that the water drop collected volume is ~0.034 mL in 30 s, corresponding to a water-collection rate of ~0.068 mL/min (Fig. 5.22a), which far exceeds that of designs lacking the JM. For v ≈ 0.5 m/s and RH ≈ 90%, the droplet can be transported over a length of ~21 mm.

Figure 5.22b illustrates such a wettability-integrated strategy of MNCS+JM, which is attributed to multigradient cooperation. A pressure difference is generated by the curvature difference of the spine, i.e., $\Delta P_{curvature} = -\int_{R2}^{R2} \frac{2\gamma}{(R+R_0)^2} \sin\theta dz$, where R is the local radius of the conical spine (R_1 and R_2 are the local radii of the spine at two opposite sides of droplet 1, and R_0 is the radius of droplet 1), γ is the water tension, θ is the half apex angle of the cone, and dz is the integral variable of the cone. The energy release force (F_e) of droplet coalescence acts in conjunction with the pressure difference, i.e., $F_p + F_e$, to overcome the resistance force (F_r) from the CAH in droplet transport. In particular, a strong wettable gradient force (Fig. 5.22c), i.e., hydrostatic pressure (F_{HP}) and hydrophobic force (F_{HF}) on the two sides of the JM with superhydrophilic and superhydrophobic properties, propels further droplet transport. These multigradient forces can effectively cooperate for droplet transport in fog harvesting.

5.3.5 Oriented-Structure-Induced Liquid Transport

Liquid unidirectional transport has cutting-edge applications ranging from fog collection and oil–water separation, to microfluidic devices (Li et al., 2017; Yang et al., 2018). Despite extensive progress, existing human-made surfaces with asymmetric wettability or micro-/nanoscale structures are still limited by complex fabrication techniques or obscure essential transport mechanisms to achieve unidirectional transport with both high speeds and large volumes (Si et al., 2018).

A liquid unidirectional transport manipulation mechanism was developed with open micro/macro dual-scale arrays to achieve long-range, rapid, and controlled transport with the aim of generating overlapping Laplace pressures as the driving force. Liquids with low surface tension can be transported spontaneously in a single preferred direction and pinned in all others without external energy input. Through digital light processing (DLP) and 3D printing technology (Li et al., 2016a; Yang et al., 2018), dual-scale arrays with parallel microgrooves can be fabricated on demand within 30 min. Figure 5.23 indicates the fabrication of dual-scale A-shaped island arrays (Si et al., 2018). Figure 5.23a,b shows a schematic diagram of the DLP 3D printing process and layer-by-layer bottom-up additive manufacturing strategy, similar to the lamellar scale on the butterfly wing. The inverted sample is pulled out of liquid resin. The precured region (red rectangle) is between the sample and screen. A light beam from a UV light-emitting diode (UV-LED) is projected onto the bottom surface of the sample through a transparent screen, and the liquid resin is UV cured to form a new layer on the sample. Then, the sample is lifted along the +Z direction with a short distance, ΔZ, for each layer. In the next circle, the cured layer becomes a substrate to form a new curing layer. Here, $\Delta\tau$ is the interval time of two continuous cycles. Figure 5.23c is the top view of the A-shaped island array model. Figure 5.23d,e shows stereomicroscope and magnitude SEM images in the side view of the A-shaped islands. The island displays an italic letter "A" morphology (red dotted line) in the top view. Figure 5.23f shows the side view of the A-shaped island array model. L is the distance between adjacent islands, α is the titled angle between the islands and the normal line, and H is the vertical height of each island. Figures 5.23g and 5.23h are low- and high-magnitude side-view SEM images of A-shaped islands, respectively. The edge of the island has a stairstep shape, and the height of a single step (h) is approximately 100 μm. Parallel microgrooves exist on the surface of the island from root to roof.

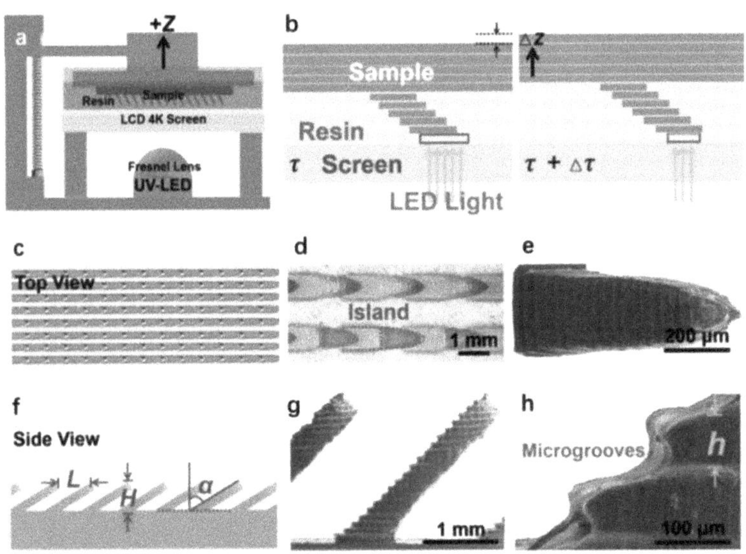

Figure 5.23 Liquid unidirectional transport on asymmetric micro-/nanoscale structured surface (Si et al., 2018). (a–b) Illustration of the fabrication of dual-scale A-shaped island arrays. (c–e) Top view of island structures. (f–h) Side view of a tilt island structure with microgrooves.

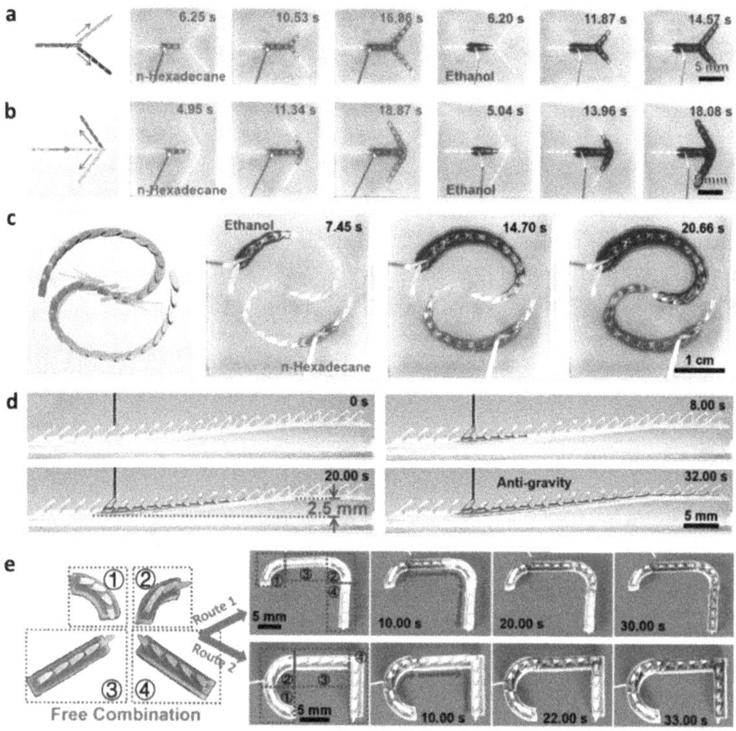

Figure 5.24 Optical images of liquid transport (Si et al., 2018). (a) In the Y style, (b) → style, (c) S style, (d) climb-up style, and (e) complex style via A-shaped island arrays.

Figure 5.24 shows liquid transport on different patterns of unidirectional A-shaped island arrays. Figure 5.24a,b shows liquid unidirectional transport on "Y"- and "→"-shaped patterns and A-shaped island arrays. Continuously depositing red-dyed n-hexadecane or blue-dyed ethanol onto the middle of the arrays, liquids are transported along the arrays with specific patterns. The liquid stream can be divided into two target directions. Figure 5.24c shows the behavior of red-dyed n-hexadecane and blue-dyed ethanol deposited on the asymmetric arc-shaped A-shaped island arrays simultaneously; two liquid streams form a Chinese-dragon-like Tai Chi symbol. Figure 5.24d shows spontaneous antigravity unidirectional transport on the A-shaped island slope. Upon continuous deposition of dyed n-hexadecane onto the bottom of the arrays, the liquid flows upward and climbs along the slope without retraction at the rear side. Within 32.00 s, liquid flows a distance of ~37.0 mm and elevates approximately 2.5 mm. Figure 5.24e shows that A-shaped island array modules can be freely combined into different microfluidics routes. Liquid can be unidirectionally transported along the target route without changing these modules.

5.3.6 Unidirectional Drop Cargo Transport

To promote drop mobility, lubricating the gap between liquid drops and solid surfaces is a facile method that has been widely exploited by nature. Inspired by lotus and rice leaves using entrapped air to "lubricate" water and *Nepenthes* pitcher plants using a slippery water layer to trap insects, a strategy for transporting drop cargoes can be achieved via the unidirectional spreading of immiscible lubricants on a peristome-mimetic surface. Figure 5.25 shows the mechanism of droplet transport for water core transport on an n-hexadecane-coated peristome-mimetic surface. Oleophilic/hydrophobic peristome-mimetic surfaces were fabricated by replicating 3D printed samples (Yu et al., 2018a). Figures 5.25a, 5.25b, and 5.25c show a diagram of the experimental setup, the wetting state of the water core on the n-hexadecane-coated peristome-mimetic substrate, and the lateral and vertical resistance forces of the water core sliding on the n-hexadecane strip, respectively. The forces in the x-axis and z-axis can be calculated from the tube deformation by detecting reflected light variations. Figure 5.25d shows the lateral and vertical resistance forces for a water drop (~2 μL) that slides on the n-hexadecane-coated peristome-mimetic surface. The measurement is performed at a constant linear stage velocity of 1.0 mm/s.

The peristome-mimetic surface, via unidirectional immiscible hexadecane spreading, can transport a wide diversity of drop cargoes over a long distance with no loss as well as controllable drop volumes and velocities, hence offering the possibility of mixing multiphase liquids and even reacting liquids. This unidirectional drop cargo transport technique is expected to find use in microfluidics, microreactors, water-harvesting systems, etc.

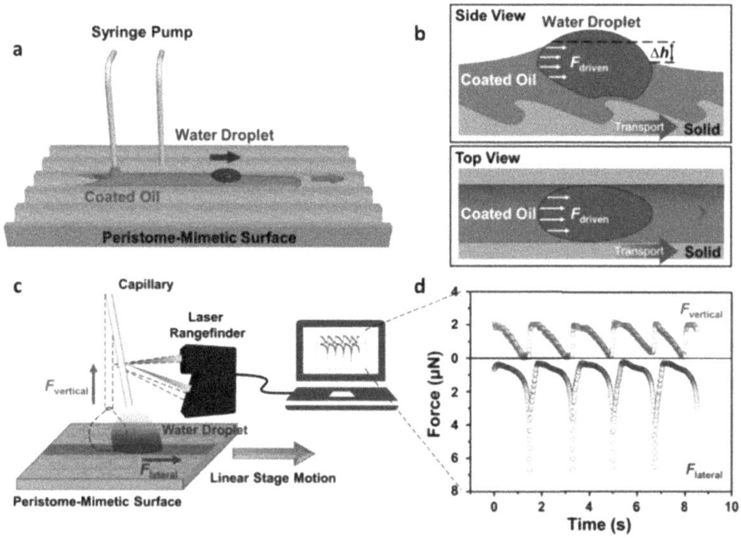

Figure 5.25 Observation on droplet transport on a peristome-mimetic surface (Yu et al., 2018a). (a) Illustration of the mechanism of water core transport on an n-hexadecane-coated peristome-mimetic surface. (b) Illustration of the driving force in the water–oil interface of the peristome-mimetic surface. (c–d) Analysis system of liquid transport (c) along with a force analysis versus time (d).

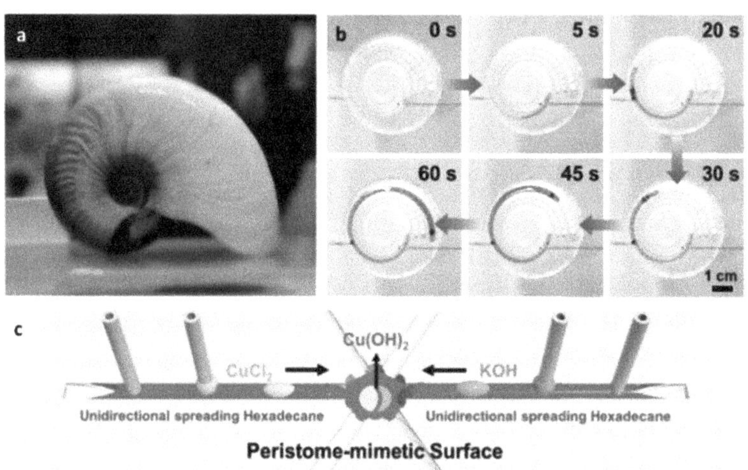

Figure 5.26 Manipulating droplet transport on peristome-mimetic surface (Yu et al., 2018a). (a) Optical image depicting the morphology of *Nautilus pompilius* according to the Archimedean spiral. (b) Droplet transport through the peristome-mimetic Archimedean-spiral microgroove. (c) Drop cargo transport for microfluidics and microreactors.

Figure 5.26 shows drop cargo transport for microfluidics and microreactors. Figure 5.26a shows the optical image depicting the morphology of *Nautilus pompilius* according to the Archimedean spiral. Figure 5.26b shows a 2 μL droplet transported through a peristome-mimetic Archimedean-spiral microgroove driven by unidirectional spreading of hexadecane without retraction. Figure 5.26c shows a microreactor processing orange ethanol droplets and pink water droplets, which contain $CuCl_2$ solute and KOH solute, respectively. The water droplets are dyed by phenolphthalein. Droplets are driven by n-hexadecane, unidirectionally spreading on the peristome-mimetic surface. Then, two droplets mix within a droplet and react to form brown $Cu(OH)_2$.

5.3.7 Time-Dependent Liquid Transport

The peristome surface, a typical physical model (Fig. 5.27a), contains surface curvatures at different scales (Wong et al., 2011). Optical microscopy and SEM are used to observe the peristome surface (Frames I–V). To the naked eye, the peristome appears as an arch-shaped ring of tissue with a radius of curvature R_1 = 5 mm in a vertical cross-sectional view, much larger than that of a water drop (~1 mm) (Frame II) (Bohn and Federle, 2004). Switching the optical view to a scanning electron micrograph view, we find that the surface of the peristome is covered by a two-order hierarchy of parallel microgrooves (Frame III) (Yu et al., 2018b). The larger microgrooves satisfy R_3 = 500 μm. Between neighboring microgrooves are microridges. The radius of microridge curvature R_2 of ~100 μm is almost one-tenth of the diameter of a water drop. A high-magnification SEM image demonstrates overlapping microcavities from smaller microgrooves (Frame VI). The microcavity slants into the surface, is enclosed at the tip, and has a sharp overhang edge at the rear side (Bohn and Federle, 2004). The radius of the overhanging edge R_4 < 1 μm (Frame V). An enclosed microcavity exists below the overhang edge, which acts as a "cage." If the impacting water does not fill the cage in time, air could be stored in the microcavities. The air entrapment resists the spreading of the precursor toward the tip of the microcavity.

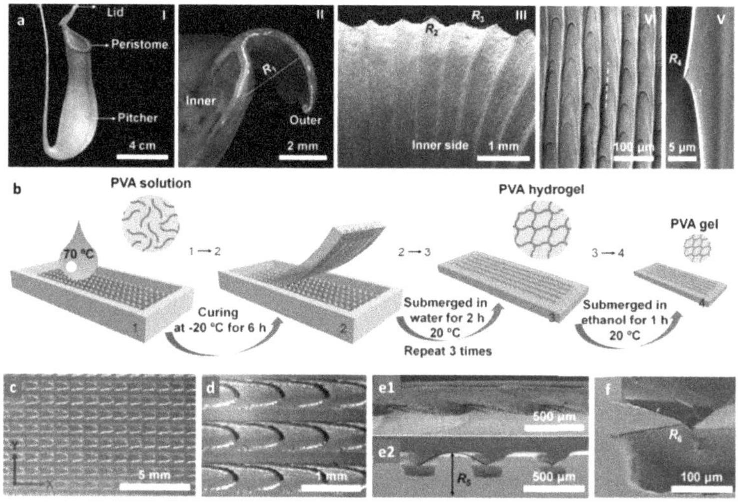

Figure 5.27 Bioinspired structured design (Yu et al., 2018b). (a) Optical and SEM images of the peristome surface with curvatures at different scales. (b) Illustration of fabrication and (c–f) characterization of peristome-mimetic surfaces with complex microstructures.

The peristome-mimetic surface can be achieved by the fabrication process of a peristome-mimetic poly(vinyl alcohol) (PVA) gel surface (Yu et al., 2018b). The mold is constructed through 3D printing. The 3D-printed mold is mounted in a tank, a hot PVA solution at 70°C is cast onto the mold (Step 1), and then the PVA-coated mold is stored in a refrigerator for 6 h (Steps 1–2). Subsequently, the PVA replica is removed from the mold (Step 2), and the peristome-mimetic flexible PVA replica is submerged into water three times every 2 h. The PVA hydrogel is finally prepared as shown in the left image of Fig. 5.27b. Figure 5.27c,d shows stereoscopic microscope images of the artificial surface. There is a microgroove distribution with a periodicity similar to that of the natural *Nepenthes* peristome, and the microgrooves consist of periodic microcavities. Figure 5.27e,f shows SEM images of the peristome-mimetic film at the top and cross-sectional views. The curvature changes as the outline of the peristome-mimetic surface varies from microridges with R_5 = 500 μm between neighboring arrayed microcavities to sharp overhangs with R_6 = 10 μm along the microcavities.

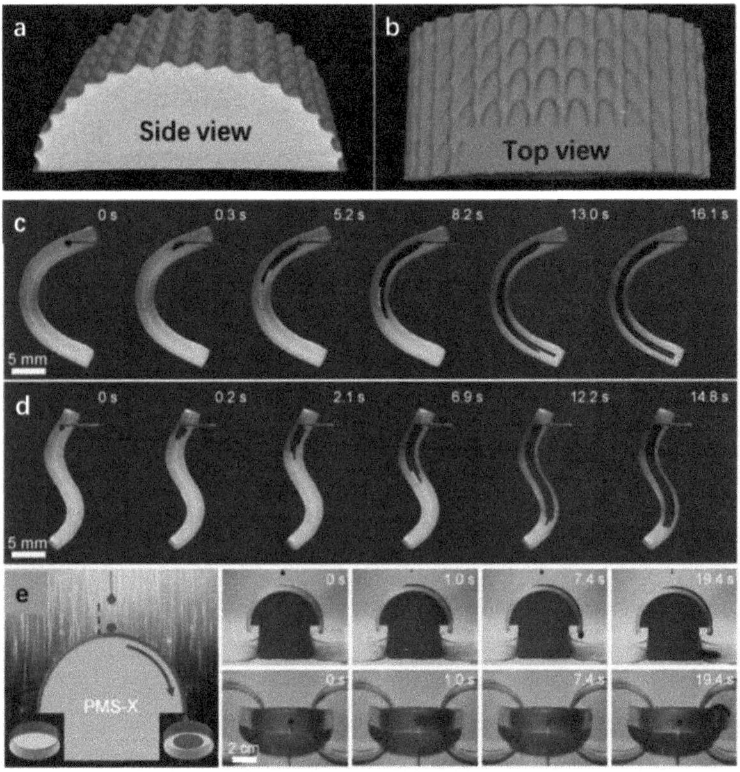

Figure 5.28 Droplet manipulation (Yu et al., 2018). (a–b) Design of the biomimetic substrate with oriented curvatures. (c–d) The liquid transport control. (e) Controlling for droplet transport of impacting drops on a 3D curved peristome substrate.

Figure 5.28 shows the transport control of impacting drops on a 3D curved peristome substrate. Figures 5.28a and 5.28b illustrate the biomimetic substrate in the side and top views, respectively. Figure 5.28b shows the C-shaped drop motion trajectory. Figure 5.28d shows the S-shaped drop motion path. The water transport path can be facilely achieved by curving the PVA substrate, which provides a design for the open microfluidic configuration. Figure 5.28e shows an illustration of the raindrop-shielding device, where the peristome-mimetic surface is covered on the curved plate. After being dropped from a height of 20.0 mm, the impacting water can be transported along the overlapping microcavities, climbing a hill against gravity and spreading toward the outside to be controlled effectively.

5.3.8 Liquid Transport in Hierarchical Microchannels

As is known, a type of insect-trapping plant (e.g., *Sarracenia*) generally grows in marshland, entrapping and digesting insects to obtain enough nutrients to survive (Chen et al., 2018). A lid (operculum) with many nectar glands emerges directly from the back of the pitcher mouth and overarches the peristome (Fig. 5.29a, left). The operculum of *Sarracenia* is generally covered by an enormous amount of trichomes oriented toward the pitcher mouth (Fig. 5.29a, right). The trichomes can capture dew and condense ambient moisture to form droplets and then quickly transport the droplets onto the peristome surface to keep the pitcher mouth wet and slippery for catching insects. In foggy environments, a thin film of water is distributed to cover the whole dry trichome. First, fine water droplets condense on the trichome surface and collect together to form several larger droplets. The white arrows indicate the transport direction of the water droplets along the dry trichome. The trichome surface becomes wet, marked by blue arrows, after these water droplets slide through. The enlarged droplet 4 moves directionally to coalesce with droplet 3 to form a larger droplet 3 + 4 and then moves to collect together with droplet 2 and droplet 1 in successive order. Droplet 6 also moves to coalesce with droplet 5 and then moves to coalesce with droplet 1 + 2 + 3 + 4. When the water droplet becomes large enough ($\alpha \approx 48°$ and $\beta \approx 65°$), the harvested water droplet is quickly transported onto the surface of the operculum and disappears immediately.

Figure 5.29b shows the ultrafast water transport process after the trichome becomes wet. A thin film of water forms on the surface of the trichome and makes the water move more rapidly, so no more large droplets form. The ultrafast transport of tiny droplet 1′ is shown in Fig. 5.29c. Figure 5.29d shows SEM images of the trichome surface showing that the trichome has unique hierarchical structural characteristics, that is, two-order ribs perfectly aligned along the gradient trichome to form approximately parallel hierarchical microchannels (Frames A, B, C, and D, corresponding to the right image of Fig. 5.28a). Low ribs gradually appear along the trichome, and the combination of high and low ribs serves as the typical structural features, in which the number of low ribs n between two neighboring high ribs H increases from 1–2 (n = 1, 2) near the tip to 4–5 (n = 4, 5) around the base.

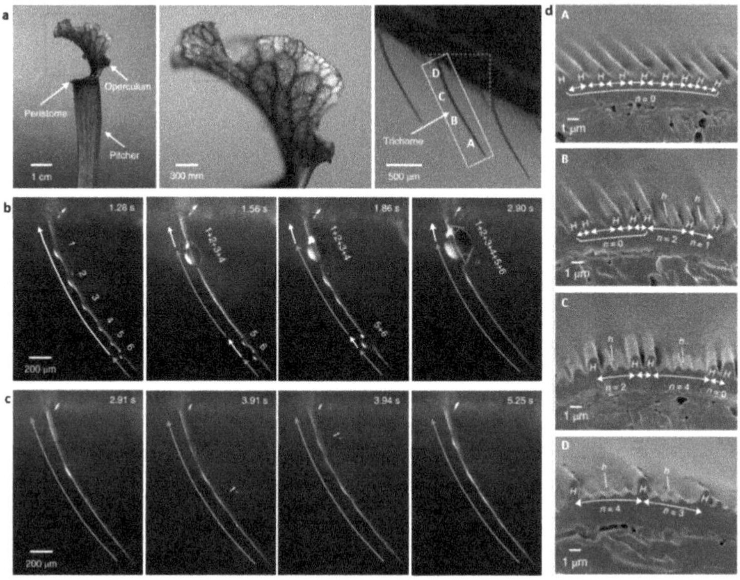

Figure 5.29 (a) Optical images of *Sarracenia*, with enormous amounts of trichomes. (b) Ultrafast liquid transport on trichomes. (c) Unique hierarchical structure of trichomes for ultrafast liquid transport (Chen et al., 2018).

Figure 5.30 shows the control of liquid transport on hierarchical microchannels. Figure 5.30a shows the water transport Mode I along dry hierarchical microchannels (N = 2). The yellow dotted lines mark the concave shape of the advancing meniscus. The transport velocity gradient appears as the water droplets fill the dry hierarchical microchannels. Figure 5.30b is the schematic diagram of the water transport in Mode I along the dry hierarchical microchannels. The advancing meniscus is concave in shape. Figure 5.30c shows the water transport in Mode II along wet smooth microchannels without low ribs. Water droplets mixed with rhodamine B or fluorescein sodium are used for water transport Modes I and II, respectively. The section images F–F and G–G, respectively, show the liquid stable states of Modes I and II without low ribs. Figure 5.30d shows the water transport in Mode II along the wet hierarchical microchannels with low ribs. The water transport in d is faster than that in a and c (see the red dotted lines). The section images J–J and K–K, respectively, show the liquid stable states of Modes I and II with low ribs. Two extra minor concave curves appear within the hierarchical microchannel (see white dotted lines).

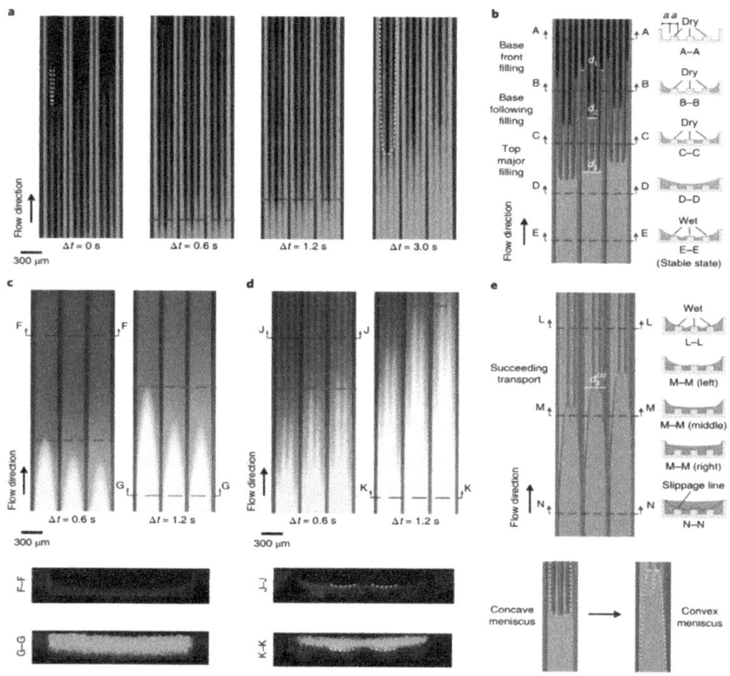

Figure 5.30 Water transport in hierarchical microchannels with different modes (Chen et al., 2018): Mode I, along dry hierarchical microchannels (a–b); Mode II, along wet smooth microchannels without low ribs (c); and Mode II along wet hierarchical microchannels with low ribs (d–e).

Figure 5.30e is a schematic diagram of water transport in Mode II along the wet hierarchical microchannels. After the microchannels become wet, the shape of the advancing meniscus changes from concave in Mode I (dry) (lower left) to convex in Mode II (wet) (lower right). Here, Δt indicates the maximum time of liquid transport.

5.4 Droplet Transport Controlled by External Actions

5.4.1 Magnet-Induced Dynamic Arrays for Microdroplet Transport

Droplet manipulation can be realized on some material-responsive surfaces, such as external temperature stimuli, chemical, light, magnet, and pneumatic mechanisms (Lin et al., 2018b). Figure 5.31a illustrates a magnetic nano-/micropillar array (MNA) that is fabricated by mixing PDMS and NdFeB microparticles via an improved soft lithography method. A PDMS mold with a negative structure array is duplicated from Si micropillar arrays (Step 1). The NdFeB/PDMS composites are poured over the PDMS mold. A magnetic field acts on the magnetic particles of the PDMS mold via the action of a permanent magnet (Step 2); thus, the magnetic microparticles can be uniformly distributed over the pillars. After pressing against the PDMS mold, curing, carefully demolding, and depositing one layer of SiO_2 nanoparticles, magnetic micropillar arrays with superhydrophobicity can be achieved (Step 3). Finally, magnetized surfaces (Step 4) can be achieved with the property of magnetic responsivity (Fig. 5.31b).

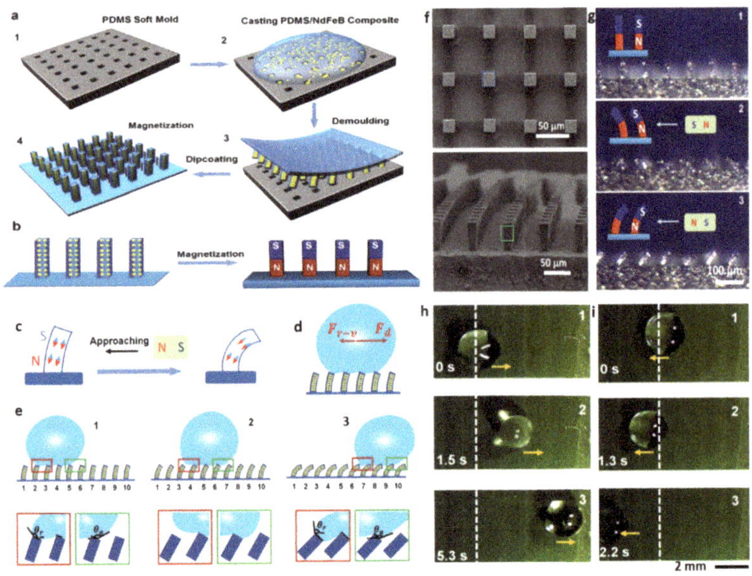

Figure 5.31 Fabrication of magnetic micropillar arrays for controlling droplet transport (Lin et al., 2018b). (a–b) Illustration of the fabrication of magnetic micropillar arrays and control of droplet transport by orientations of magnetic micropillars (c–e). (f) SEM images of magnetic micropillar arrays. (g) Optical images of tilted-angle magnetic micropillar arrays controlled by magnetic actions in various directions. (h–i) Optical images of in situ observations of the dynamic behavior of droplet transport.

Figure 5.31c–e illustrates the action of the dynamic solid–liquid–air interface in micropillars by the magnetic approach (Fig. 5.31c). The tilted pillars can generate the retention force (F_r) and the driving force (F_d) (Fig. 5.31d). When $F_d > F_r$, the droplet starts to move. The droplet is transported along the tilt-angle nano-/micropillar arrays, and there are advancing ($\theta_a = \theta_{a0} + \beta$) and receding edges ($\theta_r = \theta_{r0} - \beta$) (Fig. 5.31e). As the nano-/micropillars increase continuously, the tilt angles due to the action of the magnet increase, and the retention force decreases; thus, the droplet is driven to move directionally.

Figure 5.31f shows the micropillar arrays (top view and side view). Figure 5.31g shows the dynamic array acted by magnetic actions in directions. The droplet can be transported when the pillars are controlled into orientation at tilt angles. Figure 5.31h,i shows optical images of in situ observations of the dynamic behavior of droplet transport. When the pillars tilt toward the right (Fig. 5.31h) or left direction (Fig. 5.31i), the droplet moves correspondingly to the right (Fig. 5.31h) or left (Fig. 5.31i) direction from step to step like a walker.

5.4.2 Droplet Transport on a Photoinduced Dynamic Gradient

Photoinduced manipulation of liquids has a significantly promising future due to the advantages of contactless actions. For droplet manipulation, the design of the surface should be combined with photoresponsive materials and the control of photoirradiation.

Figure 5.32a,b illustrates the design of the photoresponsive organogel surface (POS) for droplet manipulation under near-infrared irradiation (NIR, wavelength of 808 nm). POS materials can be synthesized to achieve a surface with high photothermal efficiency and low adhesion by using appropriately prepared Fe_3O_4 nanoparticle-embedded PDMS along with oil infusion. Figure 5.32c shows the thermal images of a water droplet (~9 μL) motion on a POS (Fe_3O_4 NP content of 1 wt.%) under the action of NIR from time ~2.53 s to ~7.93 s. After the action of NIR onto the POS, a thermal gradient forms as related to the WG, which generates a driving force along the direction of the gradient for droplet motion. Figure 5.32d,e shows the time sequence images of PG (Fig. 5.32d) and glycerol (Fig. 5.32e) droplets (all are 5 μL) moving directionally by NIR in 6

s. Figure 5.32f shows a 10°-tilted POS where a light-driven droplet (~7 μL) coalesces with another droplet (~2 μL) and moves upward. Figure 5.32g shows a water droplet (~3 μL) capturing a silica sphere (~0.97 mm in diameter) and directionally moving on the POS under the action of NIR.

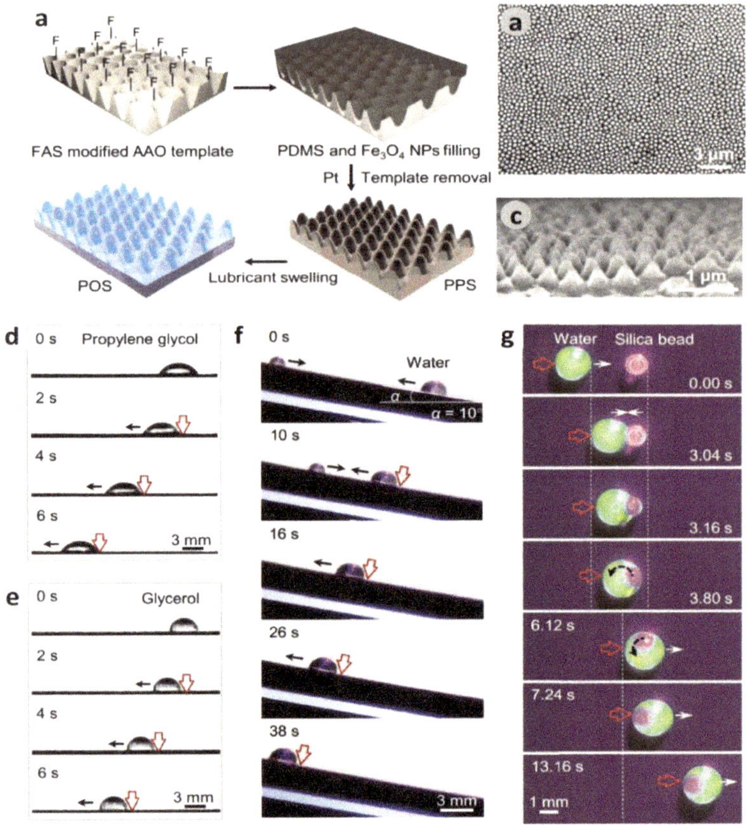

Figure 5.32 Fabrication of surface for droplet manipulation (Gao et al., 2018). (a) Illustration of the design of surface materials composed of Fe_3O_4 nanoparticles, polymer networks, and silicon oil. (b–c) SEM images of POS with a nanopapillae array. (d–g) Droplet transport of various types on light-induced gradient surfaces by action of NIR onto POS.

Droplets can execute agile light-induced motion on a POS. Figure 5.33a shows time sequence images (top view) of sequentially light-induced coalescence of droplets on a POS containing eight water droplets (numbered 1–8, volume of 3–5 μL). Following NIR, droplet

1 can be moved to coalesce with droplet 2 and both sequentially coalesce with others to form a larger droplet (containing 1 + 2 + 3 + 4 + 5 + 6 + 7 + 8) along a designed heart-shaped trajectory in 106 s (yellow dashed lines in each image indicate the transport trail of the droplet). Figure 5.33b shows the smart light-emitting circuit of droplet transport by NIR, where a circuit in a sealed glass petri dish is connected once an NIR-guided droplet (~6 μL) of saturated sodium chloride solution touches both ends of the circuit. The light-emitting diode in the circuit is turned on and gives off red light.

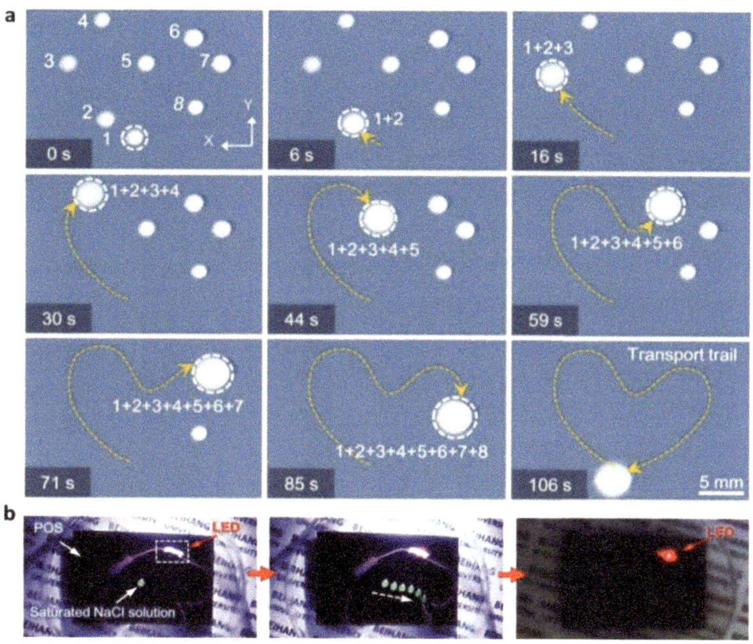

Figure 5.33 Complex transport of a droplet on surface by manipulation (Gao et al., 2018). (a) Agility of light-induced motion of droplets on a POS. Heart-shaped routes of droplets can be completed by NIR manipulation. (b) Device operation via a smart light-emitting circuit of droplet transport by NIR manipulation.

5.4.3 Magnetic Tubular Microactuators

The controlled manipulation of small amounts of liquid has attracted substantial research interest, such as microfluidics in recent years (Lonov et al., 2006; Sackmann et al., 2014; Hou et al., 2017; AbuZineh et al., 2018). Magnetic actuation is recognized as a promising method for liquid actuation based on its several advantages, including long-range action, large control forces, and weak interaction between the magnetic field and the liquid droplet. To date, most magnet-controlled liquids have depended primarily on magnetic particles added to a droplet as the actuator on an open surface. These liquids inevitably suffer from the droplets with the magnetic particles splitting or disengaging with the magnet, possibly leading to sample contamination, which severely limits the transport speed and practical applications.

Recently, a simple and reliable method for fast liquid transport was developed using asymmetric magnetic tubular microactuators (MTMAs). The MTMAs are prepared from a novel designed magnetic PDMS-based soft tube, showing asymmetric deformation in a magnetic field (Lei et al., 2018). Figure 5.34a shows the design of MTMAs for the motion of a liquid droplet of a wetting liquid confined in an MTMA driven by a magnetic field. The deformation of the upper part of the tube is induced by the magnetic field, which leads to asymmetric deformation with respect to the tube section containing the liquid droplet. As the magnet moves, this asymmetric deformation takes the shape of a dynamic conical tube, producing a Laplace pressure gradient that induces forward liquid transport. The MTMAs move droplets toward the narrower end, overcoming CAH via the asymmetric deformation of the tube induced by the magnetic field. The speed of partially wetting liquids reaches 10 cm/s, which is the maximum speed of liquid motion driven by an external-stimulus-induced capillary force in closed tubular microactuators. Various liquid (e.g., 1,2-dichloroethane, butyl alcohol, isopropanol, ethyl acetate, hexadecane, ethanol, ethanol: water (80, 70, 60), ethylene glycol, water, etc.) droplets without the addition of any sensitive magnetic particles can be manipulated in the MTMA with controllable velocity and direction and can even run uphill.

Figure 5.34b shows the side views of the corresponding instantaneous state of the liquid droplet in the obtained MTMAs. Figure 5.34c shows lateral photographs of the magnet-induced motion of an ethanol droplet in an MTMA fixed on a substrate. If the magnet moves from left to right, the ethanol droplet is propelled toward the right. If the direction of the magnet is reversed (bottom row), the direction of movement of the droplet is also reversed.

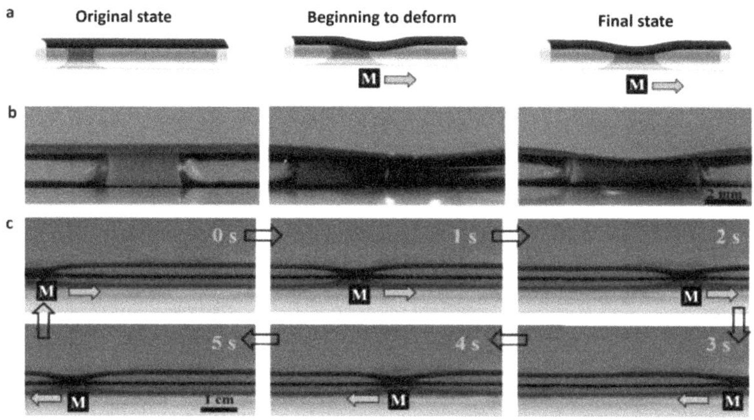

Figure 5.34 Controllable droplet transport (Lei et al., 2018). (a) Illustration of the design of MTMAs for liquid driving via channel deformation. (b) Optical images of the instantaneous state of the liquid droplet in the obtained MTMAs. (c) Optical images of in situ observation for magnet-induced motion of an ethanol droplet in an MTMA fixed on a substrate.

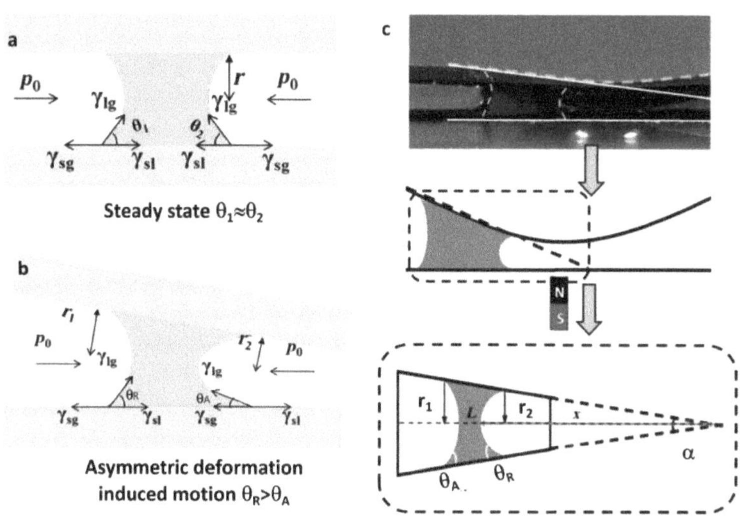

Figure 5.35 Mechanism of a liquid droplet in an MTMA driven by a magnetic field (Lei et al., 2018). (a) Steady state. (b) Asymmetric deformation-induced motion. (c) Example of analysis of driving liquid transport under the action of a magnetic force.

Figure 5.35 illustrates the mechanism of the liquid transport behavior in MTMAs. A liquid drop in an MTMA with radius r forms menisci with angles θ_1 and θ_2 (Fig. 5.35a). Without the magnetic force, the two menisci are symmetric ($\theta_1 \approx \theta_2$), and the droplet is in a stationary state of mechanical equilibrium. When a magnetic force is applied, the soft tube will deform, causing a difference in the Laplace pressure ($\theta_R > \theta_A$), which will drive the droplet toward the narrower end (Fig. 5.35b). Figure 5.35c shows that the diameter of the tube at different positions declines with increasing magnetic field (the closer the magnet is, the greater the magnetic field), whereas the diameter of the tube at different positions without the magnetic field remains nearly constant. Therefore, the tube deforms to an asymmetric cone-like geometry, which generates an adjustable capillary force to propel the liquid in the direction of the magnet movement. The curvature variation of the menisci at each end of the liquid column and apex gives rise to a difference in the Laplace pressure (ΔP) as:

$$\Delta P \approx (4\gamma\cos\theta)/\alpha x - (4\gamma\cos\theta)/\alpha(x + L) \qquad (5.12)$$

where γ is the liquid surface tension, θ is the CA of the liquid, α is the apex angle, x is the distance of the precursor from the apex, and L is the length of the precursor ($x \gg L$).

The capillary force (driving force, F_c) is described as:

$$F_c \approx \pi\alpha\gamma L\cos\theta \qquad (5.13)$$

The liquid droplet motion is resisted by not only the viscous resistance (F_V) but also the adhesive force (F_A) arising from the CAH. The adhesive force is described as:

$$F_A \approx \gamma W \Delta \cos\theta \qquad (5.14)$$

where W is the length of the advancing contact line, which is equal to πR, and R is the diameter of the MTMA. In the term $\Delta\cos\theta = \cos\theta_R - \cos\theta_A$, θ_R is the receding contact angle (RCA), and θ_A is the advancing contact angle (ACA). The viscous resistance force is given by:

$$F_V \approx 8\pi\eta L\upsilon \qquad (5.15)$$

where η is the coefficient of kinetic viscosity of the liquid and υ is the average speed of fluid motion. The forces can operate in conjunction; only when the capillary driving force is larger than the sum of the adhesive force and viscous resistance force, a partially wetting liquid can be propelled.

5.4.4 Oil Droplet Self-Transportation

Directional liquid transportation is important for a variety of biological processes and technical applications. Although surface engineering through asymmetric chemical modification or geometrical patterning facilitates effective liquid manipulation and enables water droplet self-transportation on synthetic surfaces, the self-transportation of oil droplets poses a major challenge because of their low surface tension. We consider as an example the self-transportation of oil droplets in ambient air on highly oleophobic, microtextured surfaces (Li et al., 2016b). As shown in Fig. 5.36a, Frame I, the pattern can be designed in radial undercut microstripes. Continuously decreasing the air–liquid contact fraction (f_2) radially can create an inward surface energy gradient (Fig. 5.36a, Frame II). The undercut structure with fluoropolymer coating maximizes the upward suspension force for oil droplets, preventing penetration of oil into the texture in liquid transport (Fig. 5.36a, Frame III) (Li et al., 2016b). Here, D is the stripe width, Φ is the intersection angle between two neighboring stripes, R is the radius of the central circle formed by intersection of the stripes, h is the undercut thickness, and L is the distance to the center. In addition, f_1 and f_2 are the solid and air fractions of the surface, respectively.

Different Φ and D can be obtained when designing the circular dish. Figure 5.36b shows the structural details of a surface (R = 1.3 mm). SEM is used to observe the structures of the surface, where the center of the surface shows radially arrayed microstripes from the center, along with the undercut profile (Fig. 5.36c). The stripe is locally removed by focused ion beam (FIB) milling to expose the cross-sectional morphology. Figure 5.36d,e shows the case of $\Phi = 0.625°$ and $D \approx 7$ μm.

After designing surfaces with different parameters, the droplet behaviors can be controlled effectively. Three modes of oil droplet motion on different patterns have been demonstrated. Figure 5.36f plots the observed regimes, showing three types of oil droplet behaviors depending on pattern parameters. In the green regime, oil droplets move spontaneously toward the center. In the blue regime, oil droplets remain pinned after being released. In the red regime, oil droplets exhibit outward spreading. The test liquid is hexadecane, and the droplet volume is fixed at 1 ml. Figure 5.36g shows the time-lapse images of oil droplet motion in three modes: moving toward the left, the center, and the right, observed after 0.167 s.

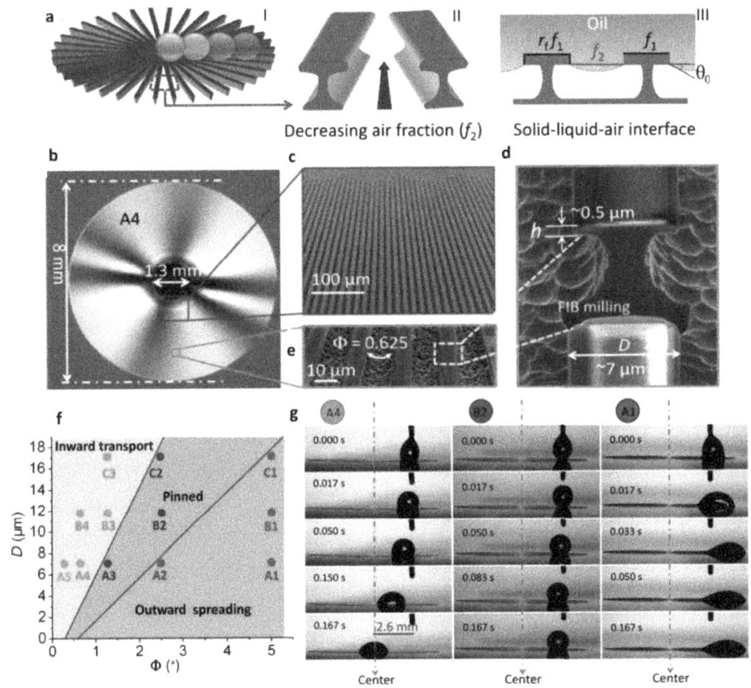

Figure 5.36 Design of surface for droplet self-transport (Li et al., 2016b). (a) Illustration of the design of a surface with radial undercut microstripes for oil droplet self-transportation. (b–d) Optical image and SEM images of the microstructure of the surface with microstripes. (f) Regimes of liquid behavior depending on pattern parameters (i.e., D, Φ). (g) Different behaviors for droplets for different structural parameters of surfaces (i.e., A4, B2, A1).

In the green group (Fig. 5.36f), on the basis of an interfacial free energy, the driving force for droplet motion on gradient surfaces scales as $F_d = -\pi r^2 \gamma d(\cos\theta)/dL$, where r is the base radius of the droplet. Thus, $-d(\cos\theta)/dL$ can be used as a characteristic parameter to evaluate the driving force of different surfaces. On the patterned area, we obtain:

$$-d(\cos\theta)/dL = [D(1 + \cos\theta_0) + 2h\cos\theta_0]/L^2\Phi \qquad (5.16)$$

The driving force $F_d \propto r^2$, whereas the hysteresis force $F_h \propto r$. In the blue group, the less steep WG is insufficient to drive the motion of oil droplets, which thus remain pinned on the surfaces.

In the red group, the upward suspension force provided by the undercut structure turns out to be insufficient for creating a robust solid–air composite interface to support oil droplets, and consequently, the oil droplets readily undergo a transition to the Wenzel state, where the liquid penetrates the texture. The breakthrough pressure P_c for oil to penetrate the texture can be quantified as:

$$P_c = 2\gamma \sin\theta_0/(L\Phi - D) \tag{5.17}$$

$P_{ref} = 2g/l_{cap}$ is used to estimate the minimum pressure of a droplet exerted on a highly hydrophobic/oleophobic surface, where $l_{cap} = \sqrt{\gamma / \rho g}$ is the capillary length of the liquid (~1.9 mm for hexadecane). P_c/P_{ref} can be used as a dimensionless parameter to evaluate the robustness of different patterns to support the Cassie state for oil droplets (see the red group, bottom). The area with the largest Φ and smallest D values thus gives much smaller P_c/P_{ref} values than the other two groups (Tuteja et al., 2007, 2008).

5.4.5 Electrochemistry-Controlled Droplet Transport

Controlling the pinning of organic droplets on solid surfaces is of fundamental and practical interest in the field of material science and engineering and has numerous applications such as surface cleaning, water treatment, and microfluidics.

Rapid in situ control of pinning and actuation of organic droplets is demonstrated on dodecylbenzenesulfonate-doped polypyrrole (PPy-(DBS)) surfaces in an aqueous environment via an electrochemical redox process (Xu et al., 2015). A dramatic change in the pinning results from the transport of DBS-molecules between the PPy(DBS) surface and the aqueous environment, as well as from a simultaneous alternation of the surface oleophobicity to organic liquids during the redox process, as shown in Fig. 5.37a–f, which illustrates the design via in situ control of the pinning of a dichloromethane (DCM) droplet on an electropolymerized PPy(DBS) surface in an aqueous electrolyte environment (0.1 M $NaNO_3$ solution) via a low-voltage (<1 V) redox process.

As illustrated in Fig. 5.37d, f, γ_{SL}, γ_{SO}, and γ_{LO} are the interfacial tensions at the interfaces of the PPy(DBS)-aqueous medium, PPy(DBS)-droplet, and droplet-aqueous medium, respectively. In addition, θ is the CA of the liquid droplet on a surface, which can be changed by the change in voltage: $\theta_{(0.6V)} < \theta_{(-0.9V)}$. This indicates that the wettability of the surface can be tuned by a voltage change from −0.9 V, i.e., from hydrophilic to hydrophobic.

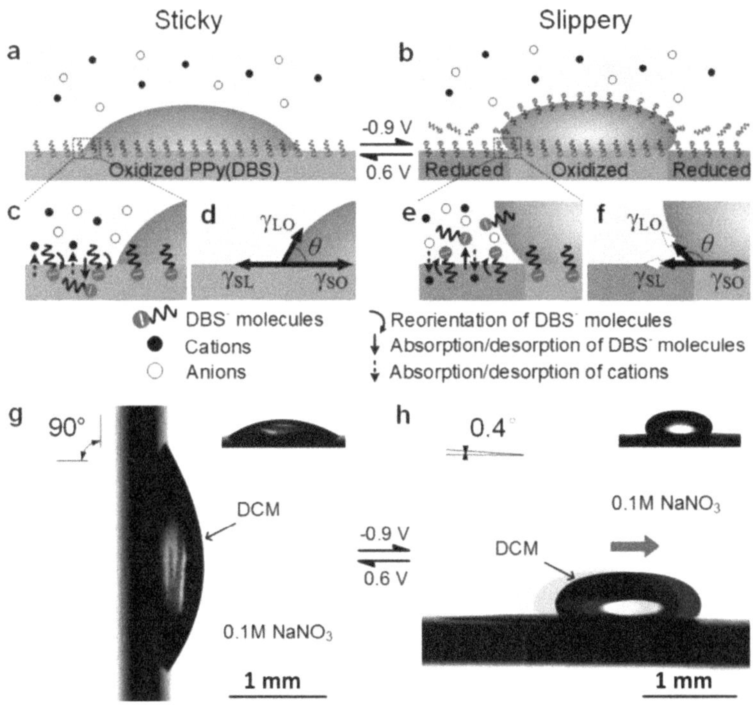

Figure 5.37 Rapid in situ control of pinning and actuation of organic droplets (Xu et al., 2015). (a–f) Illustration of wettability, i.e., sticky (a, c, d) and slippery (b, e, f) regimes controlled by an electrochemical redox process under −0.9 V (left) and 0.6 V (right). The insets illustrate the molecular features at solid–liquid interfaces. (g–h) Optical images of droplet behaviors controlled by the electrochemical redox process, i.e., sticky (g) at 0.6 V and slippery (h) at −0.9 V.

In an oxidized state, upon a voltage of 0.6 V (Fig. 5.37g), the oxidized PPy(DBS) surface exhibits a state of underwater oleophilicity and strong pinning to organic droplets, the apparent CA of a DCM droplet is ~32° on the oxidized PPy(DBS) surface, and the droplet is strongly pinned on the surface, even when the surface is tilted vertically. Upon the application of a voltage of −0.9 V (Fig. 5.37h), the PPy(DBS) surface exposed to the aqueous electrolyte is reduced and absorbs cations from the electrolyte for charge neutralization. The high slipperiness of the surface makes the droplet more mobile with less friction during movement. When the surface is at a tilting angle as low as 0.4°, the droplet slides off.

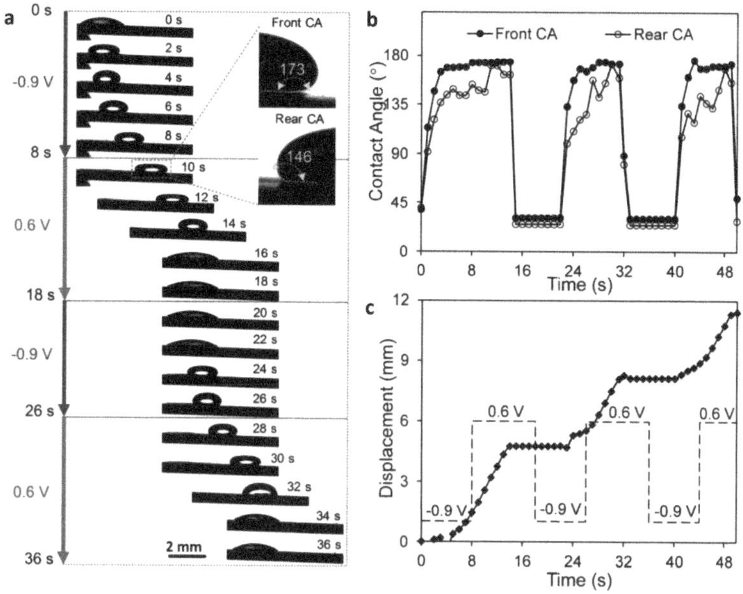

Figure 5.38 Droplet transport manipulation under different voltages (Xu et al., 2015). (a) Optical images of in situ observations of the behaviors of droplet transport in directions under control with different voltage applications. (b) The stability of contact angles at the front and rear sides of the DCM droplet. (c) The relationship of displacements of droplets versus time for voltage changes between –0.9 V and 0.6 V.

Figure 5.38 shows the controlled transportation of a DCM droplet on a PPy(DBS) surface. The DCM droplet exhibits a process of successive movement and stopping on a tilted (1.5°) PPy(DBS) surface in 0.1 M $NaNO_3$ under alternating voltages of –0.9 and 0.6 V. On the reduced PPy(DBS) surface during the movement, the front-advancing and rear-receding contact angles of the DCM droplet are 173° and 146°, respectively (Fig. 5.38a). Figure 5.38b shows the stability of the CAs at the front and rear sides of the DCM droplet. Figure 5.38c shows the displacement of the DCM droplet during transportation. The droplet's center before transportation is defined as the origin, and the displacement is calculated by the traveled distance of the droplet's center.

5.4.6 Surface Tension Transport

Phalaropes shorebirds (Fig. 5.39a) with long thin beaks feed primarily on small crustaceans and other invertebrates. By swimming in a tight circle on the water surface, they generate a vortex that draws underlying fluid and suspended prey toward the surface. By pecking on the water surface at a rate of ~1.5 Hz, the birds capture water droplets with a characteristic scale of ~2 mm between their upper and lower mandibles. The variability in bird beak morphology reflects diverse foraging strategies (Prakash et al., 2008). One such feeding mechanism in shorebirds involves surface-tension-induced transport of prey in millimetric droplets: By repeatedly opening and closing its beak in a tweezering motion, the bird moves the drop from the tip of its beak to its mouth in a stepwise ratcheting fashion. This mechanism of behavior inspires the design of a device for liquid transport. The observation in detail shows the droplet transport between beaks (Fig. 5.39b), which is related to the length and asymmetry of the CA (θ_1 on the right side, θ_2 on the left side) at the two sides of the droplet along the apex angle α (Fig. 5.39c). Figure 5.39d shows a completely wetting drop of silicone oil ($\theta_1 = \theta_2 = 0$, dynamic viscosity $\eta = 0.05$ kg/ms^2) self-propelled toward the apex of a mechanical bird beak with a constant opening angle $\alpha = 3.4°$ and uniform width of 1 mm (Fig. 5.39d).

Figure 5.39e indicates the capillary ratchet in the time sequence. Water droplet transport is generated by an opening and closing cycle of the mechanical beak. In the closing cycle, the leading contact line proceeds toward the mouth; in the opening cycle, the trailing contact line recedes toward the mouth. The result is net drop transport toward the mouth. Figure 5.39f shows a plot of the associated motion of the leading (red) and trailing (green) contact lines generated by varying the opening angle α over three cycles, where d (in millimeters) represents the distance from the beak tip to the contact line. If $\alpha_{open} - \alpha_{close}$ is sufficiently small that the leading θ_1 and trailing θ_2 CAs satisfy $\theta_r < \theta_2 < \theta_1 < \theta_a$, then the drop remains pinned. The dynamics for larger values of $\alpha_{open} - \alpha_{close}$ are best understood by considering in turn the closing and opening phases. The drop advances through a slipping ratcheting motion: In each cycle, both the leading and trailing edges of the contact lines advance and retreat. Nevertheless, due to the asymmetry in the wedge geometry, net mouthward drop motion is still achieved, albeit inefficiently. When α_{close} and α_{open} are optimally tuned, the droplet advances through a pure ratcheting motion with no slippage.

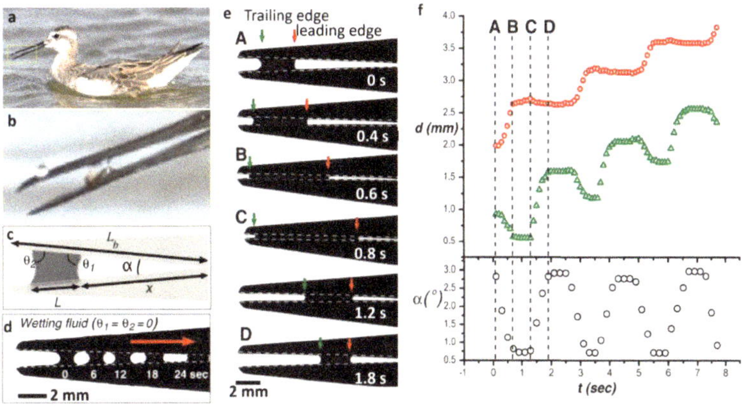

Figure 5.39 Ratchet transport of droplet by surface tension (Prakash et al., 2008). (a–b) Photograph of Phalaropes shorebirds (a) with long thin beaks feeding primarily on small crustaceans and other invertebrates (b). (c) Illustration of the mechanism for capillary ratchet transport induced by asymmetric contact angles (θ_1 on the right side, θ_2 on the left side) on the two sides of the droplet along the apex angle α. L_b is the whole length of the ratchet, L is the liquid length, and x is the distance that the liquid moves. (d) Optical image of a completely wetting drop of silicone oil ($\theta_1 = \theta_2 = 0$) self-propelled toward the apex of a mechanical bird beak. (e) Optical images of liquid transport in the time sequence via an opening–closing cycle of the mechanical beak. (f) Graph of the associated motion of the leading (red) and trailing (green) contact lines by varying the opening angle α over three cycles. Here, d (in millimeters) represents the distance from the beak tip to the contact line in (e).

5.4.7 Programmed Droplet Transport

Droplets that can achieve self-propelled transport over long ranges on solid surfaces are crucial for many applications (Wong et al., 2011). The preferential transport of droplets can be achieved by introducing topographic or chemical modulation of surface wetting gradients that break the asymmetric contact line and overcome the resistance force to move droplets along a particular direction (Lchimura et al., 2000; Daniel et al., 2001; Daniel et al., 2004 Schutzius et al., 2015; Lv et al., 2016). However, achieving fast transport is still desirable and challenging. To date, interesting work has been reported (Sun et al., 2019), i.e., high-velocity and ultralong transport of droplets is achieved on diverse substrates with surface charge density (SCD) gradients. In this work, facile water droplet printing on superamphiphobic surfaces is used to create rewritable SCD gradients without the need for additional energy input. This provides a way to program the transport of droplets on various surfaces for applications (Lagubeau et al., 2011; Vakarelski et al., 2011; Marin et al., 2013).

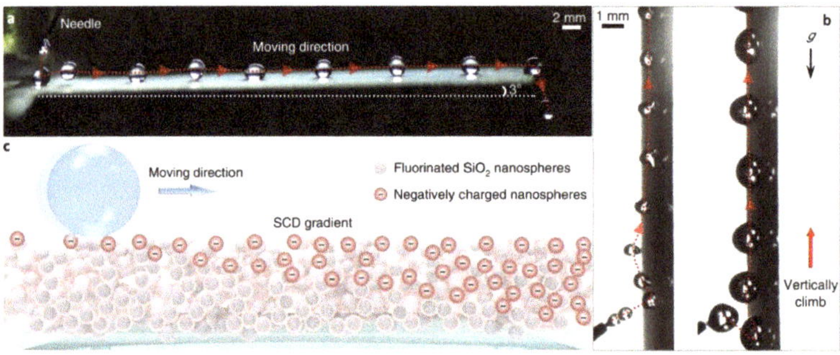

Figure 5.40 Droplet transport on a printable SCD gradient (Sun et al., 2019). (a) The time-lapse trajectory of a water droplet on an SCD gradient surface at a tilt angle of 3°. (b) The time-lapse trajectory of droplets with different radii moving upward on a vertical SCD gradient surface. (c) Illustration of droplet self-propulsion on the SCD gradient surface.

As is known, free droplets deposited on a special superamphiphobic surface exhibit spontaneous, directional motion characterized by both a high transport velocity and an ultralong transport distance. Such spectacular transport is reminiscent of the Leidenfrost phenomenon, in which droplets can migrate at a high mobility over a long distance. To achieve stable and fast transport, the strategy is capable of modulating droplet dynamics through the formation of a preferential gradient in SCD on a superamphiphobic surface. As reported previously (Sun et al., 2019), the superamphiphobic surface is constructed on a 170-μm-thick thin glass substrate coated with a 10-μm-thick nanoporous SiO_2 layer and hydrophobized by the chemical vapor deposition of 1H,1H,2H,2H-perfluorooctyltrichlorosilane (PFOTS). Furthermore, a preferential SCD gradient on the superamphiphobic surface can be established by the intricate control of the Weber number of printed water droplets (i.e., impacting droplets) along a predetermined pathway. Figure 5.40a,b shows droplet transport on an SCD gradient surface in the horizontal (Fig. 5.40a) and vertical orientations (Fig. 5.40b). Figure 5.40c illustrates droplet movement along the SCD gradient direction.

The driving force for droplet transport can be generated by an SCD gradient, as illustrated in Fig. 5.41. When a droplet impacts the superamphiphobic surface, the surface charge can act with the droplet. In this process, surface charges are generated at the contact area as a result of contact electrification (Fig. 5.41a). The surface charges are different due to the position of the air–water–solid three-phase contact line for different impact pressures, for example, a small We number or large We number, and the contact depth H is different (Fig. 5.41b). The charge density is proportional to $\rho_Q \propto We^{0.74}$, where the Weber number (We) = $\rho dv^2/\gamma$. Here, ρ, d, v, and γ are the density, diameter, impact velocity, and surface tension of the water droplet, respectively. The SCD per projected area (ρ_Q) is generated on the solid surface after the bouncing of the impinging water droplet, which can be estimated as $\rho_Q = qA_{LS}/A_D$, where q is the charge produced per unit of actual contact area, A_{LS} is the actual contact area, and A_D is the projected liquid–solid contact area. Based on a simple hydrodynamic analysis, A_{LS}/A_D is estimated: $A_{LS}/A_D \approx \frac{1}{s^2}\left[\frac{H}{2r_0}\pi r_0^2 + \pi r_0^2\right]$, where s is the average center-to-center spacing of two neighboring nanoscale posts, r_0 is the radius of the posts, and H is the liquid contact depth. H depends on the Weber number. After the droplet impacts and rebounds, a driving force of the unbalanced electric force is induced by the SCD gradient, and the droplet self-propels toward the direction with a larger SCD (Fig. 5.41c). The net electric force associated with a droplet can be expressed as $F_{Net} \sim (\varepsilon_r - 1)\varepsilon_0 kD_c$, where ε_r is the relative dielectric constant of the droplet, ε_0 is the dielectric constant of empty space, k is the SCD gradient imposed on the moving droplet, and D_c is the contact diameter. On the basis of Newton's second law, the transport velocity v scales as v ~ $(\varepsilon_r - 1)\varepsilon_0 k$; thus, the droplet transport velocity can be regulated by controlling the charge gradient k. Figure 5.41d shows a comparison of the droplet transport velocities among different surfaces with a surface free-energy gradient (e.g., wetting gradient, curvature gradient, and SCD gradient). L_t/D is used to describe the droplet transport per unit diameter, where L_t is the transport length and D is the diameter of the droplet.

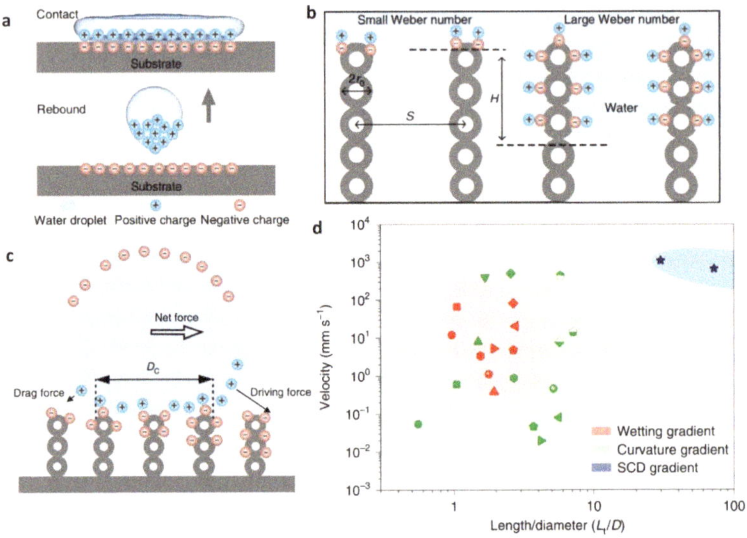

Figure 5.41 Illustration of the mechanism of droplet transport and generality of droplet transportation for various liquids (Sun et al., 2019). (a) Illustration of impacting droplet contacts with surface charge. (b) Different We numbers of droplets with different contact depths. (c) Illustration of the driving force due to the SCD gradient. D_c is the contact diameter. (d) Comparison of the droplet transport performances among different surfaces with a surface free-energy gradient. L_t is the transport length, and D is the diameter of the droplet.

Thinking Questions

1. Can you explain the significance of droplet transport in applications?
2. What are the key points for the design of materials and surfaces that can make the droplet transport in directions?
3. Can you give examples of methods and techniques for the fabrication of materials and surfaces for droplet transport in directions?
4. Which bioinspired materials and surfaces control the droplet transport?

References

AbuZineh K., L. I. Joudeh, B. Al Alwan, S. M. Hamdan, J. S. Merzaban, and S. Habuchi. Microfluidics-based super-resolution microscopy enables nanoscopic characterization of blood stem cell rolling. *Sci. Adv.* **4**, eaat5304 (2018).

Bai H., R. Sun, J. Ju, X. Yao, Y. Zheng, and L. Jiang. Large-scale fabrication of bioinspired fibers for directional water collection. *Small* **7**, 3429–3433 (2011).

Bain C. D., G. D. Burnett-Hall, and R. R. Montgomerie. Rapid motion of liquid drops. *Nature* **372**, 414–415 (1994).

Bico J. and D. Quéré. Self-propelling slugs. *J. Fluid Mech.* **467**, 101–127 (2002).

Bjelobrk N., H. L. Girard, S. B. Subramanyam, H. M. Kwon, D. Quéré, and K. K. Varanasi. Thermocapillary motion on lubricant-impregnated surfaces. *Phys. Rev. Fluids* **1**, 063902 (2016).

Bliznyuk O., H. P. Jansen, E. S. Kooij, H. J. W. Zandvliet, and B. Poelsema. Smart design of stripe-patterned gradient surfaces to control droplet motion. *Langmuir* **27**, 11238 (2011).

Bohn H. F. and W. Federle. Insect aquaplaning: Nepenthes pitcher plants capture prey with the peristome, a fully wettable water-lubricated anisotropic surface. *Proc. Natl. Acad. Sci. USA* **101**, 14138–14143 (2004).

Brochard F. Motions of droplets on solid surfaces induced by chemical or thermal gradient. *Langmuir* **5**, 432 (1989).

Chaudhury M. K. and G. M. Whitesides. How to make water run uphill. *Science* **256**, 1539–1541 (1992).

Chen H., L. Zhang, P. Zhang, D. Zhang, Z. Han, and L. Jiang. A novel bioinspired continuous unidirectional liquid spreading surface structure from the peristome surface of *Nepenthes alata*. *Small* **13**, 1601676 (2017a).

Chen H., L. Zhang, Y. Zhang, P. Zhang, D. Zhang, and L. Jiang. Uni-directional liquid spreading control on a bio-inspired surface from the peristome of *Nepenthes alata*. *J. Mater. Chem. A* **5**, 6914–6920 (2017b).

Chen H., P. Zhang, L. Zhang, H. liu, Y. Jiang, D. Zhang, Z. Han, and L. Jiang. Continuous directional water transport on the peristome surface of *Nepenthes alata*. *Nature* **532**, 85–89 (2016).

Chen H., T. Ran, Y. Gan, J. Zhou, Y. Zhang, L. Zhang, D. Zhang, and L. Jiang. Ultrafast water harvesting and transport in hierarchical microchannels. *Nat. Mater.* **17**, 935–942 (2018).

Chen Y., L. Wang, Y. Xue, L. Jiang, and Y. Zheng. Bioinspired tilt-angle fabricated structure gradient fibers: Micro-drops fast transport in a long-distance. *Sci. Rep.* **3**, 2927 (2013).

Chu K.-H., R. Xiao, and E. N. Wang. Uni-directional liquid spreading on asymmetric nanostructured surfaces. *Nat. Mater.* **9**, 413–417 (2010).

Daniel S. and M. K. Chaudhury. Rectified motion of liquid drops on gradient surfaces induced by vibration. *Langmuir* **18**, 3404 (2002).

Daniel S., M. K. Chaudhury, and J. C. Chen. Fast drop movements resulting from the phase change on a gradient surface. *Science* **291**, 633–636 (2001).

Daniel S., S. Sircar, J. Gliem, and M. K. Chaudhury. Ratcheting motion of liquid drops on gradient surfaces. *Langmuir* **20**, 4085 (2004).

Fang G., W. Li, X. Wang, and G. Qiao. Droplet motion on designed microtextured superhydrophobic surfaces with tunable wettability. *Langmuir* **24**, 11651–11660 (2008).

Feng L., S. Li, Y. Li, H. Li, L. Zhang, J. Zhai, Y. Song, B. Liu, L. Jiang, and D. Zhu. Super-hydrophobic surfaces: From natural to artificial. *Adv. Mater.* **14**, 1857–1860 (2002).

Feng S., S. Wang, L. Gao, G. Li, Y. Hou, and Y. Zheng. Controlled directional water-droplet spreading on a high-adhesion surface. *Angew. Chem. Int. Ed.* **53**, 6163–6167 (2014).

Gao C., L. Wang, Y. Lin, J. Li, Y. Liu, X. Li, S. Feng, and Y. Zheng. Droplets manipulated on photothermal organogel surfaces. *Adv. Funct. Mater.* **28**, 1803072 (2018).

Guo P., Y. Zheng, C. Liu, J. Ju, and L. Jiang. Directional shedding-off of water on natural/bio-mimetic taper-ratchet array surfaces. *Soft Matter* **8**, 1770–1775 (2012).

Hou X., Y. S. Zhang, G. Trujillo-de Santiago, M. M. Alvarez, J. Ribas, S. J. Jonas, P. S. Weiss, A. M. Andrews, J. Aizenberg, and A. Khademhosseini. Interplay between materials and microfluidics. *Nat. Rev. Mater.* **2**, 17016 (2017).

Hou Y., S. Feng, L. Dai, and Y. Zheng. Droplet manipulation on wettable gradient surfaces with micro-/nano-hierarchical structure. *Chem. Mater.* **28**, 3625–3629 (2016).

Huang J. Y., Y.-C. Lo, J. J. Niu, A. Kushima, X. Qian, L. Zhong, S. X. Mao, and J. Li. Nanowire liquid pumps. *Nat. Nanotechnol.* **8**, 277–281 (2013).

Ionov L., N. Houbenov, A. Sidorenko, M. Stamm, and S. Minko. Smart microfluidic channels. *Adv. Funct. Mater.* **16**, 1153–1160 (2006).

Ju J., H. Bai, Y. Zheng, T. Zhao, R. Fang, and L. Jiang. A multistructural and multi-functional integrated fog collection system in cactus. *Nat. Commun.* **3**, 1247–1247 (2012).

Kang E., G. S. Jeong, Y. Y. Choi, K. H. Lee, A. Khademhosseini, and S.-H. Lee. Digitally tunable physicochemical coding of material composition and topography in continuous microfibres. *Nat. Mater.* **10**, 877–883 (2011).

Kennedy R. J. Directional water-shedding properties of feathers. *Nature* **227** (1970) 736–737.

Lagubeau G., M. L. Merrer, C. Clanet, and D. Quéré. Leidenfrost on a ratchet. *Nat. Phys.* **7**, 395–398 (2011).

Lchimura K., S.-K. Oh, and M. Nakagawa. Light-driven motion of liquids on a photoresponsive surface. *Science* **288**, 1624–1626 (2000).

Lee S.-W. and P. E. Laibinis. Directed movement of liquids on patterned surfaces using noncovalent molecular adsorption. *J. Am. Chem. Soc.* **122**, 5395–5396 (2000).

Lei W., G. Hou, M. Liu, Q. Rong, Y. Xu, Y. Tian, and L. Jiang. High-speed transport of liquid droplets in magnetic tubular microactuators. *Sci. Adv.* **4**, eaau8767 (2018).

Li C., L. Wu, C. Yu, Z. Dong, and L. Jiang. Peristome-mimetic curved surface for spontaneous and directional separation of micro water-in-oil drops. *Angew. Chem., Int. Ed.* **56**, 13623–13628 (2017).

Li C., N. Li, X. Zhang, Z. Dong, H. Chen, and L. Jiang. Uni-directional transportation on peristome-mimetic surfaces for completely wetting liquids. *Angew. Chem.* **128**, 15097 (2016a).

Li J., Q. H. Qin, A. Shah, R. H. A. Ras, X. Tian, and V. Jokinen. Oil droplet self-transportation on oleophobic surfaces. *Sci. Adv.* **2**, e1600148 (2016b).

Li W. and A. Amirfazli. Superhydrophobic surfaces: Adhesive strongly to water. *Adv. Mater.* **19**, 3421 (2007).

Lin Y., Z. Hu, C. Gao, Z. Guo, C. Li, and Y. Zheng. Directional droplet spreading transport controlled on tilt-angle pillar arrays. *Adv. Mater. Interfaces* **5**, 1800962 (2018a).

Lin Y., Z. Hu, M. Zhang, T. Xu, S. Feng, L. Jiang, and Y. Zheng. Magnetically induced low adhesive direction of nano/micropillar arrays for microdroplet transport. *Adv. Funct. Mater.* **28**, 1800163 (2018b).

Liu C., J. Ju, Y. Zheng, and L. Jiang. Asymmetric Ratchet effect for directional transport of fog drops on static and dynamic butterfly wings. *ACS Nano* **8**, 1321–1329 (2014).

Liu C., Y. Xue, Y. Chen, and Y. Zheng. Effective directional self-gathering of drops on spine of cactus with splayed capillary arrays. *Sci. Rep.* **5**, 17757 (2015).

Lorenceau É. and D. Quéré. Drops on a conical wire. *J. Fluid Mech.* **510**, 29–45 (2004).

Lv J.-A., Y. Liu, J. Wei, E. Chen, L. Qin, and Y. Yu. Photocontrol of fluid slugs in liquid crystal polymer microactuators. *Nature* **537**, 179–184 (2016).

Malvadkar N. A., M. J. Hancock, K. Sekeroglu, W. J. Dressick, and M. C. Demirel. An engineered anisotropic nanofilm with unidirectional wetting properties. *Nat. Mater.* **9**, 1023–1028 (2010).

Marin A. G., W. van Hoeve, P. Garcia-Sanchez, L. Shui, Y. B. Xie, M. A. Fontelos, J. C. T. Eijkel, A. van den Berg, and D. Lohse. The microfluidic Kelvin water dropper. *Lab Chip* **13**, 4503–4506 (2013).

Prakash M., D. Quéré, and J. W. M. Bush. Surface tension transport of prey by feeding shorebirds: The capillary ratchet. *Science* **320**, 931–934 (2008).

Sackmann E. K., A. L. Fulton, and D. J. Beebe. The present and future role of microfluidics in biomedical research. *Nature* **507**, 181–189 (2014).

Schutzius T. M., S. Jung, T. Maitra, G. Graeber, M. Köhme, and D. Poulikakos. Spontaneous droplet trampolining on rigid superhydrophobic surfaces. *Nature* **527**, 82–85 (2015).

Si Y., T. Wang, C. Li, C. Yu, N. Li, C. Gao, Z. Dong, and L. Jiang. Liquids unidirectional transport on dual scale arrays. *ACS Nano* **12**, 9214–9222 (2018).

Song C., L. Zhao, W. Zhou, M. Zhang, and Y. Zheng. Bioinspired wet-assembly fibers: From nanofragments to microhumps on string in mist. *J. Mater. Chem. A* **2**, 9465–9468 (2014).

Style R. W., Y. Che, S. J. Park, B. M. Weon, J. H. Je, C. Hyland, G. K. German, M. P. Power, L. A. Wilen, J. S. Wettlaufer, and E. R. Dufresne. Patterning droplets with durotaxis. *Proc. Natl. Acad. Sci. USA* **110**, 12541–12544 (2013).

Sumino Y., N. Magome, T. Hamada, and K. Yoshikawa. Self-running droplet: Emergence of regular motion from nonequilibrium noise. *Phys. Rev. Lett.* **94**, 068301 (2005).

Sun Q., D. Wang, Y. Li, J. Zhang, S. Ye, J. Cui, L. Chen, Z. Wang, H.-J. Butt, D. Vollmer, and X. Deng. Surface charge printing for programmed droplet transport. *Nat. Mater.* **18**, 936–941 (2019).

Tian D., N. Zhang, X. Zheng, G. Hou, Y. Tian, Y. Du, L. Jiang, and S. Dou. Fast responsive and controllable liquid transport on a magnetic fluid/nanoarray composite interface. *ACS Nano* **10**, 6220 (2016).

Tian X., H. Bai, Y. Zheng, and L. Jiang. Bio-inspired heterostructured bead-on-string fibers that respond to environmental wetting. *Adv. Funct. Mater.* **21**, 1398–1402 (2011).

Tuteja A., W. Choi, J. M. Mabry, G. H. McKinley, and R. E. Cohen. Robust omniphobic surfaces. *Proc. Natl. Acad. Sci. USA* **105**, 18200–18205 (2008).

Tuteja A., W. Choi, M. Ma, J. M. Mabry, S. A. Mazzella, G. C. Rutledge, G. H. McKinley, and R. E. Cohen. Designing superoleophobic surfaces. *Science* **318**, 1618–1622 (2007).

Vakarelski I. U., J. O. Marston, D. Y. C. Chan, and S. T. Thoroddsen. Drag reduction by Leidenfrost vapor layers. *Phys. Rev. Lett.* **106**, 214501 (2011).

Wang L., F. Li, M. Kuang, M. Gao, J. Wang, Y. Huang, L. Jiang, and Y. Song. Interface manipulation for printing three-dimensional microstructures under magnetic guiding. *Small* **11**, 1900–1904 (2015b).

Wang L., L. Heng, and L. Jiang. Temperature-responsive anisotropic slippery surface for smart control of the droplet motion. *ACS Appl. Mater. Interfaces* **10**, 7442 (2018).

Wang L., M. Zhang, W. Shi, Y. Hou, C. Liu, S. Feng, Z. Guo, and Y. Zheng. Dynamic magnetic responsive wall array with droplet shedding off properties. *Sci. Rep.* **5**, 11209 (2015a).

Whitesides G. M. The origins and the future of microfluidics. *Nature* **442**, 368–373 (2006).

Wong T. S., S. H. Kang, S. K. Y. Tang, E. J. Smythe, B. D. Hatton, A. Grinthal, and J. Aizenberg. Bioinspired self-repairing slippery surfaces with pressure stable omniphobicity. *Nature* **477**, 443–447 (2011).

Wu H., R. Zhang, Y. Sun, D. Lin, Z. Sun, W. Pan, and P. Downs. Biomimetic nanofiber patterns with controlled wettability. *Soft Matter* **4**, 2429–2433 (2008).

Xu W., J. Xu, C.-H. Choi, and E.-H. Yang. In situ control of underwater-pinning of organic droplets on a surfactant-doped conjugated polymer surface. *ACS Appl. Mater. Interfaces* **7**, 25608–25617 (2015).

Xue Y., Y. Chen, T. Wang, L. Jiang, and Y. Zheng. Directional size-triggered microdroplet target transport on gradient-step fibers. *J. Mater. Chem. A* **2**, 7156–7160 (2014).

Yang J.-T., Z.-H. Yang, C.-Y. Chen, and D.-J. Yao. Conversion of surface energy and manipulation of a single droplet across micropatterned surfaces. *Langmuir* **24**, 9889–9897 (2008).

Yang Y., X. Li, X. Zheng, Z. Chen, Q. Zhou, and Y. Chen. 3D printed biomimetic super-hydrophobic structure for microdroplet manipulation and oil/water separation. *Adv. Mater.* **30**, 1704912–1704922 (2018).

Yao X., H. Bai, J. Ju, D. Zhou, J. Li, H. Zhang, B. Yang, and L. Jiang. Running droplet of interfacial chemical reaction flow. *Soft Matter* **8**, 5988–5991 (2012).

Yarin A. L., W. Liu, and D. H. Reneker. Motion of droplets along thin fibers with temperature gradient. *J. Appl. Phys.* **91**, 4751–4760 (2002).

Yu C., C. Li, C. Gao, Z. Dong, L. Wu, and L. Jiang. Time-dependent liquid transport on a biomimetic topological surface. *ACS Nano* **12**, 5149–5157 (2018b).

Yu C., L. Zhang, Y. Ru, N. Li, C. Li, C. Gao, Z. Dong, and L. Jiang. Drop cargo transfer via unidirectional lubricant spreading on peristome-mimetic surface. *ACS Nano* **12**, 11307–11315 (2018a).

Zhang J. and Y. Han. Shape-gradient composite surfaces: Water droplets move uphill. *Langmuir* **23**, 6136–6141 (2007).

Zhang M., L. Wang, Y. Hou, W. Shi, S. Feng, and Y. Zheng. Controlled smart anisotropic unidirectional spreading of droplet on a fibrous surface. *Adv. Mater.* **27**, 5057–5062 (2015).

Zhang P., H. Chen, L. Li, H. Liu, G. Liu, L. Zhang, D. Zhang, and L. Jiang. Bioinspired smart peristome surface for temperature-controlled unidirectional water spreading. *ACS Appl. Mater. Interfaces* **9**, 5645 (2017a).

Zhang P., L. Zhang, H. Chen, Z. Dong, and D. Zhang. Surfaces inspired by the nepenthes peristome for unidirectional liquid transport. *Adv. Mater.* **29**, 1702995 (2017b).

Zhao L., C. Song, M. Zhang, and Y. Zheng. Bioinspired heterostructured bead-on-string fibers via controlling the wet-assembly of nanoparticles. *Chem. Commun.* **50**, 10651–10654 (2014).

Zheng Y., D. Han, J. Zhai, and L. Jiang. In situ investigation on dynamic suspending of microdroplet on lotus leaf and gradient of wettable micro- and nanostructure from water condensation. *Appl. Phys. Lett.* **92**, 084106 (2008).

Zheng Y., H. Bai, Z. Huang, X. Tian, F.-Q. Nie, Y. Zhao, J. Zhai, and L. Jiang. Directional water collection on wetted spider silk. *Nature* **463**, 640–643 (2010).

Zheng Y., J. Cheng, C. Zhou, H. Xing, X. Wen, P. Pi, and S. Xu. Droplet motion on a shape gradient surface. *Langmuir* **33**, 4172 (2017).

Zheng Y., X. Gao, and L. Jiang. Directional adhesion of superhydrophobic butterfly wings. *Soft Matter* **3**, 178–182 (2007).

Zhou H., M. Zhang, C. Li, C. Gao, and Y. Zheng. Excellent fog-droplets collector via integrative Janus membrane and conical spine with micro/nanostructures. *Small* **14**, 1801335 (2018).

Chapter 6

Antiadhesive/Adhesive Surfaces

In this chapter, first, antiadhesive surfaces are introduced, including theories of antiadhesion on superhydrophobic surfaces and the characterization of adhesion. The solid–liquid–liquid interfaces of hydrophobic liquid-infused surfaces are introduced, such as bioinspired slippery surfaces, antiadhesion organogel materials, superhydrophilic antiadhesive surfaces, and droplet slippage on water-spreading surfaces. Low adhesion on solid–liquid–air interfaces of superhydrophobic surfaces is discussed, for example antifogging performance, fly-eye-inspired superhydrophobic nanostructures, air-trapped hollow microsphere nanocomposites, antiadhesion of coalesced droplets on superhydrophobic surfaces, including nanostructure arrays and hierarchical structured superhydrophobic surfaces, and ultra-antiadhesion with pancake bouncing on superhydrophobic surfaces. Antifouling marine surfaces are also introduced, for example biological fouling organisms and adhesion, antifouling of bioinspired engineered topographies, and vascularized fouling-release. Anti-crude-oil adhesion of polyionized hydrogels is then introduced, for example, the design of hydrogel materials and anti-crude-oil adhesion coatings. Second, adhesive surfaces are introduced. One type is wet adhesion, including the petal effect, controlled liquid splash, marine organism adhesion, and tree frog adhesion. Another type is dry adhesion, including gecko feet

Bioinspired Materials Surfaces: Nature-Inspired Micro-/Nanostructures
Yongmei Zheng

ISBN 978-981-5129-34-2 (Hardcover), 978-1-003-51388-9 (eBook)
www.jennystanford.com

with antiadhesion and adhesion, bioinspired composite microfibers, reversible adhesion, temperature-controlled adhesion, meniscus-controlled octopus-inspired adhesives, and a magnet-controlled soft millirobot.

6.1 Antiadhesive Surfaces

Surfaces with antiadhesion properties for liquids, solids, and biological objects, including bacteria, plants, land and aquatic animals, and seashells, have attracted great commercial interest. In general, the antiadhesion properties of material surfaces can be established by two routes: superhydrophobic surfaces with solid–liquid–air interfaces and hydrophobic liquid-infused surfaces with solid–liquid–liquid interfaces, which enable the development of many studies for promising applications.

For example, a lotus leaf displays a superhydrophobic surface (Barthlott and Neinhuis, 1997), along with a high contact angle hysteresis (CAH) and low CAH, and the waxy micro-/nanostructures on the surface lead to a self-cleaning effect; thus, the surface generates a low adhesion and drag reduction for fluid flow. Such low-adhesion surfaces have inspired the design of functional materials and surfaces (Lu et al., 2015). Another example is the surface of *Nepenthes* pitcher plants, producing a fluid-infused lubricating film (Wong et al., 2011) to cause insects to slide from the rim into the digestive juices. Such a liquid infusion on the surfaces of materials induces a robust antiadhesion property, which has inspired researchers to fabricate functional surfaces for antibacterial, antibiological, and anti-icing applications (Bohn and Federle, 2004). In contrast, superhydrophobic surfaces with high liquid–solid adhesion have become a new focus due to their potential applications in liquid transport, biochemical separation, in situ detection, and localized chemical reactions (Jin et al., 2005; Hong et al., 2007; Zhao et al., 2007; Winkleman et al., 2008). Inspired by biological surfaces with special wettability, superantiwetting interfaces, such as superhydrophobic (Feng et al., 2002; Feng and Jiang, 2006) and superamphiphobic surfaces in air (Li et al., 2001; Xi et al., 2008) or superoleophobic interfaces in water (Liu et al., 2009), have been designed and developed considerably in fundamental and applied research (Blossey, 2003; Sun et al., 2005; Xia and Jiang, 2008).

6.1.1 Description of Antiadhesion on Superhydrophobic Surfaces

Generally, superhydrophobic surfaces increase the contact angle (CA) of water to more than 150°. Nevertheless, the adhesion of liquid to a surface can differ considerably from that of solid to a surface (Heuer et al., 1992; Autumn et al., 2000, 2002). The superhydrophobic surface can make the droplet roll (along with low adhesion) or become pinned (along with high adhesion). In fact, whether a droplet can roll or be pinned on a superhydrophobic surface is ascribed to distinct solid–liquid contact modes (see Illustration 6.1). Usually, if a droplet is in the Wenzel state, it is in a wet contact mode, and if a droplet is in the Cassie state, it is in a composite contact mode (Feng et al., 2002; Jin et al., 2012). Depending on the interfaces, the liquid–air–solid three-phase contact line (TCL) is continuous for droplets in the Wenzel state (Wenzel, 1936) and discontinuous for droplets in the Cassie state (Cassie and Baxter, 1944). In the Cassie state, along with a discontinuous TCL, the water droplet is suspended by vapor pockets trapped on the surface (in composite contact mode). The slight solid–liquid contact greatly lowers the adhesion of the surface; thus, the droplet easily rolls off the surface. Such a surface has a self-cleaning effect. In contrast, in the Wenzel state with a continuous TCL, the completely solid–liquid contact induces strong adhesion; thus, the droplet is pinned on the surface. Both wet contact mode and composite contact mode can lead to different ultrawettability properties. In fact, in most cases, a water droplet may partially wet a superhydrophobic textured surface due to air being partially trapped in the valleys, representing an intermediate state between the Wenzel and Cassie states. It is challenging to tune the superhydrophobic states between the Wenzel state with high adhesion and the Cassie state with low adhesion, but it can be achieved by optimally designing the microstructures, especially when the microstructures are realized with variable geometric parameters that control certain solid/liquid contact modes (Aussillous and Quéré, 2001; ElSherbini and Jacobi, 2004).

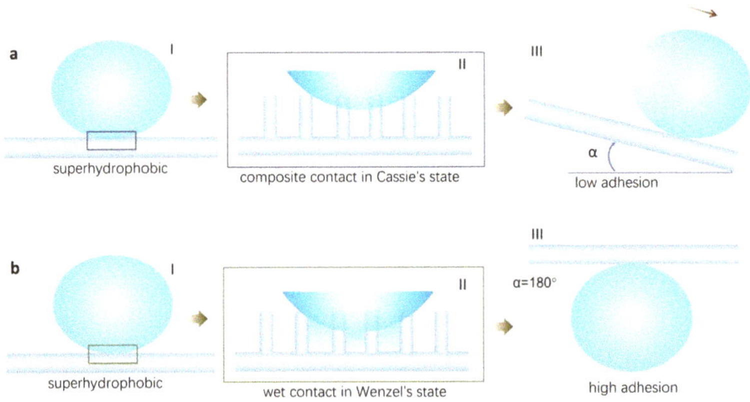

Illustration 6.1 Two types of superhydrophobic states: (a) composite contact in the Cassie state, along with low adhesion; (b) wet contact in the Wenzel state, along with high adhesion.

Based on the design of microstructured surfaces, various functions related to adhesion can be effectively achieved. Anisotropic or oriented structures have been developed to achieve directional low adhesion, leading to a low drag force for fluid transport. Furthermore, recently, superhydrophobic surfaces have been developed for anti-icing and icephobic properties with antiadhesion to ice on surfaces at low temperatures.

6.1.2 Characterization of Adhesion

A superhydrophobic surface with antiadhesion properties usually has a high CA > 150°, along with a low SA < 10°; i.e., the droplet easily rolls off along the surface, rather than becoming pinned. The adhesion can be simply characterized by the sliding angle of the droplet along the surface due to the downward force of gravity. The sliding angle can be defined as the tilt angle (α) of an inclined surface at which a droplet just slides off the surface. For surfaces with low adhesion, the sliding angle is small (Fig. 6.1a). For a superhydrophobic surface, the CA of a droplet is more than 150°, which can indicate low adhesion, i.e., $\alpha < 10°$. For surfaces with moderate or relatively higher adhesion, $10° < \alpha < 90°$ (Fig. 6.1b). If α ranges from 90° to 180°, the droplet is still pinned on the surface; thus, the surface would have the highest adhesion (Fig. 6.1c). By the balance between the downward force of gravity and surface adhesion, the adhesion of the surface can be quantitatively estimated.

To quantitatively characterize the adhesion of a surface, the CAH can be further considered. At the moment that the droplet just slides off the surface with a tilt angle of α, the advancing angle (θ_A) is defined as the CA at the front side of the droplet, and the receding angle (θ_R) is defined as the CA at the rear side of the droplet, as shown in Fig. 6.1d.

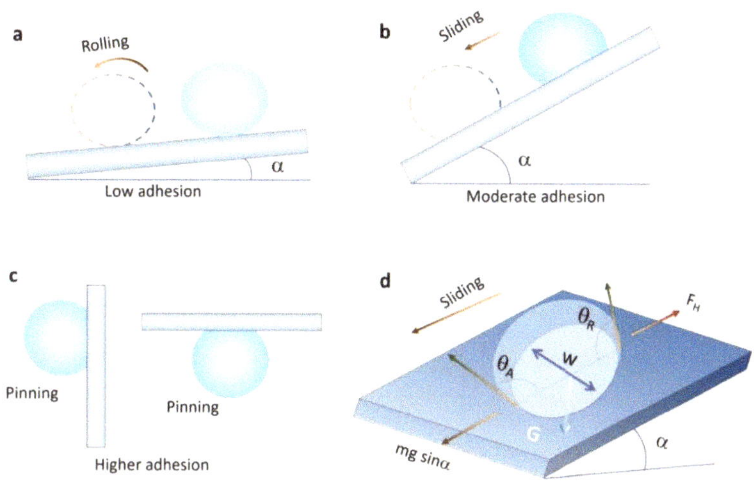

Figure 6.1 Illustration of the characterization of adhesion in different cases: (a) low adhesion, (b) moderate adhesion, and (c) high adhesion. (d) Illustration of the force balance between the downward force of gravity and contact angle hysteresis for the droplet on the surface.

According to the principle of force equilibrium, the tilt angle α can be related to the advancing angle (θ_A) and receding angle (θ_R) by the following relationship (Brandon et al., 1997; ElSherbini and Jacobi, 2004; McHale, 2009):

$$\text{mg}\sin\alpha = \text{w}\,\gamma_{LV}\,(\cos\theta_A - \cos\theta_R) \tag{6.1}$$

where w is the contact width of the droplet along the droplet sliding direction. Equation (6.1) indicates a force balance between the downward sliding force of gravity (i.e., mg sinα) and surface adhesion (or CAH). A larger α required for the droplet to slide off indicates that the droplet needs a larger energy to overcome the CAH of the surface. Considering the TCL of the droplet on the solid surface, the force of CAH (F_H) can be described as (Eral and Oh, 2013):

$$F_H = 2\pi r \gamma_{LV}(\cos\theta_A - \cos\theta_R) \quad (6.2)$$

where r is the radius of the droplet in the contact area and γ_{LV} is the liquid tension.

In addition, a small CAH is needed for surfaces with antiadhesion properties. The dynamic CA can be used to characterize the CAH and adhesion of the surface. The CAH means that the movement of the liquid is blocked due to pinning because of the TCL (Aussillous and Quéré, 2001; ElSherbini and Jacobi, 2004). Such an example can be seen on car windscreens or windows; some smaller drops become pinned to the glass, while some larger drops slide down. The method of the dynamic CA can be used to estimate the CAH. The general characterization method is used for a horizontal surface (moderate hydrophilicity or hydrophobicity) (Fig. 6.2a); the dynamic CA can be estimated by changing the CA when the droplet volume is extended, i.e., by injecting liquid into the droplet on the solid surface; thus, the droplet extends its volume to make the edge of the droplet move. When the CA reaches a critical value, this value is defined as the advancing angle (θ_A). In contrast, when some liquid is pumped away, the edge of the droplet shrinks, and then the CA reaches another threshold, which is defined as the receding angle (θ_R) (Fig. 6.2a). In this case, the dynamic CA, i.e., the advancing angle θ_A and the receding angle θ_R, both of which differ, can characterize the CAH (Eral and Oh, 2013): $\Delta\theta_{(CAH)} \approx \theta_A - \theta_R$. In general, for a superhydrophobic surface, $\Delta\theta_{(CAH)} < 10°$, and the surface is in a low adhesion state, for example, the droplet rolls off the surface of a lotus leaf, along with a self-cleaning effect.

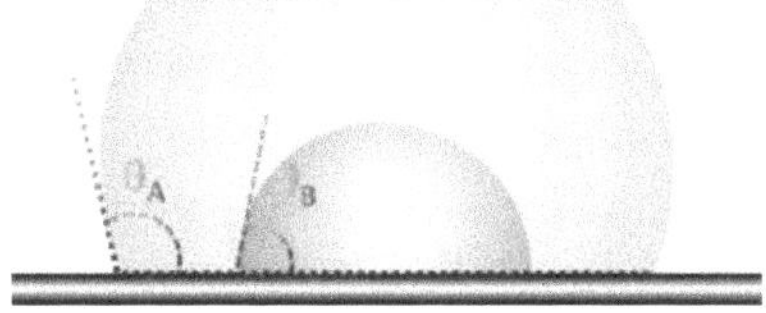

Figure 6.2 Characterization of dynamic contact angles. When the contact angle reaches the critical value, this value is defined as the advancing angle (θ_A). In contrast, when some liquid is pumped away, the edge of the droplet shrinks, and then the contact angle reaches another threshold, which is defined as the receding angle (θ_R).

6.2 Solid–Liquid–Liquid Interfaces of Hydrophobic Liquid-Infused Surfaces

6.2.1 Bioinspired Slippery Surfaces

Slippery surfaces have low adhesion properties, which have aroused increasing attention. Inspired by the liquid-infusion mechanism, a synthetic Teflon liquid-repellent surface with a slippery liquid-infused porous surface (SLIPS) has been successfully designed (Li et al., 2010). In this design, low-surface-tension perfluorinated liquids (e.g., 3 M Fluorinert FC-70, γ_B = 17.1 mN/m; or DuPont Krytox oils) are used as the as-infused liquid. The micro-/nanoporous structure locks the liquid to form a lubricating film on the surface, which has a slippery interface (Fig. 6.3a,b) that is different from the normal solid–air–liquid interface (e.g., lotus leaf). Some experiments can be used to demonstrate the antiadhesion properties. Figure 6.3c–f shows

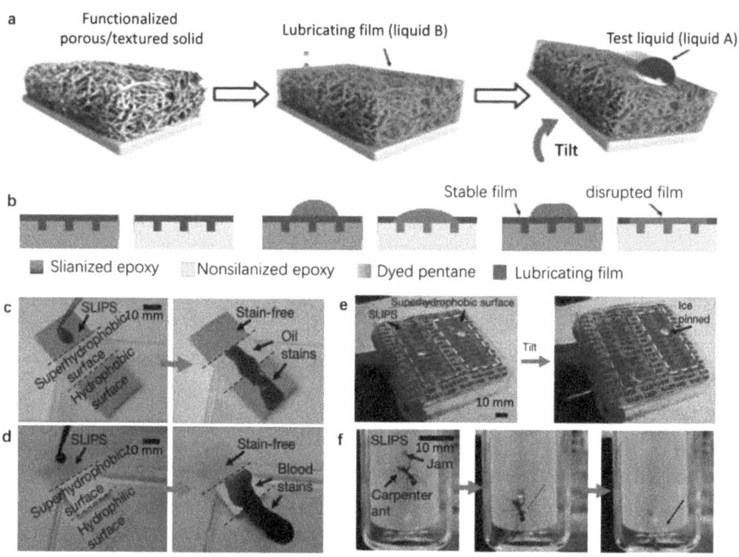

Figure 6.3 Design and application of slippery interface (Li et al., 2010). (a) Illustration of a SLIPS with antiadhesion properties. (b) Illustration of interface properties on the SLIPS. (c–d) Self-cleaning properties of the SLIPS for oil stains (c) and blood stains (d). (e) Anti-icing property of the SLIPS. (f) Ant sliding on the SLIPS.

photographs of the SLIPS effectively repelling crude oil (Fig. 6.3c), blood (Fig. 6.3d), and ice (Fig. 6.3e), as well as providing an antistick and slippery surface for insect capture (Fig. 6.3f). This surface offers another route to design surfaces with antiadhesion properties, i.e., solid–liquid–liquid interfaces, instead of solid–liquid–air interfaces. The lubricating film containing solid–liquid–liquid interfaces can serve as a self-healing coating to rapidly restore the liquid-repellent function when the surface is damaged by abrasion or impact due to surface-energy-driven capillary action.

6.2.2 Antiadhesion Organogel Materials

Organogel materials are semisolid materials due to the presence of an organic liquid phase inside the network of gelator aggregates, exhibiting low adhesion to water, blood, paint, ice, etc., which can be used to develop antiadhesive materials (Vintiloiu and Leroux, 2008; Ciriminna et al., 2015; Liu et al., 2017; Lv et al., 2017) due to their chemical or physical properties (e.g., covalent bonds, hydrogen bonding, van der Waals interactions, and π-stacking interactions) (Babu et al., 2014).

Figure 6.4a illustrates the design routes of antiadhesion organogel materials, where weakly interacting layers inhibit the adhesion of liquid or solid depositions. By the one-step procedure, crosslinked polymer networks with gelator solutions can be fabricated in situ. By the two-step procedure for antiadhesion organogels, the matrix can be prepared previously and then infused with liquids. The designability of organogels allows the selection of gelators and solvents to achieve the desired purposes (Lv et al., 2017); thus, antiadhesion properties can be achieved effectively on organogel surfaces.

As shown in Fig. 6.4b, an organogel film could be fabricated by crosslinking polyacrylate with silicone oil (viscosity of 20 sct). The organogel film exhibited a low adhesion to water, where water droplets immediately slid off the organogel surface at a tilt angle of 5° (top) and were not observed on other surfaces (bottom) (MacCallum et al., 2015). As shown in Fig. 6.4c, the surface of organogels (fluorogel-infused) greatly lowered the adhesion of blood (Liu et al., 2013). Another organogel (OG) surface with antiwax properties can be fabricated by swelling polydimethylsiloxane (PDMS) with oils, which causes extreme antiadhesion to solid wax, and the solidified paraffin particle slides off the 37°-inclined organogel surface (Fig. 6.4d) (Yao et al., 2014, 2015). These as-designed surfaces have promising prospects in the development of anti-icing surfaces (Meuler et al., 2010; Zhu et al., 2013; Urata et al., 2015; Golovin et al., 2016; Guo et al., 2019).

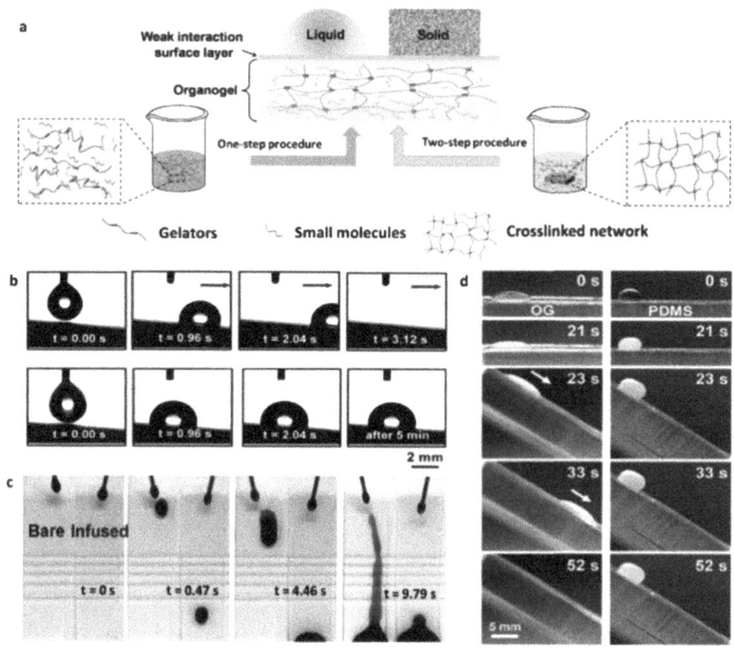

Figure 6.4 Illustration of the design of organogel (OG) materials (a) (Lv et al., 2017) with antiadhesion properties for water droplets (b), blood (c) (Liu et al., 2013), and solidified paraffin wax (d) (Yao et al., 2014).

6.2.3 Superhydrophilic Antiadhesive Surfaces

Kidney stones (KS) contain inorganic constituents such as calcium oxalate crystals (CaOx) and mainly originate from the aggregation of mineral crystals on the surface of injured renal tubules, inducing calculous diseases. Interestingly, KS do not stick on the surface of normal renal tubules (Fig. 6.5a,b) due to hydrophilic lipids and proteins infused in the nanohairs of the membrane, which induce antiadhesion properties (Zhang et al., 2018).

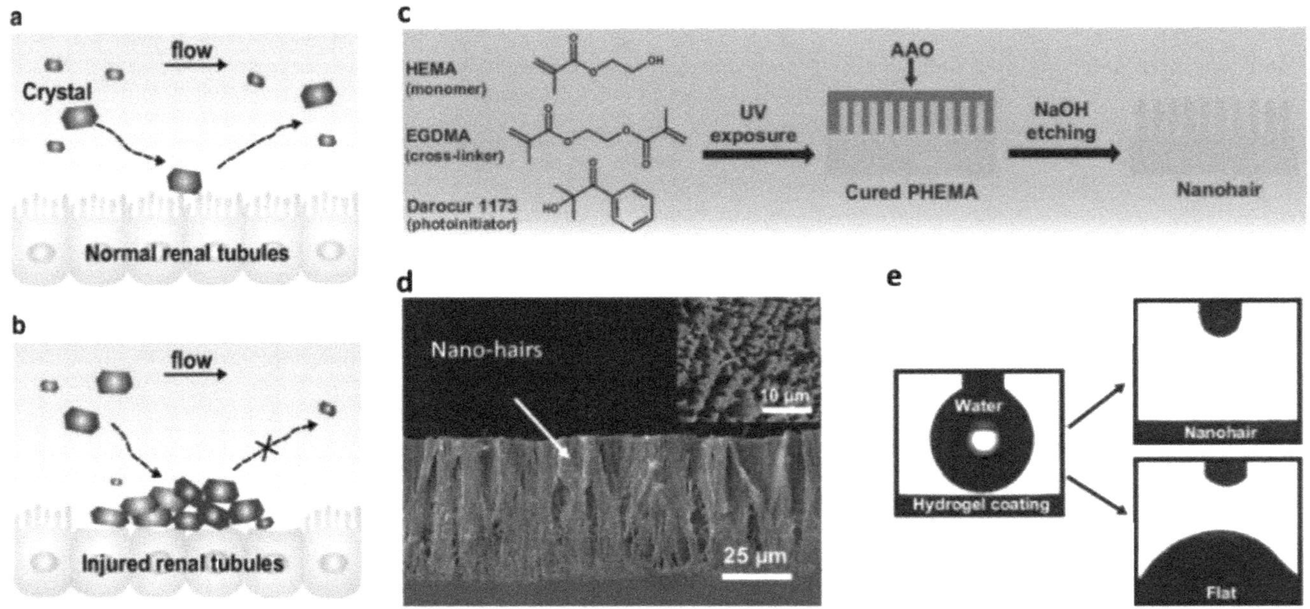

Figure 6.5 (a–b) Illustration of antiadhesion of mineral crystals on normal renal tubules (a) and adhesion on injured renal tubules (b). (c) Illustration of fabrication of the bioinspired hydrogels. (d) SEM images of bioinspired hydrogels with nanohairs. (e) Optical images of the hydrogel coating with nanohairs for antiadhesion of droplets (right top) and without nanohairs (flat) for high adhesion of droplets (right bottom) (Vervaet et al., 2010).

Based on inspiration from the unique mechanisms of antiadhesive properties, hydrophilic surfaces with nanohairs have been designed. Using integrative methods, bioinspired hydrogel coatings can be fabricated with a mixture of the monomer hydroxyethyl methacrylate (HEMA), the crosslinker ethylene glycol dimethacrylate (EGDMA), and the photoinitiator Darocur 1173 (Vervaet et al., 2010) under UV exposure; thus, poly(hydroxyethyl methacrylate) (PHEMA) nanohair hydrogel coatings can be effectively achieved, where an anodic aluminum oxide (AAO) membrane is selected as template after de-template treatment with a dilute (sodium hydroxide) NaOH solution (Fig. 6.5c). Figure 6.5d shows the nanohair morphology of the hydrogel coatings (with a height of ~51 μm and diameter of ~400 nm). The water droplet spreads with a CA of zero, indicating that the superhydrophilic surface contributes to the antiadhesion property of the hydrogel coating with nanohairs (Fig. 6.5e), in contrast to a flat surface.

6.2.4 Droplet Slippage on Water-Spreading Surfaces

There is also a special antiadhesion effect on a superhydrophilic surface when a water film forms over the surface, displaying wet-induced water repellency. As shown in Fig. 6.6a, a nanoporous structured surface with anatase titanium dioxide (TiO_2) can be obtained via a precursor solution with tetrabutylorthotitanate diluted in ethanol containing the triblock copolymer $EO_{20}PO_{70}EO_{20}$ (Pluronic P123). This surface is superhydrophilic, with water spreading to a CA of zero (inset of Fig. 6.6a). When a water droplet is gently deposited on the surface with a water-spreading film (Fig. 6.6b), interestingly, this droplet quickly runs off the surface (Frame I→II→III→VI→V) in 60–120 ms (Zheng et al., 2008).

Figure 6.6c illustrates the fast droplet roll-off process. Due to the superhydrophobicity of surfaces with abundant OH groups, water droplets spread into thin films. When a droplet contacts the water film on the superhydrophilic surface, the surface of the water droplet has a high curvature (1/R, where R is the radius of the droplet), while the water-spreading surface has a low curvature ($1/\infty$). Such a contrasting curvature at both contact regions can cause the force to convert into a shear force along the liquid–liquid interface to enable the droplet to bounce and run off. This rapid runoff of the droplet on the water film surface reveals that the contrasting liquid–liquid curvatures create ultralow adhesion for the as-deposited droplet.

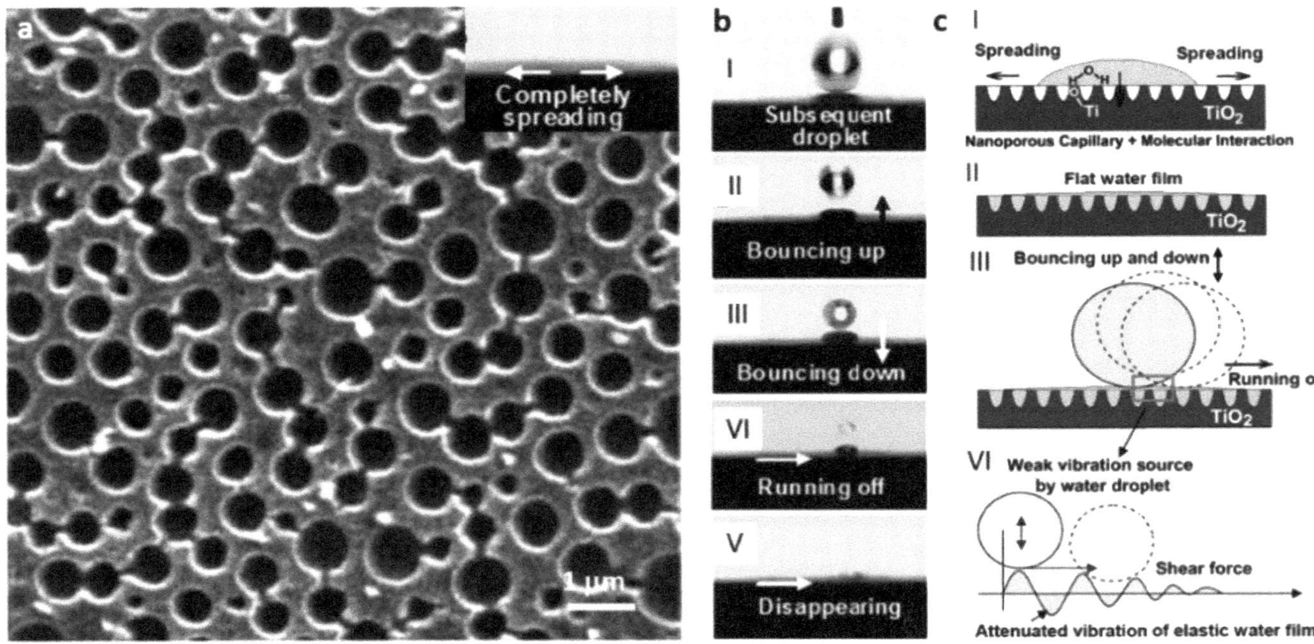

Figure 6.6 (a) SEM image of a porous, superhydrophilic surface (droplet completely spreads, inset). (b) Optical images of in situ observations of the water repellency on the water-spreading surface. (c) Illustration of extremely low adhesion on the superhydrophilic surface due to the interface curvature effect between the droplet and water film (Zheng et al., 2008).

6.3 Low Adhesion on Solid–Liquid–Air Interfaces of Superhydrophobic Surfaces

6.3.1 Antifogging Performance

Fog-droplet repellency can happen on as-designed low-adhesion superhydrophobic surfaces. As is known, in some extremely low-temperature environments, freezing fog is quickly condensed to form an ice layer on a solid surface, leading to poor optical performance on surfaces (Gao et al., 2007; Sun et al., 2014). To solve this problem, antifogging materials can be designed in two ways: via superhydrophilicity (Wang et al., 1998; Watanabe et al., 2000) and superhydrophobicity (Gao et al., 2007). Regardless of whether they are superhydrophilic or superhydrophobic, the surfaces of materials would be ineffective when surfaces are subjected to a low temperature for a long time (Howarter and Youngblood, 2007). In classical nucleation theory, a supercooled tiny water droplet condenses and grows on a surface when the as-condensed nucleation points converge or the phase transition energy is transferred to the substrates at lower temperatures (Kelton and Greer, 1988; Jung et al., 2012). Therefore, the condensation of fog droplets should be controlled first for effective design of the surfaces of materials.

6.3.1.1 Fly-eye-inspired superhydrophobic nanostructures

In nature, the eyes of some insects exhibit antifogging properties owing to their unique surface structural properties, such as the compound eyes of the green bottle fly *Lucilia sericata* (Beijing), which remain uncontaminated in dusty, miry, and moist environments. For example, bioinspired multifunctional nanostructures have been fabricated to investigate their antifogging ability (Sun et al., 2014). Figure 6.7a presents an optical image of a green bottle fly with compound eyes in a foggy chamber. The fogging droplets (diameters less than 10 μm) are generated by an ultrasonic humidifier to mimic a natural mist. The fog droplets can only condense on the body, rather than on the compound eyes, indicating the superhydrophobicity of the compound eyes. Figure 6.7b–d shows the micro- and nanostructures of the compound eye. The whole compound eye is ~5 mm in diameter. In detail, the compound eye is made up of thousands of repeating hexagonal units (~25 μm) (Fig. 6.7c), along with a nanostructure ~100 nm in size at each unit (Fig. 6.7d). It is such multilevel micro-/nanostructures that induce the superhydrophobicity of the compound eyes. In addition, the existence of wax on the surface lowers the surface energy, further increasing the superhydrophobicity of the compound eye surface.

Figure 6.7 (a) Photograph of a green bottle fly with two compound eyes. (b–d) SEM images of micro- and nanostructures of compound eyes. (e) SEM image of a bioinspired compound eye with nanostructures. (f) Optical images of antifogging properties on the bioinspired structured surface (right) in contrast to bare glass (left) (Sun et al., 2014).

Figure 6.7e shows fly-eye-inspired microspheres that were synthesized by ZnO via a "two-step" self-assembly process. The radial rhabdom-like structures are fabricated by using the triblock copolymer polyethylene oxidepolypropylene oxide-polyethylene oxide (PEO_{20}-PPO_{70}-PEO_{20}, Pluronic P123) and a ZnO precursor solution. Then, the hollow spherical structures are further assembled with a nonaqueous liquid (ethylene glycol (EG)) to nucleate the hexagonal lens structure on the surfaces. The superhydrophobicity of the bioinspired surface can be achieved by treating the surface with 1H,1H,2H,2H-perfluorooctyltriethoxysilane (PFOTES) molecules. Figure 6.7f compares the antifogging properties of the bare glass and bioinspired surface. As tiny fog droplets (<4 μL) form on both surfaces, the droplets slide off the 10°-tilted bioinspired surface during the droplet growth process, but pinning occurs on the surface of the bare glass.

6.3.1.2 Air-trapped hollow microsphere nanocomposites

Inspired by micro- and nanostructures of mosquito compound eyes, antifogging surfaces can be designed (Zhang et al., 2017a). A key design consideration is to prevent the entry of as-condensed microdroplets into the micro- and nanostructure. To date, it is still challenging to achieve robust antifogging surfaces.

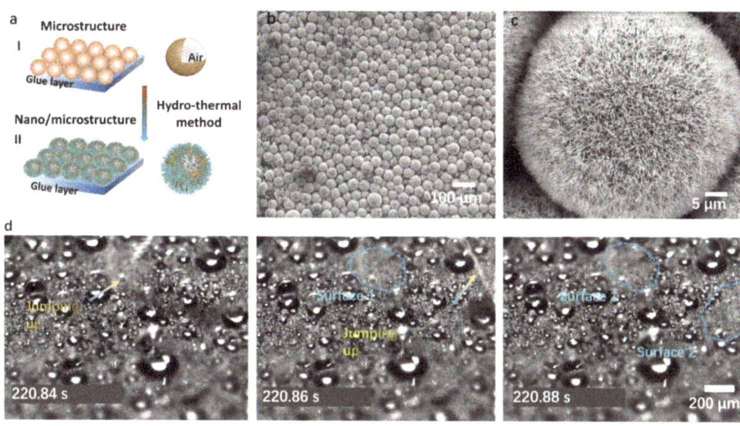

Figure 6.8 Illustration of the fabrication process of bioinspired micro- and nanostructured surfaces (a) and SEM images of surfaces (b–c) and antifogging properties (d) (Zhang et al., 2017a).

Figure 6.8a illustrates the fabrication of an antifogging film via spray and crystal growth processes. In this fabrication process, a homogeneous sol composed of soda lime borosilicate glass hollow (SGH) microspheres and ZnO seed crystals is sprayed onto the substrate surface. ZnO nanorods are grown on the surfaces of SGH spheres by the thermodynamic method. Figure 6.8b,c shows SEM images of the micro- and nanostructures of the film. The SGH microspheres are ~40 μm in size, and the ZnO nanoarrays are 50–100 nm in diameter and ~2 μm in height on average. After treatment by using FAS-17, the surface becomes superhydrophobic, thereby leading to a water CA of 163° and a rolling angle of 6° for a lower adhesion property. Figure 6.8d shows optical images in which the as-condensed droplets jumped up and shed off from the bioinspired surfaces, where the rebounding of the droplet retains the space regions on bioinspired surfaces.

The mechanism of antifogging is further illustrated in Fig. 6.9. The critical radius (r_{min}) of the as-condensed nucleus growth into droplets can be described as follows (Miljkovic et al., 2013):

$$r_{min} = \frac{2T_{sat}\sigma_{lg}}{H_{fg}\rho_l \Delta T} \tag{6.3}$$

where T_{sat} and T_{surf} are the temperatures of the saturated steam and surface, respectively, and $\Delta T = T_{sat} - T_{surf}$. σ_{lg} is the water tension, H_{fg} is the heat of transformation, and ρ_l is the water density. As r_{min} and the height and space of the ZnO nanorods are all on the order of nanometers, the tiny fog droplets (marked with 1, 2, 3, ..., 7, see Fig. 6.9a) are suspended in the Cassie state on the top of the nanostructures. As the droplets grow and coalesce, a new larger droplet (marked with 9) can jump up and shed off the surface due to coalescence-induced energy release. In addition, the driving force P_{cap} from the capillary effect for droplet repellency is $P_{cap} = P_2 - P_1$, where P_1 and P_2 are the capillary force at sites A and B, respectively (Fig. 6.9b), which is related to the curvature radius at sites A and B and can be further expressed as follows (Ju et al., 2013):

$$P_{cap} = 2\gamma\cos\theta\left(\frac{1}{r} - \frac{1}{R}\right) \tag{6.4}$$

where R is the radius at site A, r is the radius at site B, $r < R$; γ is the water tension, and θ is the apparent CA. Thus, droplets in the Cassie state would be formed on the surface with such micro-/nanostructures. As a result, the as-condensed or as-grown or as-coalesced droplets could all jump up and shed off to achieve antifogging properties of the surfaces.

Furthermore, it is understood that the air trapped by a nanostructure not only plays extremely important roles in superhydrophobicity but also has a restrained impact on droplet condensation and freezing. As shown in Fig. 6.9c, a tiny droplet maintains a similar Cassie state on top of the nanostructure, so at a lower temperature, most of the contact area is air trapped in the nanostructure supplanting the substrate. Only a small amount of energy can transfer along the nanorod to the SGH sphere. In addition, the gases in the SGH sphere are N_2 and inert gas, which have a thermal conductivity even lower than that of air. Thus, this small amount of heat can only transfer along the wall of the SGH sphere to the substrate, which uncommonly extends the heat transfer path and greatly reduces the heat transfer efficiency.

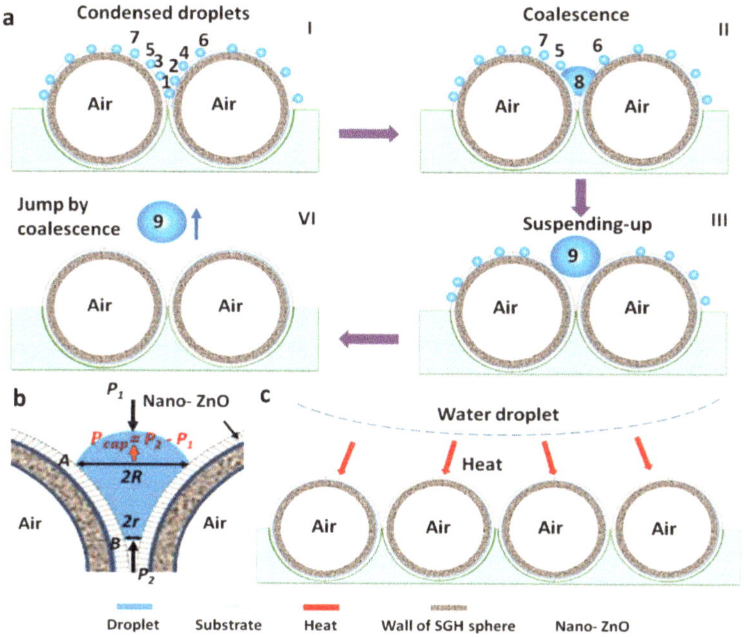

Figure 6.9 Mechanism of antifogging on a bioinspired micro-/nanostructured surface (Zhang et al., 2017a). (a) Illustration of jumping and suspending of a droplet during droplet coalescence. (b) Illustration of capillary force (P_{cap}); i.e., the droplet achieves an outward extrusion force P_{cap}, which is the difference between P_2 and P_1. (c) Illustration of heat transfer analysis from droplets to the substrate. The structures extend the heat transfer path and greatly reduce the heat transfer efficiency.

6.3.2 Antiadhesion of Coalesced Droplets on Superhydrophobic Surfaces

6.3.2.1 Nanostructure arrays

Superhydrophobic surfaces can achieve self-cleaning, displaying a "lotus effect," which facilitates the removal of contaminating particles by water droplet impact or roll-off. The mechanism is attributed to the larger CA and smaller hysteresis, which promotes the rolling motion, carrying away contaminants.

As is well known in nature, the self-propelling phenomenon can occur on some organism surfaces, including lotus leaves, water strider legs, cicada wings, etc., due to ultra-superhydrophobic properties and unique microstructures. Either nanostructures or microstructures would be important to induce the self-propelling phenomenon for the self-cleaning effect.

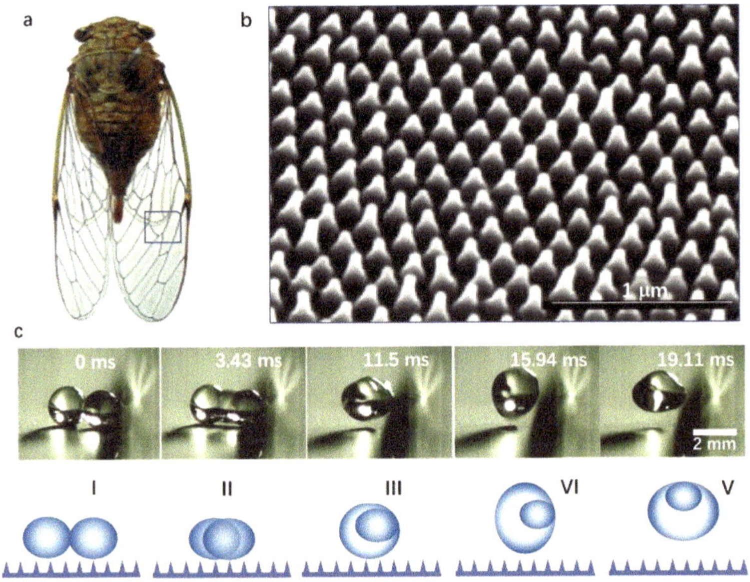

Figure 6.10 (a) Photograph of a cicada wing. (b) SEM image of the nanopapillae array on the surface of the wing. (c) Optical images (top) and illustration (bottom) of the observed in situ droplet self-propelled jumping (Wisdom et al., 2013).

With the example of a superhydrophobic surface (e.g., cicada wing) (Wisdom et al., 2013), an autonomous mechanism, i.e., droplet self-propulsion, is used to achieve self-cleaning on superhydrophobic surfaces. It is found that the contaminants are removed by self-propelled jumping and condensate-powered energy. When the superhydrophobic surface is exposed to condensing water vapor, the contaminating particles are either fully enclosed or partially covered with the liquid condensate. Figure 6.10a,b shows a cicada with its superhydrophobic wings and the nanostructured surface of a wing observed by scanning electron microscopy (SEM). This superhydrophobic surface effectively promotes the self-propelling phenomenon of condensed droplets. Figure 6.10c shows the particle removal process via the coalescence of condensed droplets, i.e., the as-grown condensed droplet coalesces with glass particles, detaches together with the particles, and subsequently jumps away from the surface. This observation demonstrates that a strong self-propelling force appears on the superhydrophobic surface of the nanostructure array.

6.3.2.2 Hierarchical structured superhydrophobic surfaces

Inspired by the wettability of creatures, surfaces with hierarchical arrays can be designed to achieve self-propelling properties (Zhang et al., 2017b). Figure 6.11a,b shows optical images of the self-propulsion of as-coalesced condensate microdrops on the surfaces, where the spaces of surfaces remain. Figure 6.11c illustrates the energy relationship by droplet coalescence.

The dynamics of as-coalesced microdrops for self-propulsion can be described as follows:

$$E_{itk} = \frac{1}{2} m V_e^2 = \Delta E_s - \Delta E_{int} - \Delta E_h \tag{6.5}$$

where $V_e = \frac{1}{5} U = \frac{1}{5}\sqrt{\frac{\sigma_{iv}}{\rho R}}$ is the average velocity, m is the mass of the droplet after coalescence, and ΔE_S is the released surface energy, which can be expressed as follows:

$$\begin{aligned} \Delta E_s &= \sigma_{lv}\Delta A_{lv} + \sigma_{sl}\Delta A_{sl} + \sigma_{sv}\Delta A_{sv} \\ &= \sigma_{lv}\pi\left(2-\sqrt[3]{4}\right)\left[2(1-\cos\theta)+\left(1-f_s-f_s\cos\theta_0\right)\sin^2\theta\right]r^2 \end{aligned} \tag{6.6}$$

ΔE_h is the gravity potential energy, as expressed by:

$$\Delta E_h = mg\Delta h = mg\left[\left(\frac{3V}{4\pi}\right)^{\frac{1}{3}} - \frac{r(3+\cos\theta)(1-\cos\theta)}{4(2+\cos\theta)}\right] \tag{6.7}$$

where E_{int} is the interfacial adhesion-induced dissipation energy; σ_{lv} is the liquid–vapor interface tension; ρ is the water density; r and R are the radii of the microdrop before and after coalescence, respectively; ΔA is the difference in surface area; and V is the volume of the merged microdrop after coalescence. θ_0 and θ are the CAs of water on smooth and rough surfaces, respectively. f_s is the solid fraction. If the water temperature is 5°C, $\sigma_{lv} = 7.5 \times 10^{-2}$ N/m, $\rho = 1.0 \times 10^3$ kg/m^3, and V = 2 μL (r = 0.79 mm), then $\Delta Es = 1 \times 10^{-7}$ J, $\Delta E_{itk} = 6.28 \times 10^{-9}$ J, and $\Delta E_h = 9 \times 10^{-9}$ J. It is inferred that $E_{int} < 8.5 \times 10^{-8}$ J would achieve self-propulsion.

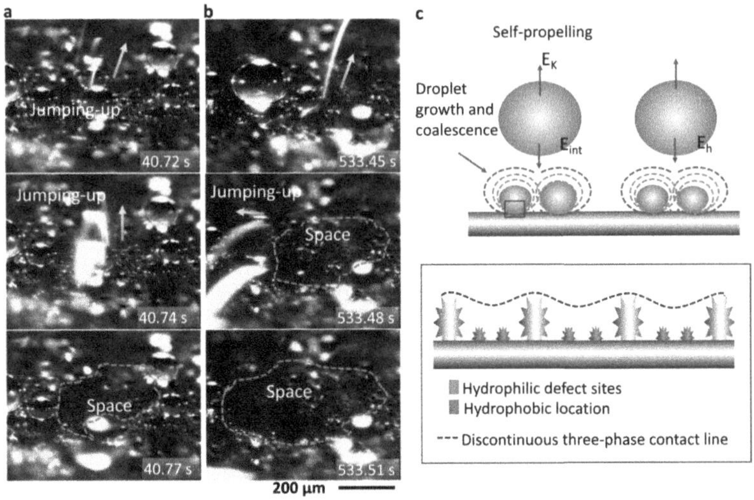

Figure 6.11 The bioinspired superhydrophobic surface for self-propulsion (a–b) and mechanism of self-propulsion due to droplet growth and coalescence along with energy generation on a superhydrophobic surface composed of micro- and nanostructures (c) (Zhang et al., 2017b).

6.3.3 Ultra-Antiadhesion with Pancake Bouncing on Superhydrophobic Surfaces

The pancake bouncing of droplets impacting at a wide range of impact velocities can be generated by designing micro-/nanotextures on a surface (Clanet et al., 2004; Liu et al., 2014). Figure 6.12a,b shows a square microscale cone array decorated with nanoflowers and treated with trichloro(1H,1H,2H,2H-perfluorooctyl)silane for superhydrophobicity (CA of water greater than 165° and low CAH less than 5°). The cones are 800 μm in height (h) and 200 μm in spacing (w), with diameters from 20 μm to 90 μm. If the unperturbed radius of the drop is r_0 = 1.45 mm or 1.10 mm and the impact velocity (v_0) ranges from 0.59 m/s to 1.72 m/s, then 7.1 < W_e < 58.5, where $W_e = \rho v^2{}_0\, r_0/\gamma$ is the Weber number, with ρ being the density and γ the surface tension of water. Figure 6.12c shows snapshots of a drop impinging on the surface at W_e = 7.1. After reaching a maximum lateral extension at 4.8 ms (Liu et al., 2014), the drop retracts at 2.9 ms and detaches from the surface at 16.5 ms (= 2.55 $\sqrt{\rho V_0^3 / \gamma}$). At a higher W_e value of 14.1, the drop retracts at 3.0 ms, and pancake bouncing occurs at 3.4 ms, along with the maximum extended length of the pancake at 4.6 ms (Fig. 6.12d). The liquid penetration is deeper, and the drop detaches from the surface (at 3.4 ms = 0.53 $\sqrt{\rho V_0^3 / \gamma}$) immediately with an ultrashort contact time. Pancake bouncing for a droplet also occurs when the droplet impacts (W_e = 31.2) on a 30°-tilted surface, as shown in Fig. 6.12e. The drop completely detaches from the surface in 3.6 ms. Figure 6.12f shows the drop impinging on the surface with nanoflows alone, which has a long contact time within 14.5 ms.

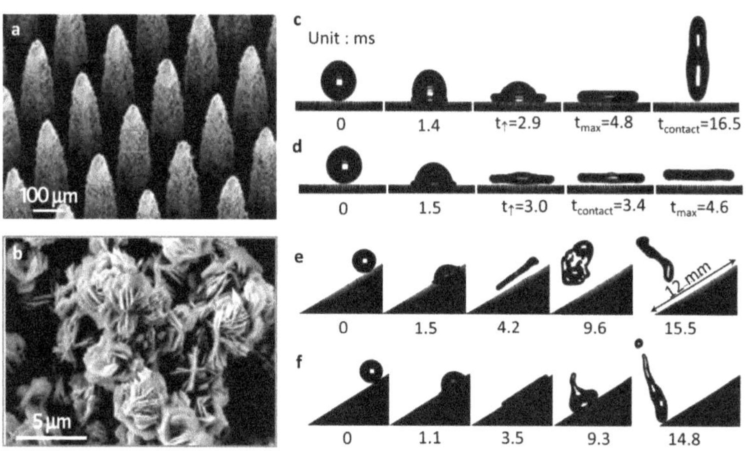

Figure 6.12 (a–b) SEM images of surfaces with cone arrays (a) and nanoflower structures on cones (b). (c–f) Optical images of in situ observations on different surfaces (horizontal (c–d) or tilted (e–f) surfaces) exhibiting pancake bouncing of impacting droplets over time (Liu et al., 2014).

6.4 Antifouling Marine Surfaces

6.4.1 Biological Fouling Organisms and Adhesion

Marine biofouling usually happens on shipping and leisure vessels, heat exchangers, oceanographic sensors, and aquaculture systems due to unwanted marine organisms, resulting in economic losses (Schultz, 2007; Jain and Bhosle, 2009; Callow and Callow, 2011; Schultz et al., 2011), such as low cruising speed (Schultz, 2007), greenhouse gas emissions, and economic costs (Jain and Bhosle, 2009; Schultz et al. 2011), as shown in Fig. 6.13, indicating the fouling by the green alga (seaweed) Ulva (image courtesy of Dr J. Lewis) (Fig. 6.13a) and barnacles (image courtesy of Dr C.D. Anderson) (Fig. 6.13b) (Callow and Callow, 2011). To date, it is still challenging to study highly effective antifouling coatings to control or deter organisms from attaching or to reduce their adhesion strength, enabling their facile removal by the shear forces from ship movement or mild mechanical cleaning devices.

Figure 6.13 Vessels fouled by marine organisms: (a) fouling by the green alga (seaweed) Ulva (image courtesy of Dr J. Lewis) and (b) barnacles (image courtesy of Dr C.D. Anderson) (Callow and Callow, 2011).

Figure 6.14 shows the diversity and size scales of a range of representative fouling organisms. Microfouling is often referred to as slime due to unicellular microorganisms such as bacteria, diatoms, and protozoa, which form a complex biofilm on the surface (Jain and Bhosle, 2009). Soft macrofouling comprises macroscopically visible algae (e.g., seaweeds) and invertebrates (e.g., soft corals, sponges, anemones, tunicates, and hydroids) (Wahl, 1989). Hard macrofouling is due to shelled invertebrates (e.g., barnacles, mussels, and tubeworms). The fouling is a dynamic process. A bacterial biofilm forms within a week by spores of macroalgae (seaweeds), fungi, and protozoa and, in turn, within several weeks by larvae of invertebrates, such as barnacles. In reality, this settlement can occur in minutes (Callow et al., 1997), a few hours (Roberts et al., 1991), or "one to several weeks" (Wahl, 1989). On the other hand, the attachment of spores and larvae can be influenced by other organisms, notably by bacterial biofilms, with positive, negative, and neutral effects (Huang and Hadfeld, 2003; Dobretsov et al., 2006; Huggett et al., 2006; Dobretsov et al., 2009). Colonization models are mostly based on studies of natural substrata, such as wood and stone or traditional antifouling coatings that rapidly foul, along with a few studies on the colonization of modern, nonbiocidal antifouling coatings by microorganisms (Molino et al., 2009a,b). The colonizing ("recruitment") stages of fouling organisms range in size from micrometers (bacteria, single-celled spores of algae, and some diatom cells) to hundreds of micrometers or even millimeters (larvae of invertebrates).

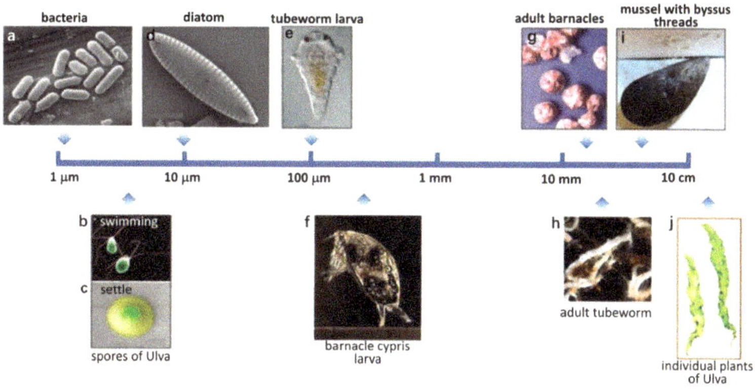

Figure 6.14 Representative fouling organisms in different size ranges: (a) bacteria, (b) swimming spores of Ulva, (c) settled spores of Ulva, (d) diatom, (e) tubeworm larva, (f) barnacle cypris larva, (g) adult barnacles, (h) adult tubeworms, (i) mussel with byssus threads, and (j) individual plants of Ulva (Callow and Callow, 2011).

In addition, the size of engineered surface topographies may deter the settlement of organisms (Smith and Callow, 2006; Aldred and

Clare, 2008). An exception is the well-characterized marine mussel foot adhesive system based on byssal threads and adhesive plaques formed from a composite of several proteins (Sagert et al., 2006). Catechol groups of the amino acid DOPA (dihydroxyphenylalanine) are oxidized to react with other catechol groups to crosslink the proteins into a strong cohesive composite. Dihydroxyphenylalanine groups also hydrogen bond to metal ions in the mineral substrata (Even et al., 2008). Therefore, to design an effective fouling-resistant coating, a major challenge is addressing the diversity of fouling organisms, which exhibit a vast range of adhesion mechanisms (including adhesives).

6.4.2 Antifouling of Bioinspired Engineered Topographies

A diverse range of mechanisms has been implicated in natural defense, including settlement-inhibiting micro- and nanotopographies, secreted bioactive molecules, sloughing surface layers, mucus secretions, and hydrolytic enzymes (Scardino and de Nys, 2011). In nature, many marine organisms are not colonized by other species, such as the skin of sharks and whales with complex surface topography, playing a role in either deterring fouling organisms from attaching or promoting their easy release. Inspired by the skin of fast-moving sharks at ~1/25th of the scale (Sharklet antifouling, Fig. 6.15a–c), the designed rough structure of the surfaces results in an 85% reduction in settlement of zoospores (motile spores) of the macroalga Ulva compared with that using smooth PDMSe. An engineered roughness index (ERI_I) can be described for various surface designs (Schumacher et al., 2007; Long et al., 2010), i.e., $ERI_I = \frac{r \times df}{1 - \varphi_S}$, where r is the Wenzel roughness factor, $(1 - \varphi_S)$ is the area fraction of the tops of features (that is, the ratio of the depressed surface area between features and the projected planar surface area), and df is the degree of freedom of spore movement. Spore settlement decreases as the ERI_I increases. Sharklet antifouling with an ERI_I of 9.5 showed a 77% reduction in settlement compared with the smooth surface.

Based on the Sharklet antifouling design for influencing the settlement of spores, nanoforce gradients are introduced to explain the effect of different feature geometries (Schumacher et al., 2008). Surfaces with nanoforce gradients ranging from 125 to 374 nN significantly reduce spore settlement, in contrast to a smooth substrate. Thus, the level of inhibition is characterized by the revised ERI model (ERI_{II}), i.e., $ERI_{II} = \frac{r \times n}{1-\varphi_s}$. Figure 6.15d shows settlement densities of spores of Ulva affected by the complexity of the microtopographical patterns, where settlement is decreased as the number (n) of features increases. The spore settlement density on the different patterns correlates well with the number of features in the ERI (ERI_{II}) model.

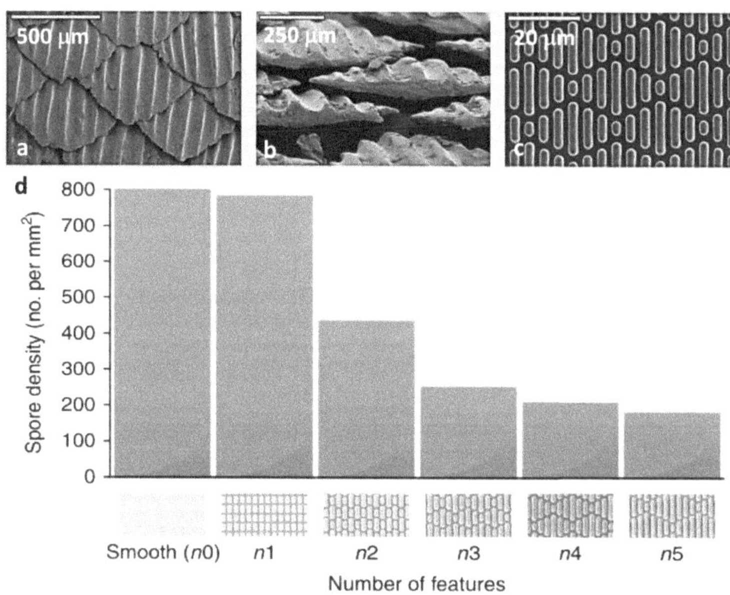

Figure 6.15 (a–c) Bioinspired topographies of PDMSe with different styles and (d) fouling for spore density versus number of features in topographies (Callow and Callow, 2011).

6.4.3 Vascularized Fouling-Release

Inspired by the long-term effectiveness of living antifouling materials, a method is developed for the self-replenishment of synthetic biofouling-release surfaces. These surfaces are created by either molding or directly embedding 3D vascular systems into PDMS and filling them with silicone oil to generate a nontoxic oil-infused material. When replenished with silicone oil from an outside source, these materials are capable of self-lubrication and continuous renewal of the interfacial fouling-release layer. Under accelerated lubricant loss conditions, the fully infused vascularized samples retain significantly more lubricant than nonvascularized samples (Howell et al., 2014).

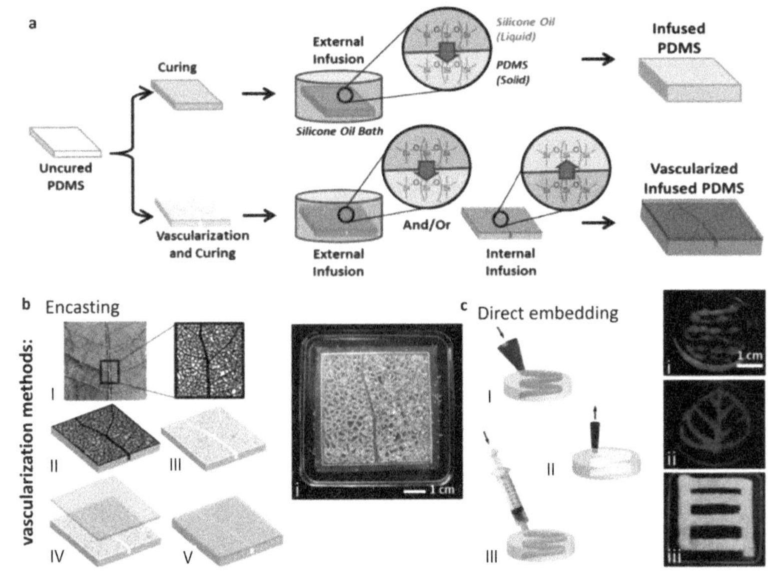

Figure 6.16 (a) Illustration of fabrication on vascularized infused surfaces. (b) Vascularization method to create vascular networks within PDMS. (c) Illustration of the direct embedding method for networks (left) and optical images (right) (Wu et al., 2011; Howell et al., 2014).

Figure 6.16a shows the schematic of the process to make either vascularized or nonvascularized silicone-oil-infused PDMS. For simple infused PDMS (upper row), cured PDMS is placed in a bath of silicone oil, which diffuses into the PDMS solid. For vascularized PDMS, the vascular pattern is created before curing. The sample is then infused externally with silicone oil in the same manner as the nonvascularized PDMS, internally through filling the vascular network, or both. Regarding methods of creating vascular networks within PDMS, as shown in Fig. 6.16b, an encased network is created using a 3D mold (Frames I–IV). The target pattern is molded (Frame I) to create the pattern in PDMS (Frame II), and then the mold is removed from the cured PDMS (Frame III); the pattern is covered with a second sheet of PDMS (Frame IV), which is chemically bonded to the pattern using O_2 plasma (Frame V). The inset shows an image of a 3D leaf vasculature network after encasing. Figure 6.16c shows the direct embedding network method that was developed (Wu et al., 2011). A pattern of 20% w/w Pluronic F127 gel at 25°C is embedded in uncured PDMS (Frame I). The PDMS is allowed to cure and cool to 4°C, and the liquid Pluronic is removed (Frame II). The channel is refilled with silicone oil (Frame III). The right insets of Fig. 6.16c show the images of fluorescently dyed silicone oil in (i) a sinusoidal channel, (ii) a leaf-shaped network, and (iii) a linear network.

Figure 6.17 The removal of *S. aureus* (a), *B. braunii* (b), *C. reinhardtii* (c), *D. salina* (d), and *N. oculata* (e) on the lubricant-infused PDMS surface. The lubricant-infused PDMS surface displays robust antiadhesion (Howwell et al., 2014).

Figure 6.17a shows the removal of glass, PDMS, and lubricant-infused PDMS from a solution of *Staphylococcus aureus* after 48 h of growth. The black arrow indicates the location of the infused PDMS sample. Figure 6.17b shows the glass slides half-coated with lubricant-infused PDMS being incubated in culture vessels containing *Botryococcus braunii*. After 12 days, the slides are slowly removed from the vessel using a dip coater attached to the slide. At t = 0 s and t = 5 s, clear attachment of the algae biofilm to the untreated glass can be seen. After t = 15 s, the boundary between the treated and untreated glass is crossed, and the biofilm begins to peel away from the surface (at t = 16 s). Finally, the intact biofilm is observed floating on the surface of the culture medium at t = 35 s (white arrow). Figure 6.17c–e shows the surface coverage of biofilm remaining on lubricant-infused PDMS versus glass control surfaces after incubation in and removal from a liquid medium containing *Chlamydomonas reinhardtii*, *Dunaliella salina*, or *Nannochloropsis oculata*. For all three species, a significant reduction in the amount of biofilm remaining on the lubricant-infused surfaces was observed.

6.5 Anti-Crude-Oil Adhesion of Polyionized Hydrogels

Antiadhesion is also important in industry. As is known, oil-spill accidents release millions of tonnes of crude oil into marine areas and generate tremendous oily wastewater every year. In addition, oily wastewater is widely produced from diverse oil-related industries, such as the food, textile, leather, steel, and metal-finishing industries. A problem is that the waste oil–water mixture kills creatures and damages people's health in a wide range of ways. Furthermore, problems with safe freshwater have been further aggravated due to oily wastewater and are expected to grow worse. Efficiently separating oil from oily water is widely recognized as a sustainable strategy. Studies that design and develop novel materials with anti-viscous-oil adhesion properties are still scarce and highly desired (Gao et al., 2016).

6.5.1 Design of Hydrogel Materials

A robust coating/membrane can be designed to display an anti-crude-oil adhesion property by using sodium polyacrylate-grafted poly(vinylidene fluoride) (PAAS-g-PVDF) via a one-step alkaline-induced phase-inversion process. As schematically shown in Fig. 6.18, poly(acrylic acid)-grafted poly(vinylidene fluoride) (PAA-g-PVDF) powder with a 7 wt.% grafting ratio of poly(acrylic acid) (PAA) is first dissolved in N,N-dimethylformamide and loaded onto solid substrates by either scraping the casting solution or immersing the substrates into the polymer solution. The substrates loaded with

PAA-g-PVDF are then immersed in a coagulation bath containing sodium hydroxide (NaOH) to carry out phase inversion. During this process, PAA-g-PVDF reacts with NaOH gradually and transforms into PAAS-g-PVDF, and the final polyionized hydrogel polymer coating is obtained. The water content in the polyionized hydrogel polymer coating is measured to be ~87 wt.%.

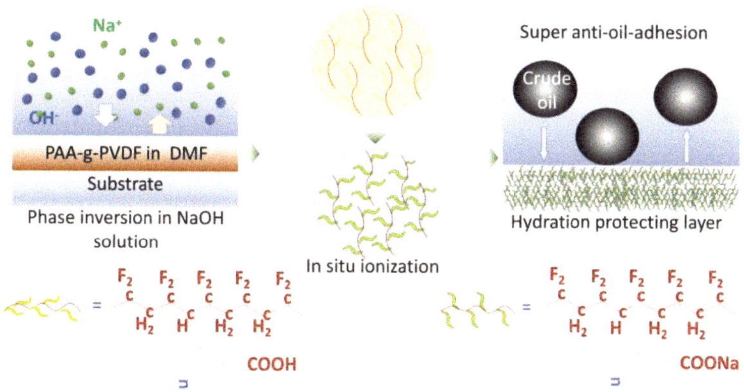

Figure 6.18 Schematic for the formation of a PAAS-g-PVDF coating/membrane for underwater super-anti-crude-oil adhesion (Gao et al., 2016).

PAAS-g-PVDF can be easily coated onto diverse substrates, including copper mesh, nickel foam, nonwoven, and polypropylene plastic plates. All the PAAS-g-PVDF-coated substrates exhibit underwater superoleophobicity, with heavy crude oil CAs above 165° and an ultralow oil adhesion force of nearly 0 μN. Among them, the PAAS-g-PVDF-coated nickel foam displays the best underwater superoleophobicity, with an underwater heavy crude oil CA reaching 172°. In addition to repelling light and heavy crude oil, PAAS-g-PVDF-coated substrates exhibit a superior oil repellency against silicone oil and heavy diesel oil, with underwater oil CAs all above 165°.

6.5.2 Anti-Crude-Oil-Adhesion Coating

The PAAS-g-PVDF materials display robust anti-crude-oil adhesion. As shown in Fig. 6.19a, the crude oil droplets display high adhesion and high CAs on PAAS-g-PVDF-coated substrates, including copper mesh, nickel foam, nonwoven and polypropylene (PP) plastic plates under water.

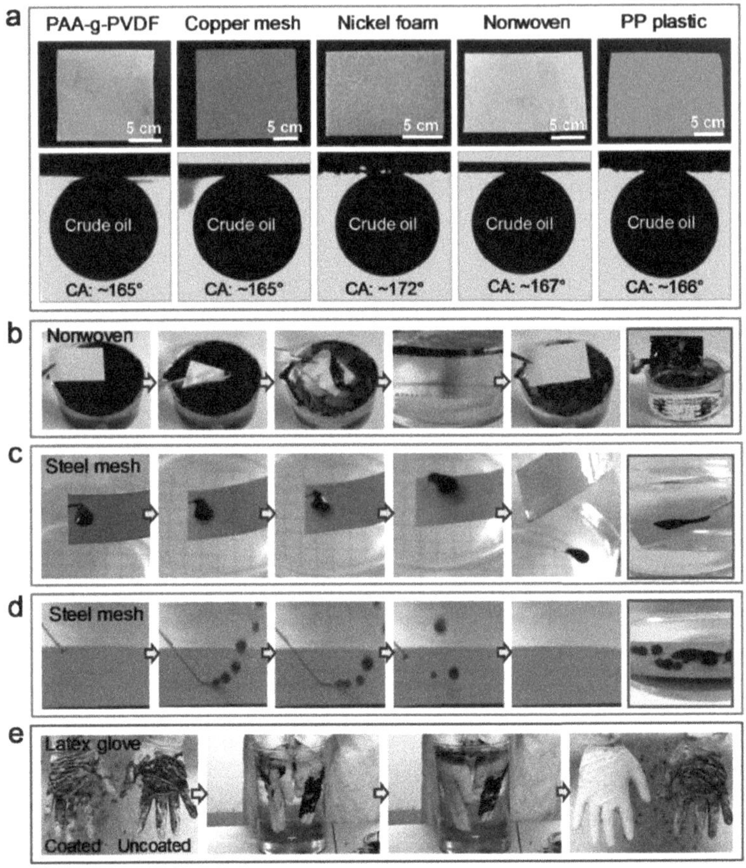

Figure 6.19 (a) Optical images of crude oil droplets with high contact angles on PAAS-g-PVDF-coated substrates, including copper mesh, nickel foam, nonwoven and polypropylene (PP) plastic plates under water. (b–e) Anti-viscous-oil-adhesion and self-cleaning performance of a PAAS-g-PVDF-coated nonwoven material (b), stainless-steel mesh (c–d), and latex glove (d) (Goa et al., 2016).

Figure 6.19b–e shows the antiadhesion and self-cleaning properties. When a PAAS-g-PVDF-coated nonwoven material is immersed and lifted repeatedly in a heavy crude oil/water mixture, the nonwoven material is free of oil adhesion and fouling and can be cleaned by washing with water (Fig. 6.19b). There are also underwater adhesion-free properties on the PAAS-g-PVDF-coated stainless-steel mesh for light and nonviscous oils (Fig. 6.19c), and soybean oil (dyed blue) can bounce off the surface of the mesh

(Fig. 6.19d). Furthermore, two latex gloves with/without PAAS-g-PVDF coating are used for comparison of the anti-crude-oil adhesion abilities (Fig. 6.19e). When immersed in water, the crude oil on the PAAS-g-PVDF-coated latex glove is immediately removed, and the glove becomes clean. However, the crude oil remains adhered to the latex glove without the PAASg-PVDF coating.

6.6 Adhesion Surfaces

6.6.1 Wet Adhesion

Wet adhesion can be generally defined as when a liquid phase exists at the interface, such as when a water droplet is pinned on a solid surface due to closed wet contact between liquid–solid interfaces, which is related to the Wenzel state of a droplet with a high liquid CA. Adhesion can be controlled by the interface effect. It is known that on some rough surfaces, water droplets may partially wet superhydrophobic textured surfaces due to air partially trapped in the valleys, which is termed a metastable state (Lafuma and Quéré, 2003; Marmur, 2004). Water drops on this type of surface in the Wenzel and Cassie states can coexist, with an energy barrier preventing a spontaneous wetting-state transition (Patankar, 2004). External stimuli, such as vibration (Bormashenko et al., 2007), electric field (Krupenkin et al., 2005; Dhindsa et al., 2006), or pressure (Bico et al., 1999; Quéré and Lafuma, 2003), can be applied to transfer drops from one state to the other. When the wetting behavior changes from the Cassie to the Wenzel state, the liquid droplet will fill the grooves of the rough surface with a dramatic increase in the adhesive force. Therefore, switching of liquid–solid adhesion can be achieved by the external-stimuli-induced wetting-state transition. In addition, the adhesion generated by marine organisms can involve special mechanisms for medical applications.

6.6.1.1 Petal effect

Biological surfaces can present special wetting adhesion and high CAs of water, for instance, the surfaces of the petal of red rose (Rosea Rehd), where the droplet can be pinned, although the surface can be rough, containing an array of micropapillae and nanowrinkles.

These hierarchical micro- and nanostructures provide sufficient roughness to allow the surface to be defined as superhydrophobic but have a high adhesive force with water, i.e., spherical droplets do not roll off and remain on the surface even when the petal is turned upside down. This phenomenon is defined as the "petal effect" (Feng et al., 2008), in contrast to the familiar "lotus effect."

Figure 6.20a shows an SEM image of a red rose petal containing a periodic array of micropapillae (average 16 μm in diameter and 7 μm in height), where the inset shows a photograph of red rose petals. Figure 6.20b shows the nanowrinkles (~730 nm in width) resulting from cuticular folds on these micropapillae. The surface reveals micro- and nanostructures, similar to those of lotus leaves. Interestingly, the droplet does not roll off the surface, even when the surface is turned upside down (Fig. 6.20d), although the droplet is spherical in shape with a water CA of 152.4° and a short contact line (Fig. 6.20c).

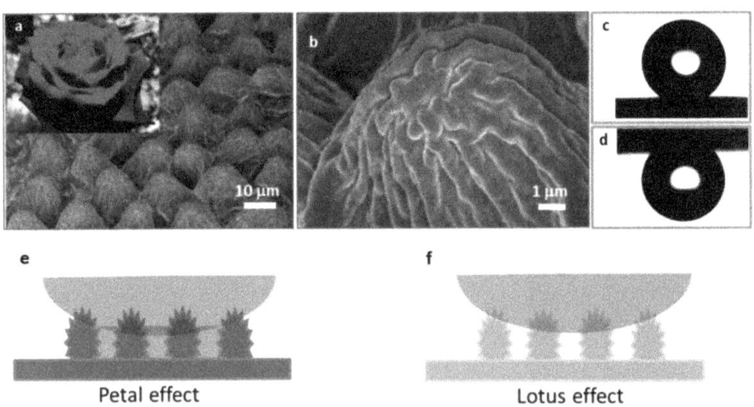

Figure 6.20 (a–b) Surface of a red rose petal composed of a periodic array of micropapillae and nanofolds on each papilla. (c–d) Optical images of the superhydrophobicity with a high water contact angle (c) and high adhesion to pin the droplet on the surface (d). (e–f) Illustration of the petal effect of droplet pinning due to the wet solid–liquid contact mode (e), different from that of the lotus effect due to the composite solid–liquid contact mode (d) (Feng et al., 2008).

This phenomenon can be understood by using the two types of superhydrophobic states on a rough surface (Lafuma and Quéré, 2003): Wenzel state and Cassie state. In the Wenzel state, the droplet is in the wet solid–liquid contact mode, and the water completely wets the solid structure of the surface where the droplet contacts, where the valleys of structures can be filled with water. The petal effect can be induced by the wet solid–liquid contact mode of the Wenzel state, along with continuous solid–liquid–air TCLs between the droplet and rough surface (Fig. 6.20e). In contrast, in the Cassie state, the droplets are suspended by the structures, where the valleys of structures contain air, and the TCL is discontinuous. The much higher amount of trapped air in the valleys of structures suspends the droplet, inducing a composite contact with the droplet on the rough surface. The composite contact mode can exist on the surface of the lotus leaf, termed the lotus effect (Fig. 6.20f). The two distinct superhydrophobic states of these surfaces thus induce different functions with high adhesion (Wenzel state) and low adhesion (Cassie state).

6.6.1.2 Controlled liquid splash

Pesticide spraying is "hard" agricultural work, where more than 50% of agrochemicals are lost because of undesired bouncing (de Ruiter et al., 2014) and splashing (Song et al., 2017) behaviors on crop leaves with "waterproof" properties. The deposition of liquid droplets on solid surfaces is of great importance to many fundamental scientific principles and technological applications, such as spraying, coating, and printing. For example, during the process of pesticide spraying, more than 50% of agrochemicals are lost because of undesired bouncing and splashing behaviors on hydrophobic or superhydrophobic leaves.

As shown in Fig. 6.21, a kind of "unavoidable" bouncing or splashing behavior is greatly inhibited or even completely eliminated on the superhydrophobic surface at varied angles by adding a small amount of a double-chain vesicle surfactant (de Ruiter et al., 2014). The in situ observations show the different behaviors when liquid droplets impact a *Brassica oleracea* L. leaf at an impact velocity of 2.53 ± 0.11 m/s, where these liquid droplets include pure water or aqueous solutions containing the same mass fraction of the micelle surfactant sodium dodecyl sulfate (SDS), trisiloxane molecules (TSs), and the vesicle surfactant sodium bis(2-ethylhexyl) sulfosuccinate [Aerosol OT (AOT)]. It is found that the water completely exhibits rebounding due to the water repellency of the leaf, but the 1% AOT droplet is completely pinned after demonstrating splashing behavior, while others (e.g., 1% SDS or 1% TS) are intermediate between water repellency and pinning.

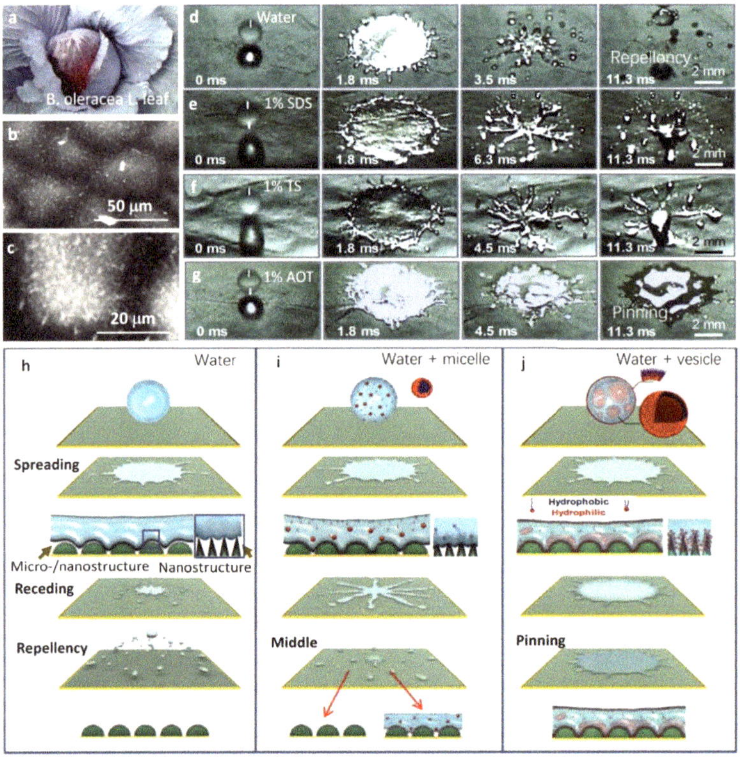

Figure 6.21 A *B. oleracea* L. leaf with microstructured/nanostructured morphology (a–c) and different splashing behaviors of different liquids: water (d), 1% SDS (e), 1% TS (f), and 1% AOT (g). (h–j) Illustration of the mechanism of splashing limitations (Song et al., 2017).

Figure 6.21h–j shows the mechanism of adhesion on solid–liquid–air interfaces. Nanoparticles such as micelles or vesicles in droplets can modulate interface tensions and change the contact mode of liquid on surfaces with micro-/nanostructures. For splashing behavior, the hydrophilic particles of the vesicle surround the nanostructures, which generates a wet contact with the surface and leads to strong wet adhesion when the composite droplet (e.g., water mixing with the vesicle) impacts the surface with micro-/nanostructures (Fig. 6.21j), in contrast to other liquids (such as micelles, Fig. 6.21i).

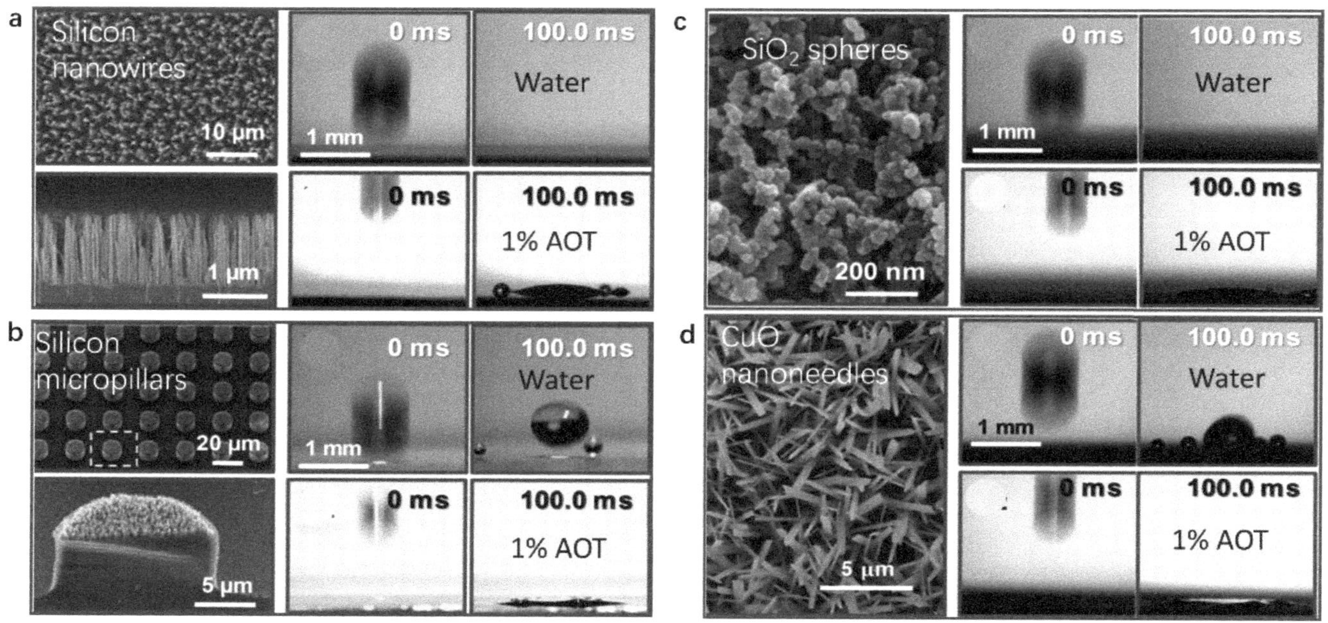

Figure 6.22 The splash inhibition of vesicle AOT generalized to other superhydrophobic surfaces: (a) silicon nanowires, (b) silicon micropillars, (c) SiO_2 spheres, and (d) CuO nanoneedles (Song et al., 2017).

Various bioinspired surfaces have been fabricated to facilitate the splash inhibition of 1% AOT droplets, as shown in Fig. 6.22; these surfaces include silicon nanowires (Fig. 6.22a), silicon micropillars covered with nanowires (Fig. 6.22b), SiO_2 spheres (Fig. 6.22c), and CuO nanoneedles (Fig. 6.22d). Figure 6.22a shows SEM images of the surface with silicon nanowires; water repellency can be observed, but the 1% AOT droplet spreads with a low CA. Figure 6.22b shows SEM images of the surface with silicon micropillars, where the nanostructures cover micropillars. The water repellency is weak, and the droplet has a high CA but does not rebound off. The 1% AOT droplet spreads strongly to form a film. Figure 6.22c shows an SEM image of surface SiO_2 spheres; water repellency is observed on this surface, and the 1% AOT droplet spreads relatively strongly, with a low CA. Figure 6.22d shows an SEM image of the surface with CuO nanoneedles, where the water droplet is pinned on the surface when the droplet impacts the surface, but the 1% AOT droplet spreads into the film. Based on these phenomena of droplet splash inhibition, the surfaces can be selected for the appropriate applications.

6.6.1.3 Marine organism adhesion

Marine organisms have evolved a diverse set of highly effective wet adhesive approaches that have inspired the design of new medical adhesives (Balkenendea et al., 2019). The underwater adhesion specialists include the mussel, sandcastle worm, and octopus (Stewat et al., 2011; Baik et al., 2017; Waite, 2017; Fears et al., 2018). Basic scientific research has elucidated highly evolved wet adhesion mechanisms in these animals, and researchers have since incorporated these chemical and physical wet adhesion strategies into medical adhesives. Much of the work in mussel-inspired adhesion has focused on understanding and mimicking the unusual adhesive proteins found in the terminal plaque of the mussel byssus (Waite et al., 2005). Sandcastle worms employ coacervate-forming protein adhesives to construct their sand grain dwellings (Stewart et al., 2011). Cephalopods, on the other hand, rely on anatomical features in the form of suction cups to adhere robustly onto wet surfaces (Baik et al., 2017).

Mussels secrete protein-based byssal threads to tether themselves to diverse underwater surfaces, including rocks, ships, and other organisms (Waite et al., 2005). At the distal end of these threads, adhesive proteins form an adhesive plaque that securely anchors the thread (and thus the mussel) to the surface, allowing the mussel to withstand the high shear forces of waves (Lin et al., 2007; Wei et al., 2013; DeMartini et al., 2017). The mussel's adhesive plaque is arguably the most well-studied marine bioadhesive (Fig. 6.23a,b).

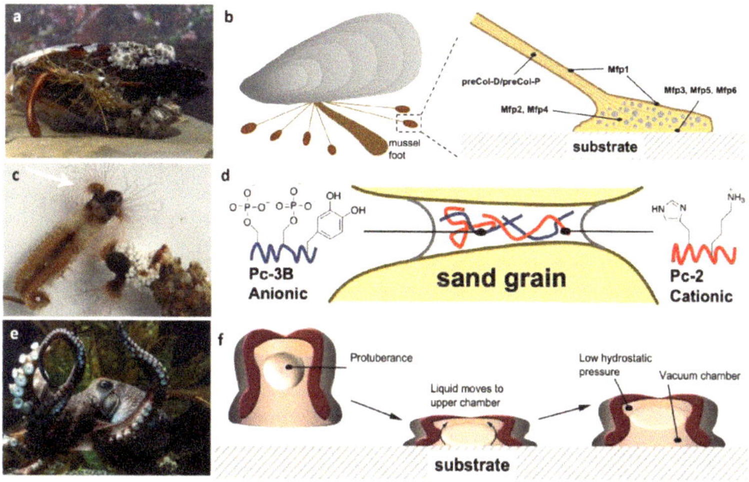

Figure 6.23 (a–b) Mussels strongly attach to underwater surfaces with a protein-based plaque and can withstand the strong forces of crashing waves. (c–d) The sandcastle worm builds tubular dwellings of sand particles cemented together with adhesive proteins. Photograph of a sandcastle worm (left, arrow) and a worm inside a tubular dwelling under construction. The (lighter) white granules contain the coacervated and uncured adhesive. (e–f) A common New Zealand octopus uses suction cups to reversibly adhere to diverse surfaces and can use these attachments to move around its environment (Balkenendea et al., 2019).

Inspiration from the sandcastle worm is shown in Fig. 6.23c,d. The sandcastle worm connects sand grains to build tubular dwellings. The load-bearing bioadhesive that connects sand particles consists of a complex coacervate of highly phosphorylated anionic proteins (Pc-3B) and cationic proteins (Pc-2). Octopus-inspired tissue adhesive patches have also been made. As shown in Fig. 6.23e,f, it is hypothesized that *Octopus vulgaris* muscular hydrostats reversibly adhere to surfaces by compressing suction cups, causing the liquid to flow toward the upper chamber above the protuberance, resulting in a low hydrostatic pressure. A vacuum is created in the lower chamber (Baik et al., 2017). A medical example is a simplified biomimetic suction cup patch for wet tissue adhesion to a porcine heart (Chen and Yang, 2017).

6.6.1.4 Tree frog adhesion

The adhesive mechanisms of climbing animals have clear implications for biomimetics. For example, these animals adhere well to many surfaces, the adhesion is reversible so that the adhesive is reusable, and sticking only occurs when required (Autumn, 2007). Additionally, these surfaces can be self-cleaning so that they are resistant to failure through the accumulation of dirt particles (Hansen and Autumn, 2005; Crawford et al., 2012; Hu et al., 2012).

The adhesion mechanism of tree and torrent frogs is quite different from that of geckos (Autumn et al., 2000, 2002; Hawkes et al., 2015). Tree frogs mainly adhere by wet adhesion, as there is a thin layer of fluid between the toe pad and the adhering substrate. The adhesive forces are thought to be capillarity- and viscosity-dependent hydrodynamic forces (Emerson and Diehl, 1980; Hanna and Barnes, 1991), but a contribution by van der Waals forces cannot be excluded (Langowski et al., 2018). Tree frog toe pads also generate friction forces. Indeed, the maximum friction forces may exceed those arising from adhesion (Endlein et al., 2017). The tree frog adhesive mechanism provides good adhesion and friction on soft wet surfaces. As a result, such a surface with wet adhesion is likely to have many important applications, particularly in the field of medicine (Meng et al., 2019).

Figure 6.24a,b shows photographs of a frog living in trees or shrubs (tree frog such as *Hypsiboas boans*). *H. boans* is the largest tree frog found in Trinidad (West Indies), with a snout/vent length of ~100 mm (Fig. 6.24a). This frog has just landed on a bamboo stem and has slipped downward, leaving behind some of the watery fluid that forms its adhesive joint (Fig. 6.24b). An SEM observation shows the toe pad surrounded on three sides by a groove (black line) (Fig. 6.24c) and polygonal (mostly hexagonal) epithelial cells surrounded by deep channels (Fig. 6.24d). The actual epithelial cells (the outermost layer) have a complex structure, for instance including nanopillars and a fluid secreted by the toe pads, so the tree frog toe pads produce higher friction, along with wet adhesion force (capillary forces, viscous forces, viscous-poroelastic adhesion, etc.) (Meng et al., 2019).

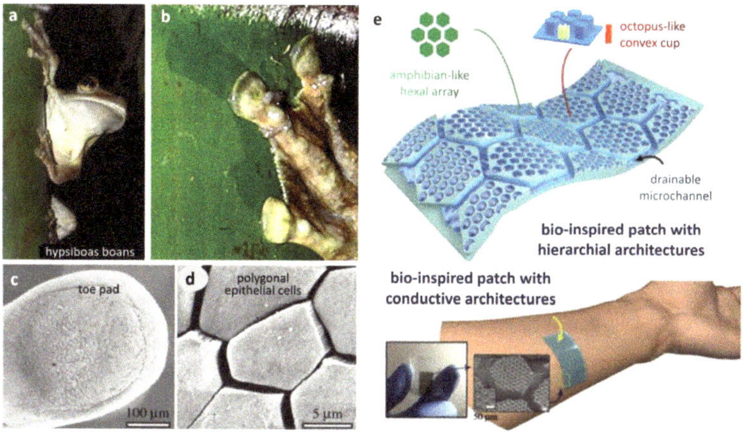

Figure 6.24 (a–b) Photograph of the tree frog *H. boans*. (b) Optical image of the watery fluid of adhesion on a bamboo stem when the frog slipped downward. (c–d) SEM images of the toe pad composed of micropatterns. (e) Illustration of the design of bioinspired tree frog adhesion for biomimetic skin patch applications (Meng et al., 2019).

Although the mimicking of tree frog wet adhesion is still in its infancy, there have already been some attempts to make use of it in the areas of bionic robots (Nguyen et al., 2018), soft tissue engineering (Tsipenyuk and Varenberg, 2014), etc. Figure 6.24e shows a biomimetic skin patch inspired by the hexagonal pattern of tree frog toe pads with the convex cups of octopus suckers (Kim et al., 2019). The patch exhibits outstanding adhesion performance on human skin, even in a flowing water environment. In addition, the patch can serve as a flexible electrode sensitive enough to receive biosignals on wet skin, even under motion.

6.6.2 Dry Adhesion

Dry adhesion can be generally defined when the solid phase contacts a solid surface, i.e., solid–solid contact, and is not related to the liquid phase. The adhesion can be determined by the solid–solid contact force, such as drag force and friction force.

6.6.2.1 Gecko feet with antiadhesion and adhesion

Geckos can climb on almost any complex surface because their feet contain microsetae split into hundreds of nanoscale spatulae (Autumn et al., 2000, 2002; Hansen and Autumn, 2005). Gecko setae arrays achieve a water CA of ~160°, indicating a superhydrophobic surface, to carry out the solid–solid adhesion mechanism (Northen et al., 2010).

Figure 6.25 (a) Photograph of a gecko, as well as the superhydrophobicity of gecko feet along with a high adhesive ability toward water (insets). (b–d) SEM images of the structural hierarchy of gecko feet composed of nanohairs (Liu et al., 2012).

Figure 6.25a shows a photograph of a gecko (*Gekko subpalmatus* Gunther) under anesthesia. Both insets in Fig. 6.25a indicate that gecko feet possess superhydrophobicity and a high adhesive ability toward water. Gecko feet have a structural hierarchy (Fig. 6.25b), containing millions of well-aligned microscopic keratinous setae (20–70 μm in length and 3–7 μm in diameter) (Fig. 6.25c). The setae are further composed of nanoscale spatulae (100–200 nm in diameter) (Fig. 6.25d) (Liu et al., 2012). The adhesive force of the gecko foot is wide, ranging from 10–60 μN, as tested by a high-sensitivity microelectromechanical balance system. This wide range is partly attributed to the conformational changes in the surface proteins (e.g., hydrophilic–lyophilic balance) in gecko setae, which is demonstrated when setae are exposed to water, inducing a change in the setal array (Liu et al., 2012).

Figure 6.26a shows the typical force–distance curves before and after a water droplet (4 μL) contacts gecko setae. The surface of setae exhibits a moving speed of 0.01 mm/s to contact the water droplet and then move downward. The force is increased gradually until it reaches its maximum, and then the droplet is pulled away from the surface, and the contact force is sharply reduced to zero (see process 1→2→3). In addition to rough structures that induce the superhydrophobicity of gecko feet (Hansen and Autumn, 2005), rough structures exhibit two tunable solid–liquid contact states to achieve high wetting adhesive properties: the default state in which the spatulae minimally contact water (Fig. 6.26b) and the adhered state in which the spatulae maximally contact water (Fig. 6.26c). The main adhesive force is the result of high-density nanopillars contacting the water.

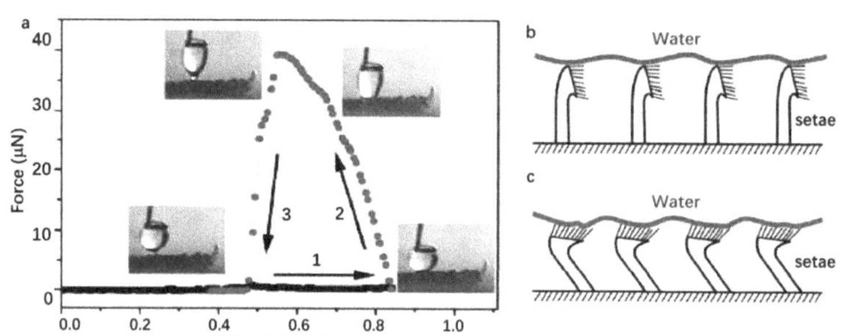

Figure 6.26 Force–distance curves recorded before and after a water droplet (4 μL) contacts a gecko foot (a) and a schematic of two typical contact states between water droplets and gecko setae: (b) default state and (c) adhered state (Liu et al., 2012).

The feet of geckos retain self-cleaning properties, similar to that of plant surfaces. The billions of integrated spatulae maximize van der Waals interactions to generate large adhesive and shear forces, which depend on the self-cleaning ability of the feet (Autumn et al., 2002). The particles contacting the unloaded surface may wash away easily in the presence of water due to the unique, flexible micro- and nanostructures on the gecko feet.

The detachment or attachment to a surface is dependent on the oriented anisotropy of the flexible array of setae. The anisotropic adhesion can be tested. When dragged against the natural path, the setal arrays are compressed with nonadherence (Fig. 6.27a), along with an average friction force, $F_{II} = mg\cos\alpha^* \approx 7.5$ mN, for an average normal (compressive) force, $F_{\perp} = mg\sin\alpha^* \approx 25.0$ mN, and the friction coefficient μ ($= F_{II}/F_{\perp}$) ≈ 0.31, where m is the total mass (body mass plus added mass of backpack), g is the gravitational acceleration, α^* is the critical detachment angle, i.e., the peel-off angle over vertical at which detachment occurs, defined as $\alpha^* = \tan^{-1}(F_{\perp}/F_{II})$. When dragged along their natural path (along the oriented direction), the setal arrays compressed initially and then adhered, resulting in normal tensile forces (Fig. 6.27b). In arrays dragged along the oriented direction (Fig. 6.27c,d), the average friction force is $F_{II} \approx 74.6$ mN. The average adhesive force is $F_{\perp} \approx 34.8$ mN. The adhesive force was a linear function of the shear force: $F_{\perp} = -0.487F_{II} + 0.002$ ($R^2 = 0.89$). The critical detachment angle is $\alpha^* \approx 24.6°$ on average (Autumn et al., 2006).

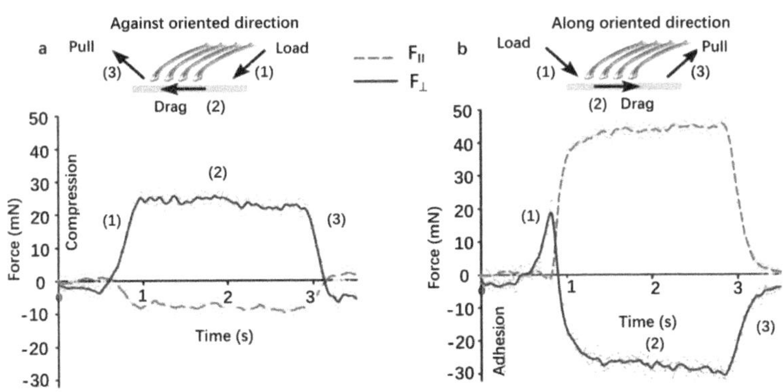

Figure 6.27 (a, c) Illustration of the anisotropic force test method against the oriented direction (a) and along the oriented direction (c). (b, d) Curves of shear and normal forces in isolated gecko setal arrays on a glass surface, corresponding to (a) and (c), respectively (Autumn et al., 2006).

6.6.2.2 Bioinspired composite microfibers

Biological systems for sensors can be used in various manners, for example chemical adhesives (Autumn et al., 2006; Mahdavi et al., 2008), adhesion by smart nanohairs (Cui et al., 2012; Aksak et al., 2014; Drotlef et al., 2014; Marvi et al., 2015; Hennebert, et al., 2014), and microneedle arrays with swellable tips (Yang et al., 2013) and soft skin (Drotlef et al., 2017) by means of complex surface topographies. Inspired by biological mechanisms, soft and stretchable skin-adhesive microfibers can be designed by using PDMS decorated with mushroom-shaped vinylsiloxane (VS) tips (Fig. 6.28a).

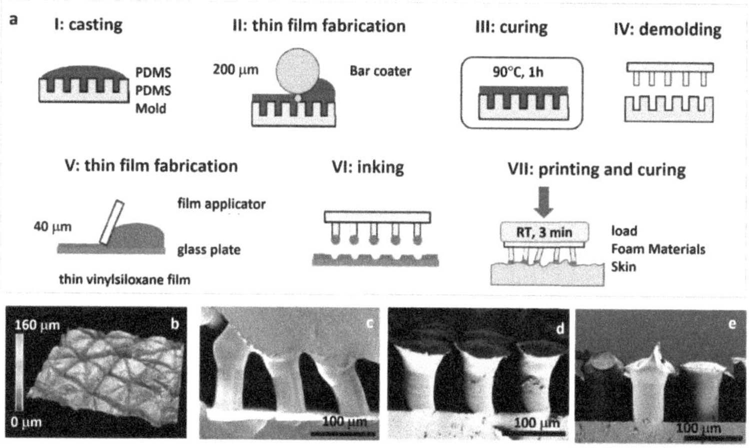

Figure 6.28 (a) Illustration of the fabrication process of skin-adhesive films via casting (i), thin film fabrication (ii), curing (iii), demolding (iv), thin film fabrication again (v), inking (vi), and printing and curing (vii). (b–e) 3D laser-scanning microscopy image (b) and SEM images (c–e) of the composite microfiber adhesive film attached to a skin replica (Drotlef et al., 2017).

Figure 6.28b shows 3D laser-scanning microscopy images of the skin surface in the forearm. The skin surface is composed of island-like planar areas separated by interconnected microgrooves with different roughnesses of 0.4 μm and 104.2 μm. Figure 6.28c–e illustrates SEM images of the composite microfiber adhesive film attached to a skin replica, showing conformal interfacing between the adhesive film and skin replica.

The integrated skin-adhesive sensors were highly flexible and could be easily attached onto the skin (Fig. 6.29a). Figure 6.29b illustrates the electromechanical behavior of a strain-sensor prototype under repeated stretching–releasing cycles (strains from 0.1% to 1%, $\Delta R/R_0$ indicates the extent of the response signal).

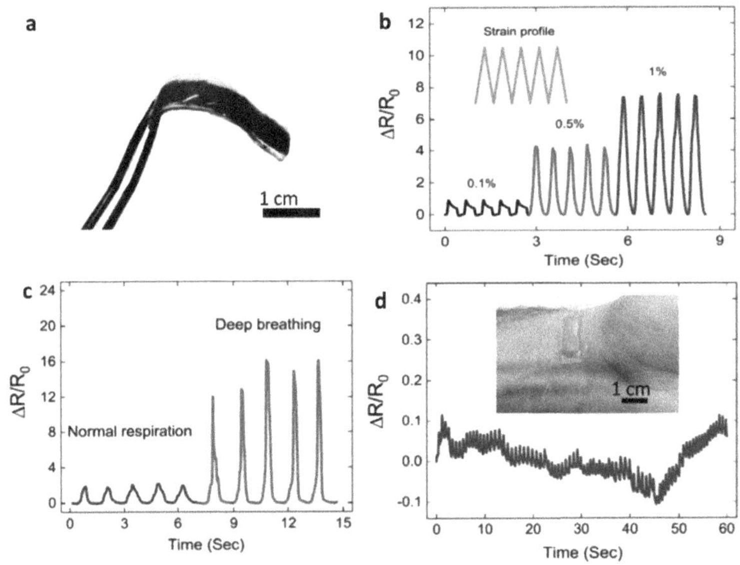

Figure 6.29 (a) Photograph of integrated skin-adhesive sensors. (b) Illustration of the electromechanical behavior of a strain-sensor prototype under repeated stretching–releasing cycles. (c) Illustration of the response of a microfibrillar skin-adhesive strain sensor attached to the human chest area. (d) Illustration of a record of the blood-flow pulse over 1 min and wearable skin-adhesive strain sensors for healthcare applications (Drotlef et al., 2017).

Figure 6.29c shows the response of a microfibrillar skin-adhesive strain sensor attached to the human chest area, which would be significant to help early diagnosis of several diseases, including asthma, hearth failure, embolism, and so on (Ramires et al., 2013; Lee et al., 2014). Figure 6.29d illustrates a record of the blood flow pulse over 1 min, showing the heart rate with a frequency of 84 beats per minute. The inset of Fig. 6.29d depicts a microfibrillar skin-adhesive strain sensor on the radial artery of the wrist, which can record the blood flow pressure with high sensitivity (Prabha et al., 2014).

6.6.2.3 Gradient micropillars with optimal adhesion

For biologically inspired dry adhesives, the fibrillar structure of the surface requires sufficient flexibility to form contacts and high rigidity to maintain stability. This fundamental conflict has greatly hindered the advance of synthetic adhesives toward mass-scale and practical applications, where the adhesion is desired to be simultaneously strong, durable, directional, and adaptive to roughness.

Figure 6.30 illustrates the fabrication and design of slanted functional gradient pillars (s-FGPs). The material gradient of the s-FGPs mimics the gradient tarsal setae of the ladybird beetle (Peisker et al., 2013). In this fabrication, a silicon master with patterned angled cavities (Fig. 6.30a) (leaning angle, 20°; diameter, 2.5 μm; AR, 8) is first filled with homogeneous, UV-curable polyurethane acrylate (PUA)-based nanocomposites (Fig. 6.30b). The composites comprise 15 vol.% superparamagnetic iron oxide nanoparticles coated with functionalized silica shells ($Fe_3O_4@SiO_2$, total diameter of ~20 nm) as the reinforcements. The magnetic-responsive nanoreinforcements inside the mold cavities are then redistributed by applying a magnetic field gradient along the axial direction of the cavities (Fig. 6.30c). The s-FGPs are finally obtained by solidifying the redistributed nanocomposite (Fig. 6.30d) and peeling them off from the mold (Fig. 6.30e). Figure 6.30f illustrates the slanted functional gradient pillar array, i.e., s-FGPs (Wang, 2018).

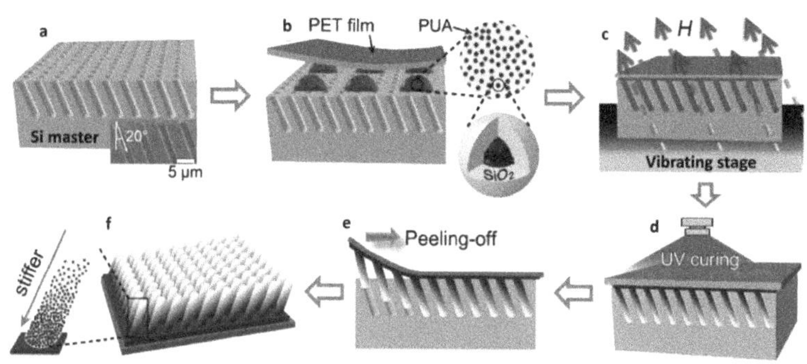

Figure 6.30 Schematic for the fabrication of slanted functional gradient pillars (s-FGPs) via the following processes: Si master with a pattern of angled cavities (a)→filling with PUA and $Fe_3O_4@SiO_2$ nanocomposites (b)→applying a magnetic field and vibrating stage (c)→UV curing (d)→peeling-off (e)→achievement of the gradient tilted pillar array (f) (Wang, 2018).

Surfaces with various types of structures are used to compare the shear adhesion, including the vertical and slanted compliant pillars (i.e., v-CPs, s-CPs), stiff pillars (i.e., v-SPs and s-SPs), and functional gradient pillars (i.e., v-FGPs and s-FGPs) (see SEM images in insets of Fig. 6.31a). The shear adhesion can be compared on these surfaces (Fig. 6.31a), where the s-FGPs have the highest shear adhesion, followed by the v-FGPs. Such s-FGPs achieve both desired geometric (slanted and high AR) and material properties; for instance, the elastic modulus gradually increases from pillar tip to base, indicating the adaptive adhesion of the s-FGPs measured against a rough steel substrate. The right-top color inset in Fig. 6.31a shows a 3D surface profile of the substrate within an area of 60 × 60 μm^2 obtained by a scanning probe microscope.

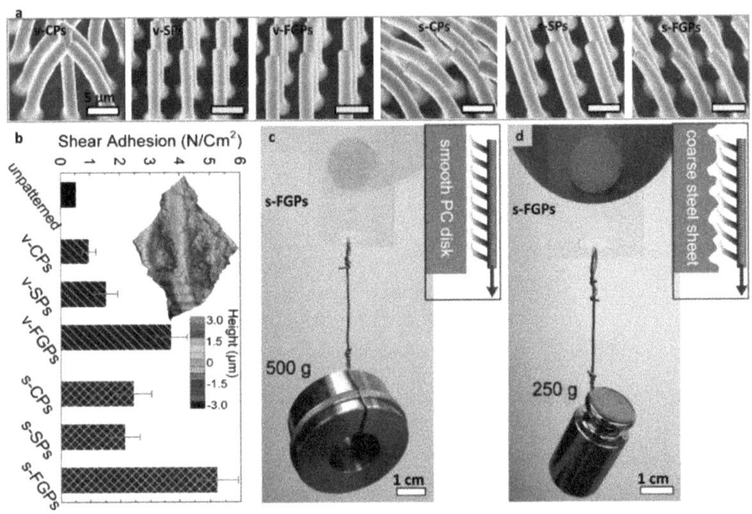

Figure 6.31 (a) Adaptive adhesion of surfaces with various types of structures: v-CPs, v-SPs, v-FGPs, s-CPs, s-SPs, and s-FGPs (insets from left to right). (b–c) Photographs of the s-FGPs for high adhesion against a smooth PC disk (b) and coarse steel sheet (c) as substrates (Wang, 2018).

Figure 6.31b,c shows the comparison of vertically adhering performance between a smooth polycarbonate disk supporting a 500 g weight and a coarse steel plate supporting a 250 g weight for a 1 cm^2 s-FGPs sample, where the right insets show the schematic diagrams of the attachment to surfaces.

6.6.3 Reversible Adhesion

6.6.3.1 Temperature-controlled adhesion

The design of smart surfaces with switchable adhesive properties in a wet environment has remained a challenge in adhesion science and materials engineering. In nature, many marine organisms such as mussels (Lee et al., 2006, 2007a, 2011), sandcastle worms, and barnacles orchestrate a powerful wet adhesion mechanism through an elegant blend of catechol chemistry (Shao and Stewart, 2010; Brennan et al., 2017), polyelectrolyte complexes (Zhao et al., 2016), and supramolecular architectures (Ahn et al., 2013; Heinzmann et al., 2016). Over the past decade, it has been shown that the remarkable wet adhesion tenacity can be mimicked by the use of 3,4-dihydroxy-L-phenylalanine (DOPA) groups (Lee et al., 2007b; Ahn et al., 2015; North et al., 2017). Extensive progress has been made in the understanding of basic mechanisms, and the design of artificial adhesives has served multifunctional applications such as vessel and heart (Lang et al., 2014), blood (Shin et al., 2017), tissue (Yang et al., 2013; Rose et al., 2014), human-protein-based sealant (Annabi et al., 2017), and others (Kim et al., 2009; Maier et al., 2015; Li et al., 2017). As is well known, in the dry condition, reversible interfacial adhesion can be achieved by an external stimulus (Michal et al., 2016), including pH, light, temperature, humidity, magnetic force, modulation of the bulk material properties (Reddy et al., 2007; Shafiq et al., 2012; Wang et al., 2016; Ye et al., 2016; Kizilkan et al., 2017), etc. However, in a wet environment, attaining reversible adhesion remains extremely elusive due to the complex interfacial interactions (Boesel et al., 2010; Zeng et al., 2010), in particular, producing a robust and smart underwater adhesive with tunable, reversible modulation for a wide range of substrates is still greatly challenging (Zhong et al., 2014; Zhao et al., 2017).

Recently, such wet adhesives have been designed based on two copolymers: a mussel-inspired guest-adhesive copolymer and a thermoresponsive host copolymer (Zhao et al., 2017). Figure 6.32a illustrates the bioinspired wet adhesive synthesis process, which involves the creation of an adhesive guest copolymer pDOPA-AD-MEA on a clean Si substrate using the dip-coating method, followed by self-assembly of the host copolymer pNIPAM-CD using host–guest molecular recognition. Figure 6.32b illustrates the temperature-responsive properties of the adhesion mechanism. The lower critical solution temperature (LCST) is measured to be ~35°C. It is found that there is low interfacial adhesion when the local temperature of the adhesive is 25°C (<LCST) (top of Fig. 6.32b). At high temperatures of 40°C (>LCST) (bottom of Fig. 6.32b), the collapse of pNIPAM-CD chains is also associated with the formation of numerous agglomerates during the phase transition. As a result, the adhesive group is exposed, leading to high interfacial adhesion.

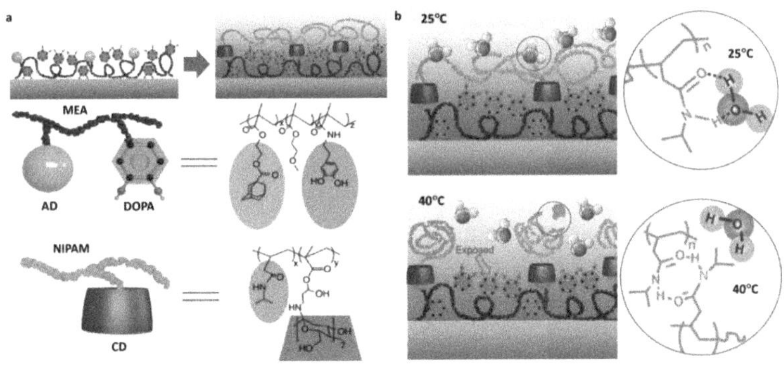

Figure 6.32 (a) Illustration of the bioinspired wet adhesive synthesis process. (b) Illustration of the temperature response to adhesion. At a low temperature of 25°C (<LCST), low interfacial adhesion occurs (top), and at a high temperature of 40°C (>LCST), high interfacial adhesion occurs (bottom) (Zhao et al., 2017).

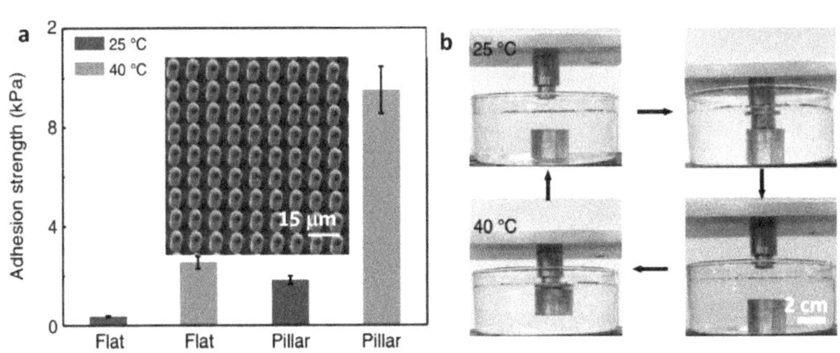

Figure 6.33 (a) Comparison of adhesion strength on different surfaces: flat and pillar (at 25°C and 40°C). At 40°C, the surface generates a higher adhesion strength. (b) Photographs of reversible adhesion changes between 25°C (nonadhesion) and 40°C (adhesion) for an object (Zhao et al., 2017).

Figure 6.33 shows the versatility of the reversible wet adhesive. Figure 6.33a depicts the combination of the wet adhesive with rough PDMS pillar arrays, which increase the adhesion strength. Figure 6.33b shows a photograph of the as-fabricated adhesive for the well-controlled lifting of a solid object (weight = 200 g) in a water environment.

6.6.3.2 Meniscus-controlled octopus-inspired adhesives

The adhesion capabilities of various skin architectures found in nature can generate remarkable physical interactions with their engaged surfaces. Among them, octopus suckers have unique hierarchical structures for reversible adhesion under dry and wet conditions to develop the application (Baik et al., 2018). Based on nanoscale hierarchical structures, without any chemical parts, some bioinspired adhesion strategies are known to induce mechanical interlocking or molecular interactions (Laulicht et al., 2012; Yang et al., 2013; Choi et al., 2016). These architectures have been applied to develop reversible and repeatable skin-adhesive patches, which serve in wound protection, smart skin-attachable devices, etc. (Pang et al., 2015; Drotlef et al., 2017; Park et al., 2017).

Figure 6.34a shows a photograph of suction cups of *Octopus vulgaris* (the most common species of octopus), along with a disk-like acetabular roof, adopting the grooved surfaces encircled by an extruded rim of epithelium around the edge of the sucker. The columnar rim, in particular, plays a key role in the high adaptability to various surface morphologies under wet conditions to completely lock the suction chamber, while the wrinkled surface in the acetabular roof maximizes the contact area during attachment (Tramacere et al., 2014, 2015). Figure 6.34b,c illustrates the fabrication of solution processing of the μ-SC array by controlling capillarity in hole pattern mode, where it generates external pressure via the underlying controlled meniscus (stage I) and modulates the volume of trapped air (stage II). Figure 6.34d–f shows an illustration of meniscus-controlled octopus-inspired adhesives (Fig. 6.34d), a photograph of the adhesive (Fig. 6.34e), and SEM images of bioinspired octopus-like surfaces (Fig. 6.34f).

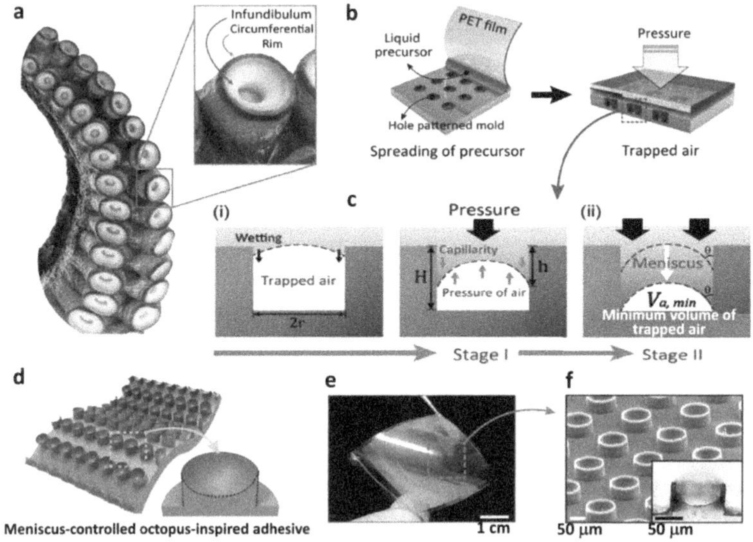

Figure 6.34 (a) Photographs of the structure of an octopus sucker (*Octopus vulgaris*) with infundibular and circumferential rims. (b) Illustration of the fabrication of bioinspired structures. (c) Illustration of the solid–liquid interfaces with trapped air. Under pressure action, the capillary meniscus tunes the volume of trapped air to control adhesion. (d) Illustration of the meniscus-controlled octopus-inspired adhesive. (e) Photograph of the bioinspired adhesive. (f) SEM image of a bioinspired octopus-like structure. The inset indicates the meniscus-controlled interfaces (Baik et al., 2018).

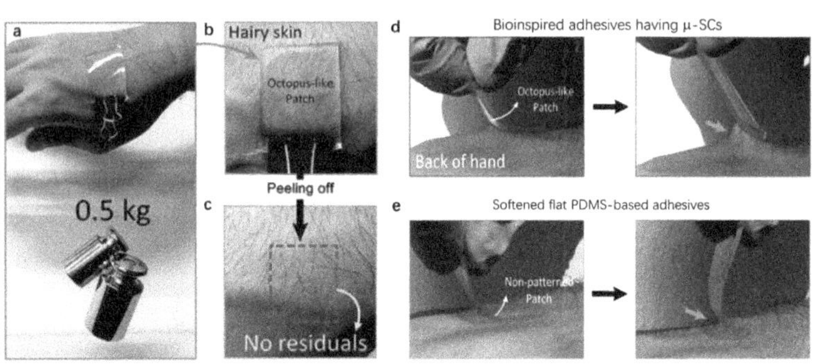

Figure 6.35 (a–c) Excellent bioinspired octopus-like patch with strong adhesions on skin. (d–e) Effects of bioinspired adhesives containing μ-SCs (d) and softened flat PDMS-based adhesives (e) (Baik et al., 2018).

Figure 6.35 shows the application of the bioinspired adhesive (2 × 2 cm^2) fully supporting a 0.5 kg weight on hairy skin (Fig. 6.35a–c) without any sticky residuals after detachment (Fig. 6.35c). Figure 6.35d,e shows photographs of an octopus-inspired adhesive on the hairy skin of a volunteer's hand, indicating the higher peel strength of the bioinspired octopus-like patch (i.e., bioinspired adhesives containing μ-SCs, Fig. 6.35d) in contrast to the nonpatterned patch (i.e., softened flat PDMS-based adhesives, Fig. 6.35e).

6.6.3.3 Magnet-controlled soft millirobot

By means of adhesion, developing untethered millirobots is of interest for emerging applications in various industrial and biomedical settings, robotics science, and engineering (Kim et al., 2013; Rus and Tolley, 2015; Miriyrv et al., 2017). Various soft micro- or millirobots have been proposed based on different actuation mechanisms. Legs and/or feet are commonly found in many living animals, including both land animals (ants, dogs, cheetahs, etc.) and ocean animals (starfish, octopi, etc.), after billions of years of evolution. These legs lift the animal's body from the ground on demand, leading to smaller body friction with the ground, higher degrees of freedom in locomotion, less energy cost, and enhanced obstacle-crossing ability. Legged animals usually demonstrate great adaptability to complex terrain and can probably access virtually 100% of earth's land surface.

Inspired by the behaviors of these creatures, a milliscale (height ~1 mm) soft robot decorated with tapered feet structures has been designed. Under the trigger of an external magnetic field, this robot can achieve combined discontinuous and continuous locomotion, and the motion of the soft legs is reminiscent of human walking (Lu et al., 2018). Figure 6.36a,b shows an optical image of the as-developed soft robot with a flexible tapered feet structure that exhibits superior adaptability, fast speed, heavy load-carrying ability, and excellent ability to move over obstacles. On the basis of the structural topographies of hundreds of legged animals, the length of their legs (L) is normally 1~2 times larger than that of their foot-to-foot spacing (S). The thickness of the body, foot length, and foot-to-foot spacing of the robot are designed as ~150 μm, ~650 μm, and ~600 μm, respectively. The tip angle of the tapered feet is ~15°, and Young's modulus of the as-fabricated soft body is measured to be ~2 MPa.

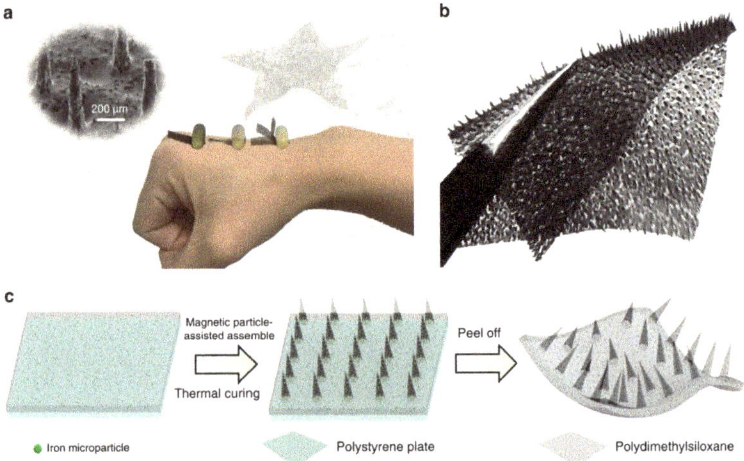

Figure 6.36 Multilegged millirobot (a) with flexible tapered feet (b) and its fabrication (c) (Lu et al., 2018).

An individual foot's performance in one gait cycle can be effectively controlled. Figure 6.36c illustrates the methods to fabricate a multilegged millirobot with flexible tapered feet using a modified magnetic-particle-assisted molding approach (Timonen et al., 2010). A mixture containing PDMS prepolymer and iron microparticles is deposited on a polystyrene plate via spin coating. Then, a magnet bar is placed underneath to induce the formation of tapered feet under the guidance of a magnetic field. After solidification by thermal curing for 1 h at 80°C, the as-fabricated film is peeled off the plate and cut into a rectangular shape.

Figure 6.37a–d shows optical images of an individual foot's deformation in one gait cycle, which can be divided into four typical stages: controlled dorsiflexion (CD), power plantar flexion (PPF), swing state (SS), and controlled plantar flexion (CPF). Figure 6.37e–g shows robot locomotion under harsh conditions. For locomotion on a wet surface with a liquid film, the robot can move with an average speed of 0.5 mm/s under a drive frequency of 1 Hz owing to its small contact area with the underlying substrate and its hydrophobic tapered feet (Fig. 6.37e). In addition, the robot can locomote with a load 100 times its own weight and move 8 mm in 45 s (Fig. 6.37f), and it can also cross a steep obstacle with a height ~10 times higher than its own leg (taking 113.1 s). The robot's posture can be easily adjusted to climb obstacles via the discontinuous flap-wave mode and continuous inverted-pendulum motion locomotion mode (Fig. 6.37g).

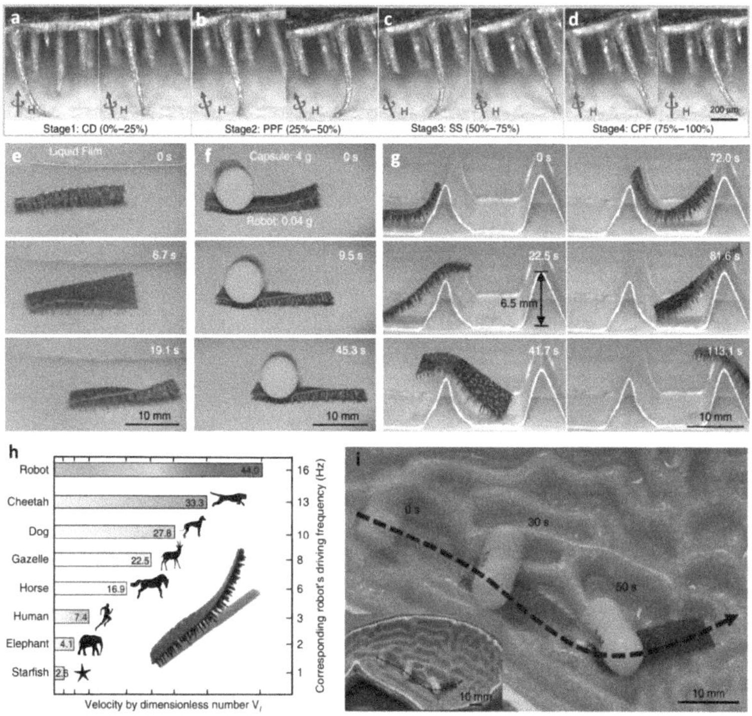

Figure 6.37 Robot locomotion with complex styles via the action of an external magnetic field (a) and optical images (e–h) and comparison of robot driving frequency (i) (Lu et al., 2018).

Figure 6.37h shows a comparison of the normalized speed between the soft robot and other animals. If a dimensionless number, V_l, i.e., the locomotion distance in one second relative to the lower limb (leg or foot) length, is defined to quantify the locomotion efficiency of various living organisms, the maximum velocities of humans and cheetahs are 10.4 m/s and 33.3 m/s, respectively, corresponding to V_l values of 7.4 and 33.3, respectively. In contrast, the soft robot can easily reach the maximum speed of humans and cheetahs at swing frequencies of 3 Hz and 13 Hz, respectively. Figure 6.37i shows the case of drug transport in a stomach model under wet conditions in 50 s. The complex internal structure of the stomach is 1.5–6.8 mm in depth and 2.4–6.2 mm in width (2–10 orders of magnitude larger than the leg length). The robot can move 32 mm in 50 s in this harsh in-vivo simulated environment while carrying a medical tablet (~91.4 mg) that is twice as heavy as itself. This development opens up promising applications.

Thinking Questions

1. Can you explain the significance of antiadhesion in applications?
2. What are the key points of designing materials and surfaces that can achieve antiadhesion properties?
3. Can you give examples of methods and techniques for the fabrication of materials and surfaces with antiadhesion?
4. Do you know some bioinspired materials and surfaces with antiadhesion? Give examples of their applications.
5. Which bioinspired materials and surfaces achieve adhesion performance?
6. Can you give examples of materials and surfaces with adhesion for applications?

References

Ahn B. K., S. Das, R. Linstadt, Y. Kaufman, N. R. Martinez-Rodriguez, R. Mirshafian, E. Kesselman, Y. Talmon, B. H. Lipshutz, J. N. Israelachvili, and J. H. Waite. High-performance mussel-inspired adhesives of reduced complexity. *Nat. Commun.* **6**, 8663 (2015).

Ahn Y., Y. Jang, N. Selvapalam, G. Yun, and K. Kim. Supramolecular Velcro for reversible underwater adhesion. *Angew. Chem.* **125**, 3222–3226 (2013).

Aksak B., K. Sahin, and M. Sitti. The optimal shape of elastomer mushroom-like fibers for high and robust adhesion. *Beilstein J. Nanotechnol.* **5**, 630 (2014).

Aldred N. and A. S. Clare. Mini-review: The adhesive strategies of cyprids and development of barnacle-resistant marine coatings. *Biofouling* **24**, 351–363 (2008).

Annabi N., Y. N. Zhang, A. Assmann, E. S. Sani, G. Cheng, A. D. Lassaletta, A. Vegh, B. Dehghani, G. U. Ruiz-Esparza, X. Wang, S. Gangadharan, A. S. Weiss, and A. Khademhosseini. Engineering a highly elastic human protein–based sealant for surgical applications. *Sci. Transl. Med.* **9**, eaai7466 (2017).

Aussillous P. and D. Quéré. Liquid marbles. *Nature* **411**, 924 (2001).

Autumn K., A. Dittmore, D. Santos, M. Spenko, and M. Cutkosky. Frictional adhesion: A new angle on gecko attachment. *J. Exp. Biol.* **209**, 3569–3579 (2006).

Autumn K. Gecko adhesion: Structure, function, and applications. *MRS Bull.* **32**, 473–478 (2007).

Autumn K., M. Sitti, Y. A. Liang, A. M. Peattie, W. R. Hansen, and S. Sponberg. Evidence for van der Waals adhesion in gecko setae. *Proc. Natl Acad. Sci. USA* **99**, 12252–12256 (2002).

Autumn K., Y. A. Liang, S. T. Hsieh, W. Zesch, W. P. Chan, T. W. Kenny, and R. Fearing. Adhesive force of a single gecko foot-hair. *Nature* **405**, 681–685 (2000).

Babu S. S., V. K. Praveen, and A. Ajayaghosh. Functional π-gelators and their applications. *Chem. Rev.* **114**, 1973 (2014).

Baik S., D. W. Kim, Y. Park, T. J. Lee, S. Ho Bhang, and C. Pang. A wet-tolerant adhesive patch inspired by protuberances in suction cups of octopi. *Nature* **546**, 396–400 (2017).

Baik S., J. Kim, H. J. Lee, T. H. Lee, and C. Pang. Highly adaptable and biocompatible octopus-like adhesive patches with meniscus-controlled unfoldable 3D microtips for underwater surface and hairy skin. *Adv. Sci.* **5**, 1800100 (2018).

Balkenendea D. W. R., S. M. Winklera, and P. B. Messersmith. Marine-inspired polymers in medical adhesion. *Eur. Polym. J.* **116**, 134–143 (2019).

Barthlott W. and C. Neinhuis. Purity of the sacred lotus, or escape from contamination in biological surfaces. *Planta* **202**, 1 (1997).

Bico J., C. Marzolin, and D. Quéré. Pearl drops. *Europhys. Lett.* **47**, 220 (1999).

Blossey R. Self-cleaning surfaces: Virtual realities. *Nat. Mater.* **2**, 301–306 (2003).

Boesel L. F., C. Greiner, E. Arzt, and C. A. Del. Gecko-inspired surfaces: A path to strong and reversible dry adhesives. *Adv. Mater.* **22**, 2125–2137 (2010).

Bohn H. F. and W. Federle. Insect aquaplaning: Nepenthes pitcher plants capture prey with the peristome, a fully wettable water-lubricated anisotropic surface. *Proc. Natl Acad. Sci.* **101**, 14138–14143 (2004).

Bormashenko E., R. Pogreb, G. Whyman, and M. Erlich. Resonance Cassie–Wenzel wetting transition for horizontally vibrated drops deposited on a rough surface. *Langmuir* **23**, 12217 (2007).

Brandon S., A. Wachs, and A. Marmur. Simulated contact angle hysteresis of a three-dimensional drop on a chemically heterogeneous surface: A numerical example. *J. Colloid Interface Sci.* **191**, 110 (1997).

Brennan M. J., B. F. Kilbride, J. J. Wilker, and J. C. Liu. A bioinspired elastin-based protein for a cytocompatible underwater adhesive. *Biomaterials* **124**, 116–125 (2017).

Callow J. A. and M. E. Callow. Trends in the development of environmentally friendly fouling-resistant marine coatings. *Nat. Commun.* **2**, 244 (2011).

Callow M. E., J. A. Callow, J. D. Pickett-Heaps, and R. Wetherbee. Primary adhesion of Enteromorpha (Chlorophyta, Ulvales) propagules: Quantitative settlement studies and video microscopy. *J. Phycol.* **33**, 938–947 (1997).

Cassie A. B. D. and S. Baxter. Wettability of porous surfaces. *Trans. Faraday Soc.* **40**, 546 (1944).

Chen Y. C. and H. Yang. Octopus-inspired assembly of nanosucker arrays for dry/wet adhesion. *ACS Nano* **11**, 5332–5338 (2017).

Choi M. K., O. K. Park, C. Choi, S. Qiao, R. Ghaffari, J. Kim, D. J. Lee, M. Kim, W. Hyun, and S. J. Kim. Cephalopod-inspired miniaturized suction cups for smart medical skin. *Adv. Healthcare Mater.* **5**, 80 (2016).

Ciriminna R., F. V. Bright, and M. Pagliaro. Ecofriendly antifouling marine coatings. *ACS Sustainable Chem. Eng.* **3**, 559 (2015).

Clanet C., C. Beguin, D. Richard, and D. Quere. Maximal deformation of an impacting drop. *J. Fluid Mech.* **517**, 199–208 (2004).

Crawford N., T. Endlein, and W. J. P. Barnes. Self-cleaning in tree frog toe pads: A mechanism for recovering from contamination without the need for grooming. *J. Exp. Biol.* **215**, 3965–3972 (2012).

Cui J., D. M. Drotlef, I. Larraza, J. P. Fernández-Blázquez, L. F. Boesel, C. Ohm, M. Mezger, R. Zentel, and A. del Campo. Bioinspired actuated adhesive patterns of liquid crystalline elastomers. *Adv. Mater.* **24**, 4601 (2012).

de Ruiter J., R. Lagraauw, D. van den Ende, and F. Mugele. Wettability-independent bouncing on flat surfaces mediated by thin air films. *Nat. Phys.* **11**, 48–53 (2014).

DeMartini D. G., J. M. Errico, S. Sjoestroem, A. Fenster, and J. H. Waite. A cohort of new adhesive proteins identified from transcriptomic analysis of mussel foot glands. *J. R. Soc. Interface* **14**, 20170151 (2017).

Dhindsa M. S., N. R. Smith, J. Heikenfeld, P. D. Rack, J. D. Fowlkes, M. J. Doktycz, A. V. Melechko, and M. L. Simpson. Reversible electrowetting of vertically aligned superhydrophobic carbon nanofibers. *Langmuir* **22**, 9030 (2006).

Dobretsov S., H.-U. Dahms, and P.-Y. Qian. Inhibition of biofouling by marine microorganisms and their metabolites. *Biofouling* **22**, 43–54 (2006).

Dobretsov S., M. Teplitski, and V. Paul. Mini-review: Quorum sensing in the marine environment and its relationship to biofouling. *Biofouling* **25**, 413–427 (2009).

Drotlef D. M., P. Blümler, and A. del Campo. Magnetically actuated patterns for bioinspired reversible adhesion (dry and wet). *Adv. Mater.* **26**, 775 (2014).

Drotlef D.-M., M. Amjadi, M. Yunusa, and M. Sitti. Bioinspired composite microfibers for skin adhesion and signal amplification of wearable sensors. *Adv. Mater.* **29**, 1701353 (2017).

ElSherbini A. and A. Jacobi. Critical contact angles for liquid drops on inclined surfaces. *J. Colloid Interface Sci.* **273**, 556 (2004).

Emerson S. B. and D. Diehl. Toe pad morphology and mechanisms of sticking in frogs. *Biol. J. Linn. Soc.* **13**, 199–216 (1980).

Endlein T., A. Ji, S. Yuan, I. Hill, H. Wang, W. J. P. Barnes, Z. Dai, and M. Sitti. The use of clamping grips and friction pads by tree frogs for climbing curved surfaces. *Proc. R. Soc. B* **284**, 20162867 (2017).

Eral H. and J. Oh. Contact angle hysteresis: A review of fundamentals and applications. *Colloid Polym. Sci.* **291**, 247 (2013).

Even M. A., J. Wang, and Z. Chen. Structural information of mussel adhesive protein Mefp-3 acquired at various polymer/Mefp-3 solution interfaces. *Langmuir* **24**, 5795–5801 (2008).

Fears K. P., B. Orihuela, D. Rittschof, and K. J. Wahl. Acorn barnacles secrete phase-separating fluid to clear surfaces ahead of cement deposition. *Adv. Sci.* **5**, 1700762 (2018).

Feng L., S. H. Li, Y. S. Li, H. J. Li, L. J. Zhang, J. Zhai, Y. L. Song, B. Q. Liu, L. Jiang, and D. B. Zhu. Super-hydrophobic surfaces: From natural to artificial. *Adv. Mater.* **14**, 1857–1860 (2002).

Feng L., Y. Zhang, J. Xi, Y. Zhu, N. Wang, F. Xia, and L. Jiang. Petal effect: A superhydrophobic state with high adhesive force. *Langmuir* **24**, 4114–4119 (2008).

Feng X. J. and L. Jiang. Design and creation of superwetting/antiwetting surfaces. *Adv. Mater.* **18**, 3063–3078 (2006).

Gao S., J. Sun, P. Liu, F. Zhang, W. Zhang, S. Yuan, J. Li, and J. Jin. A robust polyionized hydrogel with an unprecedented underwater anti-crude-oil-adhesion property. *Adv. Mater.* **28**, 5307–5314 (2016).

Gao X., X. Yan, X. Yao, L. Xu, K. Zhang, B. Yang, and L. Jiang. The dry-style antifogging properties of mosquito compound eyes and artificial analogues prepared by soft lithography. *Adv. Mater.* **19**, 2213 (2007).

Golovin K., S. P. Kobaku, D. H. Lee, E. T. DiLoreto, J. M. Mabry, and A. Tuteja. Designing durable icephobic surfaces. *Sci. Adv.* **2**, e1501496 (2016).

Guo Z., C. Gao, J. Li, Y. Liu, and Y. Zheng. Antiicing properties of bioinspired liquid-infused double-layer surface with internal wetting transport ability. *Adv. Mater. Interfaces* **6**, 1900244 (2019).

Hanna G. and W. J. P. Barnes. Adhesion and detachment of the toe pads of tree frogs. *J. Exp. Biol.* **155**, 103–125 (1991).

Hansen W. R. and K. Autumn. Evidence for self-cleaning in gecko setae. *Proc. Natl Acad. Sci. USA* **102**, 385–389 (2005).

Hawkes E. W., E. V. Eason, D. L. Christensen, and M. R. Cutkosky. Human climbing with efficiently scaled gecko-inspired dry adhesives. *J. R. Soc. Interface* **12**, 20140675 (2015).

Heinzmann C., C. Weder, and L. M. Espinosa. Supramolecular polymer adhesives: Advanced materials inspired by nature. *Chem. Soc. Rev.* **45**, 342–358 (2016).

Hennebert E., R. Wattiez, M. Demeuldre, P. Ladurner, D. S. Hwang, J. H. Waite, and P. Flammang. Sea star tenacity mediated by a protein that fragments, then aggregates. *Proc. Natl. Acad. Sci. USA* **111**, 6317 (2014).

Heuer A. H., D. J. Fink, V. J. Laraia, J. L. Arias, P. D. Calvert, K. Kendall, G. L. Messing, J. Blackwell, P. C. Rieke, D. H. Thompson, A. P. Wheeler, A. Veis, and A. I. Caplan. Stability of tetragonal ZrO_2 particles in ceramic matrices. *Science* **255**, 1098 (1992).

Hong X., X. F. Gao, and L. Jiang. Application of superhydrophobic surface with high adhesive force in no lost transport of superparamagnetic microdroplet. *J. Am. Chem. Soc.* **129**, 1478–1479 (2007).

Howarter J. A. and J. P. Youngblood. Self-cleaning and anti-fog surfaces via stimuli-responsive polymer brushes. *Adv. Mater.* **19**, 3838–3843 (2007).

Howell C., T. L. Vu, J. J. Lin, S. Kolle, N. Juthani, E. Watson, J. C. Weaver, J. Alvarenga, and J. Aizenberg. Self-replenishing vascularized fouling-release surfaces. *ACS Appl. Mater. Interfaces* **6**, 13299–13307 (2014).

Hu S., S. Lopez, P. H. Niewiarowski, and Z. Xia. Dynamic self-cleaning in gecko setae via digital hyperextension. *J. R. Soc. Interface* **9**, 2781–2790 (2012).

Huang S. and M. G. Hadfeld. Composition and density of bacterial biofilms determine larval settlement of the polycheate *Hydroides elegans. Mar. Ecol. Prog. Ser.* **260**, 161–172 (2003).

Huggett M. J., J. E. Williamson, R. de Nys, S. Kjelleberg, and P. D. Steinberg. Larval settlement of the common Australian sea urchin *Heliocidaris erythrogramma* in response to bacteria from the surface of coralline algae. *Oecologia* **149**, 604–619 (2006).

Jain A. and N. B. Bhosle. Biochemical composition of the marine conditioning film: Implications for bacterial adhesion. *Biofouling* **25**, 13–19 (2009).

Jin M. H., X. J. Feng, L. Feng, T. L. Sun, J. Zhai, T. J. Li, and L. Jiang. Superhydrophobic aligned polystyrene nanotube films with high adhesive force. *Adv. Mater.* **17**, 1977–1981 (2005).

Jin X., S. Yang, Z. Li, K. S. Liu, and L. Jiang. Bio-inspired special wetting surfaces via self-assembly. *Sci. China Chem.* **55**, 2327 (2012).

Ju J., K. Xiao, X. Yao, H. Bai, and L. Jiang. Bioinspired conical copper wire with gradient wettability for continuous and efficient fog collection. *Adv. Mater.* **25**, 5937–5942 (2013).

Jung S., M. K. Tiwari, and D. Poulikakos. Frost halos from supercooled water droplets. *Proc. Natl. Acad. Sci. USA.* **109**, 16073–16078 (2012).

Kelton K. F. and A. L. Greer. Test of classical nucleation theory in a condensed system. *Phys. Rev. B: Condens. Matter Mater. Phys.* **38**, 10089–10092 (1988).

Kim D. W., S. Baik, H. Min, S. Chun, H. J. Lee, K. H. Kim, J. Y. Lee, and C. Pang. Highly permeable skin patch with conductive hierarchical architectures inspired by amphibians and octopi for omnidirectionally enhanced wet adhesion. *Adv. Funct. Mater.* **29**, 201807614 (2019).

Kim S., C. Laschi, and B. Trimmer. Soft robotics: A bioinspired evolution in robotics. *Trends Biotechnol.* **31**, 287–294 (2013).

Kim T. I., H. E. Jeong, K. Y. Suh, and H. H. Lee. Stooped nanohairs: Geometry-controllable, unidirectional, reversible, and robust gecko-like dry adhesive. *Adv. Mater.* **21**, 2276–2281 (2009).

Kizilkan E., J. Strueben, A. Staubitz, and S. N. Gorb. Bioinspired photocontrollable microstructured transport device. *Sci. Robot.* **2**, eaak9454 (2017).

Krupenkin T., J. A. Taylor, P. Kolodner, and M. Hodes. Electrically tunable superhydrophobic nanostructured surfaces. *Bell Labs Tech. J.* **10**, 161 (2005).

Lafuma A. and D. Quéré. Superhydrophobic states. *Nat. Mater.* **2**, 457 (2003).

Lang N., M. J. Pereira, Y. Lee, I. Friehs, N. V. Vasilyev, E. N. Feins, K. Ablasser, E. D. O'Cearbhaill, C. Xu, A. Fabozzo, R. Padera, S. Wasserman, F. Freudenthal, L. S. Ferreira, R. Langer, J. M. Karp, and P. J. del Nido. A blood-resistant surgical glue for minimally invasive repair of vessels and heart defects. *Sci. Transl. Med.* **6**, 218ra6–218ra6 (2014).

Langowski J. K. A., D. Dodou, M. Kamperman, and J. L. van Leeuwen. Tree frog attachment: Mechanisms, challenges, and perspectives. *Front. Zool.* **2018**, 15–32 (2018).

Laulicht B., R. Langer, and J. M. Karp. Quick-release medical tape. *Proc. Natl. Acad. Sci. USA* **109**, 18803 (2012).

Lee B. P., P. B. Messersmith, J. N. Israelachvili, and J. H. Waite. Mussel-inspired adhesives and coatings. *Annu. Rev. Mater. Res.* **41**, 99–132 (2011).

Lee H., B. P. Lee, and P. B. Messersmith. A reversible wet/dry adhesive inspired by mussels and geckos. *Nature* **448**, 338–341 (2007a).

Lee H., N. F. Scherer, and P. B. Messersmith. Single-molecule mechanics of mussel adhesion. *Proc. Natl. Acad. Sci. USA* **103**, 12999–13003 (2006).

Lee Y. S., P. N. Pathirana, C. L. Steinfort, and T. Caelli. Monitoring and analysis of respiratory patterns using microwave Doppler radar. *IEEE J. Transl. Eng. Health Med.* **2**, 1 (2014).

Lee, H., S. M. Dellatore, W. M. Miller, and P. B. Messersmith. Mussel-inspired surface chemistry for multifunctional coatings. *Science* **318**, 426–430 (2007b).

Li H. J., X. B. Wang, Y. L. Song, Y. Q. Liu, Q. S. Li, L. Jiang, and D. B. Zhu. Super "amphiphobic" aligned carbon nanotube films. *Angew. Chem., Int. Ed.* **40**, 1743–1746 (2001).

Li J., A. D. Celiz, J. Yang, Q. Yang, I. Wamala, W. Whyte, B. R. Seo, N. V. Vasilyev, J. J. Vlassak, Z. Suo, and D. J. Mooney. Tough adhesives for diverse wet surfaces. *Science* **357**, 378–381 (2017).

Li Y., L. Li, and L. J. G. Sun. Bioinspired self-healing superhydrophobic coatings. *Angew. Chem. Int. Ed. Engl.* **49**, 6129–6133 (2010).

Lin Q., D. Gourdon, C. Sun, N. Holten-Andersen, T. H. Anderson, J. H. Waite, and J. N. Israelachvili. Adhesion mechanisms of the mussel foot proteins mfp-1 and mfp-3. *Proc. Natl. Acad. Sci. USA* **104**, 3782–3786 (2007).

Liu H., P. Zhang, M. Liu, S. Wang, and L. Jiang. Organogel-based thin films for self-cleaning on various surfaces. *Adv. Mater.* **25**, 4477 (2013).

Liu K., J. Du, J. Wu, and L. Jiang. Superhydrophobic gecko feet with high adhesive forces towards water and their bio-inspired materials. *Nanoscale* **4**, 768 (2012).

Liu M. J., S. T. Wang, Z. X. Wei, Y. L. Song, and L. Jiang. Bioinspired design of a superoleophobic and low adhesive water/solid interface. *Adv. Mater.* **21**, 665–669 (2009).

Liu P., H. Zhang, W. He, H. Li, J. Jiang, M. Liu, H. Sun, M. He, J. Cui, L. Jiang, and X. Yao. Development of "Liquid-like" copolymer nanocoatings for reactive oil-repellent surface. *ACS Nano* **11**, 2248 (2017).

Liu Y., L. Moevius, X. Xu, T. Qian, J. M. Yeomans, and Z. Wang. Pancake bouncing on superhydrophobic surfaces. *Nat. Phys.* **10**, 515–519 (2014).

Long C. J., J. F. Schumacher, P. A. C. Robinson 2nd, J. A. Finlay, M. E. Callow, J. A. Callow, and A. B. Brennan. A model that predicts the attachment behavior of *Ulva linza* zoospores on surface topography. *Biofouling* **26**, 411–419 (2010).

Lu H., M. Zhang, Y. Yang, Q. Huang, T. Fukuda, Z. Wang, and Y. Shen. A bioinspired multilegged soft millirobot that functions in both dry and wet conditions. *Nat. Commun.* **9**, 3944 (2018).

Lu Y., S. Sathasivam, J. Song, C. R. Crick, C. J. Carmalt, and I. P. Parkin. Robust self-cleaning surfaces that function when exposed to either air or oil. *Science* **347**, 1132 (2015).

Lv J., X. Yao, Y. Zheng, J. Wang, and L. Jiang. Antiadhesion organogel materials: From liquid to solid. *Adv. Mater.* **29**, 1703032 (2017).

MacCallum N., C. Howell, P. Kim, D. Sun, R. Friedlander, J. Ranisau, O. Ahanotu, J. J. Lin, A. Vena, B. Hatton, T.-S. Wong, and J. Aizenberg. Liquid-infused silicone as a biofouling-free medical material. *ACS Biomater. Sci. Eng.* **1**, 43 (2015).

Mahdavi A., L. Ferreira, C. Sundback, J. W. Nichol, E. P. Chan, D. J. Carter, C. J. Bettinger, S. Patanavanich, L. Chignozha, and E. Ben-Joseph. A biodegradable and biocompatible gecko-inspired tissue adhesive. *Proc. Natl. Acad. Sci. USA* **105**, 2307 (2008).

Maier G. P., M. V. Rapp, J. H. Waite, J. N. Israelachvili, and A. Butler. Adaptive synergy between catechol and lysine promotes wet adhesion by surface salt displacement. *Science* **349**, 628–632 (2015).

Marmur A. Lotus effect: Superhydrophobicity metastability. *Langmuir* **20**, 3517 (2004).

Marvi H., S. Song, and M. Sitti. Experimental investigation of optimal adhesion of mushroom-like elastomer microfibrillar adhesives. *Langmuir* **31**, 10119 (2015).

McHale C., M. I. Newton, and N. J. Shirtcliffe. Dynamic wetting and spreading and the role of topography. *J. Phys.: Condens. Matter* **21**, 464122 (2009).

Meng F., Q. Liu, X. Wang, D. Tan, L. Xue, and W. J. P. Barnes. Tree frog adhesion biomimetics: Opportunities for the development of new, smart adhesives that adhere under wet conditions. *Phil. Trans. R. Soc. A* **377**, 20190131 (2019).

Meuler A. J., G. H. McKinley, and R. E. Cohen. Exploiting topographical texture to impart icephobicity. *ACS Nano* **4**, 7048 (2010).

Michal B. T., E. J. Spencer, and S. J. Rowan. Stimuli-responsive reversible two-level adhesion from a structurally dynamic shape-memory polymer. *ACS Appl. Mater. Interfaces* **8**, 11041–11049 (2016).

Miljkovic N., R. Enright, Y. Nam, K. Lopez, N. Dou, J. Sack, and E. N. Wang. Jumping-droplet-enhanced condensation on scalable superhydrophobic nanostructured surfaces. *Nano Lett.* **13**, 179–187 (2013).

Miriyev A., K. Stack, and H. Lipson. Soft material for soft actuators. *Nat. Commun.* **8**, 596 (2017).

Molino P. J., E. Campbell, and R. Wetherbee. Development of the initial diatom microfouling layer on antifouling and fouling-release coatings in temperate and tropical Australia. *Biofouling* **25**, 685–694 (2009a).

Molino P. J., S. Childs, M. R. E. Hubbard, J. M. Carey, M. A. Burgman, and R. Wetherbee. Development of primary bacterial microfouling layer on antifouling and fouling release coatings in temperate and tropical environments in Eastern Australia. *Biofouling* **25**, 149–162 (2009b).

Nguyen P. V., N. V. Huynh, T. T. Phan, and V. A. Ho. Soft grasping with wet adhesion: Preliminary evaluation, in *2018 IEEE International Conference on Soft Robotics (Robosoft)*, Livorno, Italy, April 24–28, 2018, pp. 418–422.

North M., C. A. Grosso, and J. J. Wilker. High strength underwater bonding with polymer mimics of mussel adhesive proteins. *ACS Appl. Mater. Interfaces* **9**, 7866–7872 (2017).

Northen M. T., C. Greiner, E. Arzt, and K. L. Turner. A gecko-inspired reversible adhesive. *Adv. Mater.* **20**, 3905–3909 (2010).

Pang C., J. H. Koo, A. Nguyen, J. M. Caves, M. G. Kim, A. Chortos, K. Kim, P. J. Wang, J. B. H. Tok, and Z. Bao. Highly skin-conformal microhairy sensor for pulse signal amplification. *Adv. Mater.* **27**, 634 (2015).

Park Y., J. Shim, S. Jeong, G. R. Yi, H. Chae, J. W. Bae, S. O. Kim, and C. Pang. Microtopography-guided conductive patterns of liquid-driven graphene nanoplatelet networks for stretchable and skin-conformal sensor array. *Adv. Mater.* **29**, 1606453 (2017).

Patankar N. A. Transition between superhydrophobic states on rough surfaces. *Langmuir* **20**, 7097 (2004).

Peisker H., J. Michels, and S. N. Gorb. Evidence for a material gradient in the adhesive tarsal setae of the ladybird beetle *Coccinella septempunctata*. *Nat. Commun.* **4**, 1–7 (2013).

Prabha V., P. Buhutkar, M. Bhutkar, and G. Sivagami. A comparative study of arterial stiffness indices between smokers & non-smokers. *Int. J. Med. Res. Health Sci.* **7**, 207–212 (2014).

Quéré D., A. Lafuma, and J. Bico. Slippy and sticky microtextured solids. *Nanotechnology* **14**, 1109 (2003).

Ramirez J.-M., C. S. Ward, and J. L. Neul. Breathing challenges in Rett syndrome: Lessons learned from humans and animal models. *Respir. Physiol. Neurobiol.* **189**, 280 (2013).

Reddy S., E. Arzt, and A. Campo. Bioinspired surfaces with switchable adhesion. *Adv. Mater.* **19**, 3833–3837 (2007).

Roberts D., D. Rittschof, E. Holm, and A. R. Schmidt. Factors influencing initial larval settlement: Temporal, spatial and surface molecular components. *J. Exp. Marine Biol. Ecol.* **150**, 203–221 (1991).

Rose S., A. Prevoteau, P. Elzière, D. Hourdet, A. Marcellan, and L. Leibler. Nanoparticle solutions as adhesives for gels and biological tissues. *Nature* **505**, 382–385 (2014).

Rus D. and M. T. Tolley. Design, fabrication and control of soft robots. *Nature* **521**, 467 (2015).

Sagert J., C. Sun, and J. H. Waite. Chemical subtleties of mussel and polychaete holdfasts, in *Biological Adhesives* (A. M. Smith, J. A. Callow, eds), Springer, 2006, pp. 125–144.

Scardino A. J. and R. de Nys. Biomimetic models and bioinspired surfaces for fouling control. *Biofouling* **27**, 73–86. (2011).

Schultz M. P. Effects of coating roughness and biofouling on ship resistance and powering. *Biofouling* **23**, 331–341 (2007).

Schultz M. P., J. A. Bendick, E. R. Holm, and W. M. Hertel. Economic impact of biofouling on a naval surface ship. *Biofouling* **27**, 87–98 (2011).

Schumacher J. F., M. L. Carman, T. G. Estes, A. W. Feinberg, L. H. Wilson, M. E. Callow, J. A. Callow, J. A. Finlay, and A. B. Brennan. Engineered antifouling microtopographies-effect of feature size, geometry and roughness on settlement of zoospores of the green alga Ulva. *Biofouling* **23**, 55–62 (2007).

Schumacher J. F., C. J. Long, M. E. Callow, J. A. Finlay, J. A. Callow, and A. B. Brennan. Engineered nanoforce gradients for inhibition of settlement (attachment) of swimming algal spores. *Langmuir* **24**, 4931–4937 (2008).

Shafiq Z., J. Cui, L. Pastor-Pérez, V. S. Miguel, R. A. Gropeanu, C. Serrano, and A. del Campo. Bioinspired underwater bonding and debonding on demand. *Angew. Chem.* **124**, 4408–4411 (2012).

Shao H. and R. J. Stewart. Biomimetic underwater adhesives with environmentally triggered setting mechanisms. *Adv. Mater.* **22**, 729–733 (2010).

Shin M., S.-G. Park, B.-C. Oh, K. Kim, S. Jo, M. S. Lee, S. S. Oh, S.-H. Hong, E.-C. Shin, K.-S. Kim, S.-W. Kang, and H. Lee. Complete prevention of blood loss with self-sealing haemostatic needles. *Nat. Mater.* **16**, 147–152 (2017).

Smith A. M. and J. A. Callow. *Biological Adhesives*, Springer, 2006.

Song M., J. Ju, S. Luo, Y. Han, Z. Dong, Y. Wang, Z. Gu, L. Zhang, R. Hao, and L. Jiang. Controlling liquid splash on superhydrophobic surfaces by a vesicle surfactant. *Sci. Adv.* **3**, e1602188 (2017).

Stewart R. J., C. S. Wang, and H. Shao. Complex coacervates as a foundation for synthetic underwater adhesives. *Adv. Colloid Interface Sci.* **167**, 85–93 (2011).

Sun T. L., L. Feng, X. F. Gao, and L. Jiang. Bioinspired surfaces with special wettability. *Acc. Chem. Res.* **38**, 644–652 (2005).

Sun Z., T. Liao, K. Liu, L. Jiang, J. H. Kim, and S. X. Dou. Fly-eye inspired superhydrophobic anti-fogging inorganic nanostructures. *Small* **10**, 3001–3006 (2014).

Timonen J. V., C. Johans, K. Kontturi, A. Walther, O. Ikkala, and R. H. A. Ras. A facile template-free approach to magnetodriven, multifunctional artificial cilia. *ACS Appl. Mater. Interfaces* **2**, 2226–2230 (2010).

Tramacere F., E. Appel, B. Mazzolai, and S. N. Gorb. Hairy suckers: the surface microstructure and its possible functional significance in the Octopus vulgaris sucker. *Beilstein J. Nanotechnol.* **5**, 561 (2014).

Tramacere F., M. Follador, N. Pugno, and B. Mazzolai. Bioinspiration. *Biomimetics* **10**, 035004 (2015).

Tsipenyuk A. and M. Varenberg. Use of biomimetic hexagonal surface texture in friction against lubricated skin. *J. R. Soc. Interface* **11**, 20140113 (2014).

Urata C., G. J. Dunderdale, M. W. England, and A. Hozumi. Self-lubricating organogels (SLUGs) with exceptional syneresis-induced anti-sticking properties against viscous emulsions and ices. *J. Mater. Chem. A* **3**, 12626 (2015).

Vervaet B. A., A. Verhulst, M. E. De Broe, and P. C. D'Haese. The tubular epithelium in the initiation and course of intratubular nephrocalcinosis. *Urol. Res.* **38**, 249–256 (2010).

Vintiloiu A. and J.-C. Leroux. Organogels and their use in drug delivery: A review. *J. Controlled Release* **125**, 179 (2008).

Wahl M. Marine epibiosis 1. Fouling and antifouling-some basic aspects. *Marine Ecol. Prog. Ser.* **58**, 175–189 (1989).

Waite J.H. Mussel adhesion: Essential footwork. *J. Exp. Biol.* **220**, 517–530 (2017).

Waite J. H., N. H. Andersen, S. Jewhurst, and C. J. Sun. Mussel adhesion: Finding the tricks worth mimicking. *J. Adhes.* **81**, 297–317 (2005).

Wang R., K. Hashimoto, A. Fujishima, M. Chikuni, E. Kojima, A. Kitamura, M. Shimohigoshi, and T. Watanabe. Photogeneration of highly amphiphilic TiO_2 surfaces. *Adv. Mater.* **10**, 135–138 (1998).

Wang Y., H. Tian, J. Shao, D. Sameoto, X. Li, L. Wang, H. Hu, Y. Ding, and B. Lu. Switchable dry adhesion with step-like micropillars and controllable interfacial contact. *ACS Appl. Mater. Interfaces* **8**, 10029–10037 (2016).

Wang Z. Slanted functional gradient micropillars for optimal bioinspired dry adhesion. *ACS Nano* **12**, 1273–1284 (2018).

Watanabe T., S. Fukayama, M. Miyauchi, A. Fujishima, and K. Hashimoto. Photocatalytic activity and photo-induced wettability conversion of TiO_2 thin film prepared by sol-gel process on a soda-lime glass. *J. Sol-Gel Sci. Technol.* **19**, 71–76 (2000).

Wei W., J. Yu, C. Broomell, J. N. Israelachvili, and J. H. Waite. Hydrophobic enhancement of Dopa-mediated adhesion in a mussel foot protein. *J. Am. Chem. Soc.* **135**, 377–383 (2013).

Wenzel R. N. Resistance of solid surfaces to wetting by water. *Ind. Eng. Chem.* **28**, 988 (1936).

Winkleman A., G. Gotesman, A. Yoffe, and R. Naaman. Immobilizing a drop of water: Fabricating highly hydrophobic surfaces that pin water droplets. *Nano Lett.* **8**, 1241–1245 (2008).

Wisdom K. M., J. A. Watson, X. Qu, F. Liu, G. S. Watson, and C. Chen. Self-cleaning of superhydrophobic surfaces by self-propelled jumping condensate. *PNAS*, **110**, 7992–7997 (2013).

Wong T.-S., S. H. Kang, S. K. Y. Tang, E. J. Smythe, B. D. Hatton, A. Grinthal, and J. Aizenberg. Bioinspired self-repairing slippery surfaces with pressure-stable omniphobicity. *Nature*, **477**, 443–447 (2011).

Wu W., A. DeConinck, and J. A. Lewis. Omnidirectional printing of 3D microvascular networks. *Adv. Mater.* **23**, H178–H183 (2011).

Xi J. M., L. Feng, and L. Jiang. A general approach for fabrication of superhydrophobic and superamphiphobic surfaces. *Appl. Phys. Lett.* **92**, 053102 (2008).

Xia F. and L. Jiang. Bio-inspired, smart, multiscale interfacial materials. *Adv. Mater.* **20**, 2842–2858 (2008).

Yang S. Y., E. D. O'Cearbhaill, G. C. Sisk, K. M. Park, W. K. Cho, M. Villiger, B. E. Bouma, B. Pomahac, and J. M. Karp. A bio-inspired swellable microneedle adhesive for mechanical interlocking with tissue. *Nat. Commun.* **4**, 1702 (2013).

Yao X., S. S. Dunn, P. Kim, M. Duffy, J. Alvarenga, and J. Aizenberg. Fluorogel elastomers with tunable transparency, elasticity, shape-memory, and antifouling properties. *Angew. Chem. Int. Ed.* **53**, 4418 (2014).

Yao X., S. Wu, L. Chen, J. Ju, Z. Gu, M. Liu, J. Wang, and L. Jiang. Self-replenishable anti-waxing organogel materials. *Angew. Chem.* **127**, 9103 (2015).

Ye Z., G. Z. Lum, S. Song, S. Rich, and M. Sitti. Phase change of gallium enables highly reversible and switchable adhesion. *Adv. Mater.* **28**, 5088–5092 (2016).

Zeng H., D. S. Hwang, J. N. Israelachvili, and J. H. Waite. Strong reversible Fe^{3+}-mediated bridging between dopa-containing protein films in water. *Proc. Natl Acad. Sci. USA* **107**, 12850–12853 (2010).

Zhang M., L. Wang, S. Feng, and Y. Zheng. A strategy of antifogging: Air-trapped hollow microsphere nanocomposites. *Chem. Mater.* **29**, 2899–2905 (2017a).

Zhang S., J. Huang, Y. Tang, S. Li, M. Ge, Z. Chen, K. Zhang, and Y. Lai. Understanding the role of dynamic wettability for condensate microdrop self-propelling based on designed superhydrophobic TiO_2 nanostructures. *Small* **13**, 1600687 (2017b).

Zhang T., Y. Wang, F. Zhang, X. Chen, G. Hu, J. Meng, and S. Wang. Bio-inspired superhydrophilic coatings with high anti-adhesion against mineral scales. *NPG Asia Materials* **10**, e471 (2018).

Zhao N., Q. D. Xie, X. Kuang, S. Q. Wang, Y. F. Li, X. Y. Lu, S. X. Tan, J. Shen, X. L. Zhang, Y. Zhang, J. Xu, and C. C. Han. A novel ultra-hydrophobic surface: Statically non-wetting but dynamically non-sliding. *Adv. Funct. Mater.* **17**, 2739–2745 (2007).

Zhao Q., D. W. Lee, B. K. Ahn, S. Seo, Y. Kaufman, J. N. Israelachvili, and J. H. Waite. Underwater contact adhesion and microarchitecture in polyelectrolyte complexes actuated by solvent exchange. *Nat. Mater.* **15**, 407–412 (2016).

Zhao Y., Y. Wu, L. Wang, M. Zhang, X. Chen, M. Liu, J. Fan, J. Liu, F. Zhou, and Z. Wang. Bio-inspired reversible underwater adhesive. *Nature Commun.* **8**, 2218 (2017).

Zheng Y., D. Han, J. Zhai, and L. Jiang. In situ investigation on dynamic suspending of microdroplet on lotus leaf and gradient of wettable micro- and nanostructure from water condensation. *Appl. Phys. Lett.* **92**, 084106 (2008).

Zhong C., T. Gurry, A. A. Cheng, J. Downey, Z. Deng, C. M. Stultz, and T. K. Lu. Strong underwater adhesives made by self-assembling multiprotein nanofibres. *Nat. Nanotechnol.* **9**, 858–866 (2014).

Zhu L., J. Xue, Y. Wang, Q. Chen, J. Ding, and Q. Wang. Ice-phobic coatings based on silicon-oil-infused polydimethylsiloxane. *ACS Appl. Mater. Interfaces* **5**, 4053–4062 (2013).

Chapter 7

Outlook

In this chapter, advances in bioinspired material surfaces are introduced, including the significance, development, and multidisciplinary applications of bioinspired material surfaces. In addition, increasing research trends are introduced, such as worldwide attention, promising future applications, and an increase in bioinspired research.

7.1 Advances in Bioinspired Material Surfaces

7.1.1 Significance

Since natural creatures, such as animals and plants, have evolved to exhibit surfaces and materials with smart properties, including self-cleaning, self-repair, energy conservation, drag reduction, wet/dry adhesion, and so on, these functional surfaces and materials have inspired scientists to create products, for example, artificial fibers similar to the thread of spider silk, fishing nets originating from cobwebs, and self-cleaning coatings based on the effect of lotus leaves.

Based on the understanding of biological features and functions, the concepts of "bioinspired," "biomimetic," or "bionics" have

Bioinspired Materials Surfaces: Nature-Inspired Micro-/Nanostructures
Yongmei Zheng

ISBN 978-981-5129-34-2 (Hardcover), 978-1-003-51388-9 (eBook)
www.jennystanford.com

subsequently appeared (Vincent et al., 2006; Esomba, 2012), revealing hot research topics in some fields. In particular, the results of studies may even go beyond these biological features, although the concepts are rooted in biology. Such bioinspired studies can supplement the scientific research process to design novel engineering systems and modern devices.

As is well known, the mysteries of biological systems in the natural world are comprehensive, rich and colorful, and inexhaustible, offering knowledge from biological molecules, biological micro-/nanoscales, biological responsive processes, etc. Bioinspired materials are novel smart materials that can be created by scientists and researchers after learning from biological features in biological systems, including surface physical structures, material chemistry components, geometry and shapes at the micro- and nanolevel, environment-responsive properties, biological processing, biological mechanical properties, etc. Scientists and researchers can design artificial materials and achieve creature-like properties of materials by learning biological principles. It is important to understand biological phenomena, reveal the physical and chemical mechanisms, and establish the relationships between material properties and functional properties, such as mechanically adaptive nanocomposites inspired by sea cucumbers, structural colors of butterfly wings, water-harvesting structures similar to the surface of beetle backs, spider silk and cactus spines, self-cleaning coatings inspired by lotus leaves, and so on.

7.1.2 Development

With the increasing demand for science and technology in applications, bioinspired materials have been rapidly developed. Novel ideas can be advanced by understanding the mysteries of natural creatures after evolving over billions of years. To date, bioinspired materials have been developed at the nanolevel, microlevel, and macrolevel, which has led to the development of innovative and intelligent materials that can be used in devices, engineering, industry, etc.

A classic example is the development of dirt- and water-repellent paint (coating) based on the lotus effect; the surface of lotus leaves has water-repellent, antistick properties, which can be applied to buildings, windows, devices, etc. Examples of bionics in engineering include the hulls of boats imitating the thick skin of dolphins; sonar, radar, and medical ultrasound imaging imitating animal echolocation; artificial neurons; artificial neural networks; and so on (Vincent et al., 2006). By learning how the wing structure of the blue morpho butterfly reflects light, a tag that can read through water and on metal (Clark et al., 1999) has been created, as well as new nanosensors to detect explosives (Odum, 2003). By mimicking ecological functions, technoecosystems involve the coupling of ecological processes to technological processes, resulting in the creation of a self-regulating hybrid system. The structure and energy dynamics of ecosystems are analogous to the energy flow between components of an electrical circuit (Videler, 2011). Inspired by piers and boat hulls, medical adhesives involving glue and tiny nanohairs are being developed, the mechanism of which allows mussels to stick to rocks and enables bioadhesive gels for blood vessels (Videler et al., 2004; Kastrup et al., 2012). In addition, new aircraft designs need far greater agility. By learning about biological systems and understanding the relationships between the multiscale structure and macroscopic functional properties, scientists and researchers are using novel ways to design bioinspired materials based on different aspects (Xia and Jiang, 2008; Jiang, 2014) from the synthesis of suitable target molecules to the construction of architectural structures by selecting organic and inorganic components and assembling at nano- and microscales via methods and technologies, including chemical vapor deposition (Liu et al., 2013), self-assembly (Yao et al., 2010; Li et al., 2012), lithography (Pokroy et al., 2009; Zhao et al., 2012), etc.

7.1.3 Multidisciplinary Approaches

Scientists know well that biological functional materials have robust structural properties; for instance, nacre and bone are strong, rough biological materials due to their composite multiscale architecture comprising ceramic platelets aligned with a small fraction of polymers (Ortiz and Boyce, 2008); lotus leaves possess micro- and nanoscale hierarchical structures on the surfaces, consisting of micropapillae covered with waxy crystalline nanohairs, which generate unique self-cleaning properties (Furstner et al., 2005; Bormashenko et al., 2006). The glaringly beautiful wings of morpho butterflies are composed of periodic hierarchical multilayer microstructures that create structural colors, representing the origin of photonic crystals (Parker and Townley, 2007). These multilevel and multiform structures in biological material surfaces provide endless inspiration to scientists, who seek new ways to design novel materials. The cross-discipline efforts related to the design of materials can be promoted in developing many applications. Based on the principles from biological systems, including processes, organizational design, and strategic thinking, biomimetics has been extended to a range of industries in defense, packaging, and business services (Bhushan, 2009) for formulating business strategy.

Since the discovery of unique structures, properties, theories, and performance in biological systems, scientists have been looking for novel design ideas as well as new theories and methods to solve practical problems. By learning from biological systems, such as bioinorganic materials (biomineralization); bioinspired, multiscale structured materials (chiral morphologies); bionanomaterials (bionanoparticles); hybrid organic/inorganic implant materials (bone-like composites); and smart biomaterials, inspired research has begun on many fronts in conjunction with multidisciplinary research, including life science, matter science, physics and chemistry, materials, information science, engineering and so on, thus serving human beings well.

7.2 Increasing Research Trends

7.2.1 Worldwide Attention

In the late 21st century, long-term foundational research on bionics began worldwide. The United States prioritized the development of diverse fields, including advanced manufacturing, materials, and military equipment. Germany put significant attention on and investment into nanotechnology, electronic technology, fullerene and carbon materials, photonics, bioinspired materials, biological sensors, and so on. In the United Kingdom, a national strategy for science was developed by the House of Commons Science and Technology Committee after the impact of the 1993 White Paper,

"Realising our potential: a strategy for science, engineering and technology." Other countries (e.g., Japan, Russia, and Korea) have established long-term plans to conduct foundational research on advanced manufacturing, materials, biological technology, high-performance computers, and communication. In 2003, a science conference in Xiangshan, China, focused on the theme of bionic science, including the following topics: bioinspired structures and forces; bioinspired materials and microsystems; bioinspired functional devices and controls; bioinspired molecules, artificial intelligence, and cognition; and bionic research and application. The International Society of Bionic Engineering (ISBE), an educational, nonprofit, nonpolitical organization, was formed in 2010 in China to foster the exchange of information on bionic engineering research, development, and application. The society was launched by 15 countries, including China, the United Kingdom, Germany, the United States, Russia, Canada, Japan, South Korea, India, Thailand, Austria, Romania, the Netherlands, Belgium, and Brazil. Events and meetings have been held every year and all over the world, including the Biomimetic Workshop 2013, 2014, International Workshop on Bionic Engineering (IWBE2015–2018), Nature-Inspired Engineering Conference (2019), International Conference on Nature Inspired Surface Engineering (NISE)-2019, and so on.

7.2.2 Promising Future for Applications

Benefiting from nanoscience and nanotechnology, human beings can see a beautiful and interesting microcosmic world of biological material surfaces. For instance, natural surfaces of biological materials, such as lotus leaves (Zheng et al., 2008), cactus spines (Ju et al., 2012), spider silk (Zheng et al., 2010), pitcher plants (Chen et al., 2016), water strider legs (Gao and Jiang, 2014), butterfly wings (Zheng et al., 2007), and beetle backs (Parker and Lawrence, 2001), are endowed with various micro- and nanostructures in conjunction with smart biological functions in nature. Some ideas for designing novel materials are rooted in natural biology in the constituents of biological systems, such as bioinorganic materials (biomineralization), multiscale structured materials (chiral morphologies), bionanomaterials (bionanoparticles), hybrid organic–inorganic implant materials (bone-like composites), and smart biomaterials. In particular, many biological materials can dynamically alter the physicochemical properties to respond to changes in environmental conditions. Thus, the surfaces can be triggered to control the interfacial properties at the solid–water interface, such as ion channels, directional fluid motions, and controllable adhesion for biological functions, for instance, the self-cleaning effect of lotus leaves, the anisotropic dewetting of butterfly wings or rice leaves, the superhydrophobic forces of water strider legs, the controllable attachment of geckos, the fog-harvesting abilities of spider silk and cactus spines and beetle backs, etc. The understanding of biological systems enables scientists and researchers to advance novel thoughts, developing a scientific way to fabricate the materials and surfaces for real-world applications, including biomimetic fins, actively moving polymers, neural memory devices, smart micro-/nanocontainers for drug delivery, various biosensors, dual-/multi-responsive materials, self-cleaning buildings, windows, and clothes, and water harvesting from industry, forest, seaside, and so on. Figure 7.1 illustrates some typical examples and applications (Sun et al., 2005; Feng and Jiang, 2006; Liu et al., 2010a; Su et al., 2016).

Studies on bioinspired materials have been extended to a wide range of research fields due to the design of artificial materials with various scopes, such as molecular/supramolecular, functional, or structural materials, including responsive materials with changes in properties in response to an external stimulus (light, electronic, magnetic, thermal, etc.) or hierarchical structured materials at micro- and nanolevels. These bioinspired materials can have promising applications, including novel adhesives and paints, buildings, consumer products, bioinspired robots, microdevices, electrons, etc., which involve multi- and interdisciplinary fields such as chemistry, engineering, physics, materials science, biology, and medicine.

To achieve bioinspired materials, multidisciplinary approaches are necessary, based on multilevel structural models in biological systems, material design and fabrication via nanotechnology, physics, and chemistry. Scientists have opened the possibility to design bioinspired materials by understanding biological surfaces. For example, a fluid-controlling surface can be designed to mimic the micro- and nanostructures on the superhydrophobic wings of the butterfly *Morpho aega* with directional adhesion, i.e., a droplet easily rolls off the surface of wings along one direction but is pinned tightly against rolling in the opposite direction. Through coordinating stimuli-responsive materials and appropriate surface-geometry structures, materials can be developed by micromanufacturing techniques to realize a microratchet similar to the microscales of a butterfly wing, and nanoparticles can be synthesized as coatings; thus, the surface can act with reversible transitions between a low-adhesive rolling state and a high-adhesive pinning state for water droplets, controlled by temperature as well as magnetic and electric fields.

Significance of bioinspired materials for promising applications

biological surfaces	functions	promising applications
Lotus leaves	Self-cleaning	Building, texture, window
Rose leaves	Adhesion	Droplet carrier as machine hand
Butterfly wings	Structural color	Photonic crystal and photo devices
	Water repellency	Droplet transport or low adhesion
Composite eyes	Antifogging	Windshields of vehicles
Water strider legs	Superhydrophobic force	Movers on water
Spider silks Beetle backs	Fog harvesting	Water collection system and devices
Seashells, bones, teeth	Toughness, High mechanical strength	Aerospace materials or biomaterials
Moth eyes	Antireflection	Photo devices
Shark skin	Antidrag	Shipping, boats, watercraft, vessels
Gecko setae	High adhesion force	Climber on walls, or high buildings
Pitcher plants	Directional liquid transport	Microfluidics
Goose feathers	Flexible fiber	Droplet transport

Figure 7.1 List of bioinspired examples for human applications.

Bioinspired materials can be realized for many additional functions, such as superhydrophobic (Feng et al., 2002) and superamphiphobic (Li et al., 2001; Xi et al., 2008) surfaces in air and superoleophobic interfaces in water (Liu et al., 2009), which can be achieved by nanotechnology, electrochemistry, electrospinning, and so on (Blossey, 2003; Sun et al., 2005; Xia et al., 2008). In particular, superhydrophobic surfaces with high liquid–solid adhesion (contact angle greater than 150°) can have potential applications, such as liquid transport, biochemical separation, in situ detection, localized chemical reactions (Jin et al., 2005; Zhao et al., 2007; Cho and Choi, 2008; Winkleman et al., 2008), a "mechanical hand" for reversible no-loss transport of microdroplets (Feng et al., 2008), dynamically tuning liquid–solid adhesion (Song et al., 2005; Boduroglu et al., 2007; Hong et al., 2007; Balu et al., 2008; Lai et al., 2008; Huang et al., 2009), and stimuli-responsive switchable wetting (Zhao et al., 2006; Cheng et al., 2008; Li et al., 2009). Multidisciplinary approaches are necessary to extend the application scope of bioinspired materials (Liu et al., 2010b; Kreder et al., 2016; Su et al., 2016; Zhang et al., 2017), and typical examples are illustrated in Fig. 7.2.

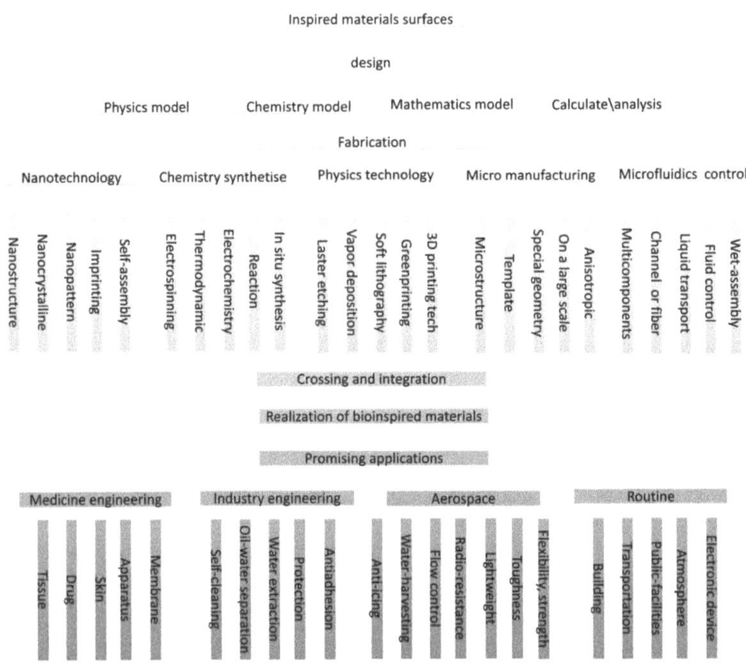

Figure 7.2 Typical examples of multidisciplinary bioinspired materials and surfaces.

7.2.3 Increase in Bioinspired Research

Many studies have demonstrated that bioinspired materials and surfaces are significant in the development of novel materials, functions and devices, and promising future applications. With the development of science and technology, research on bioinspired materials has become a hot topic in many fields, exhibiting increasing trends (Fig. 7.3) and an increasing citation frequency of publications in recent years (Fig. 7.4).

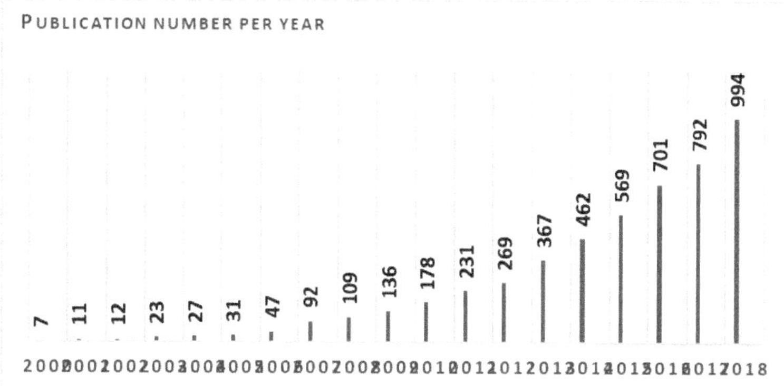

Figure 7.3 Publications with the keyword "bioinspired" have increased in recent years (from Web of Science, July 14, 2019).

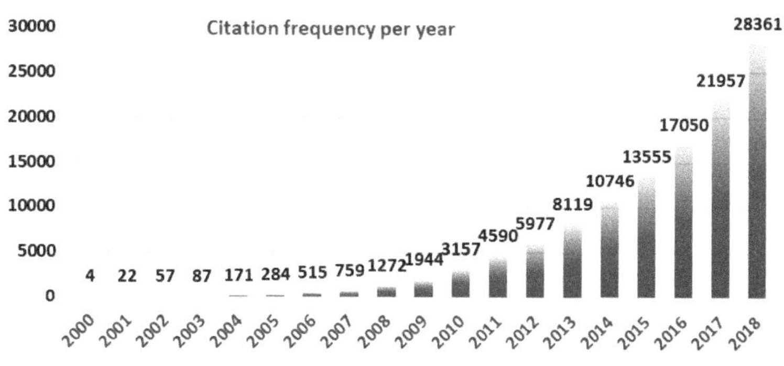

Figure 7.4 Publication citation frequency per year (from Web of Science, July 14, 2019).

Based on the keywords of "bioinspired or wettability," there is frequently an increase in the number of publications in high-level journals, such as *Nature*, *Science*, and *Nature Materials* (Fig. 7.5), and there are increasing citation frequencies of these publications

(Fig. 7.6), which indicates hot research areas in terms of bioinspired themes and subjects (Fig. 7.7–7.9).

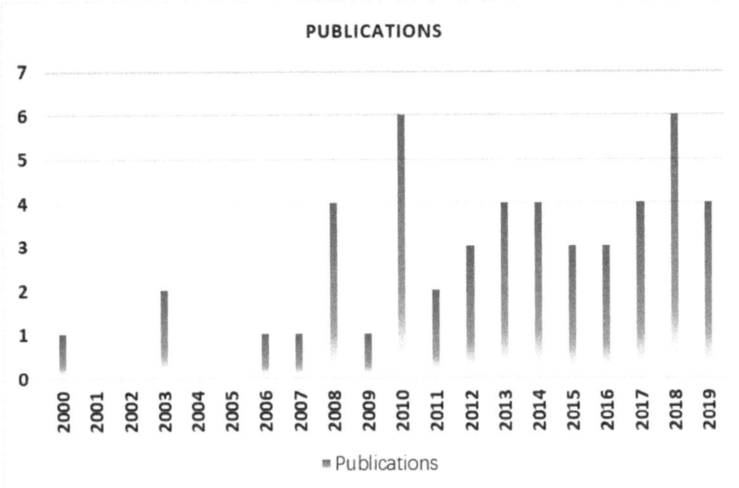

Figure 7.5 The numbers of publications per year in *Nature*, *Science*, and *Nature Materials* (from the Web of Science, November 11, 2019).

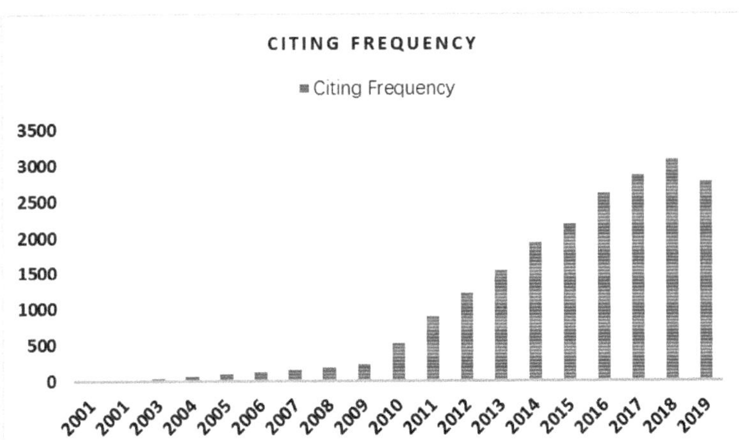

Figure 7.6 Increase in the citing frequency of publications in *Nature*, *Science*, and *Nature Materials* per year (from the Web of Science, November 11, 2019).

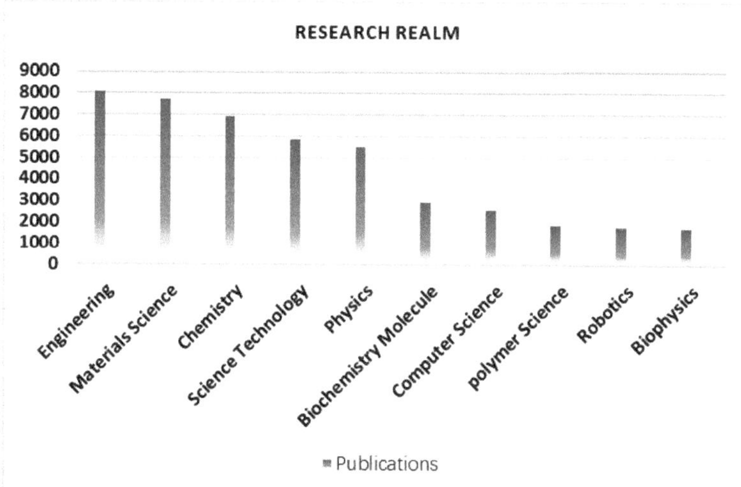

Figure 7.7 Publications in research realms on "bioinspired" during 2014–22 (from the Web of Science, July 19, 2023).

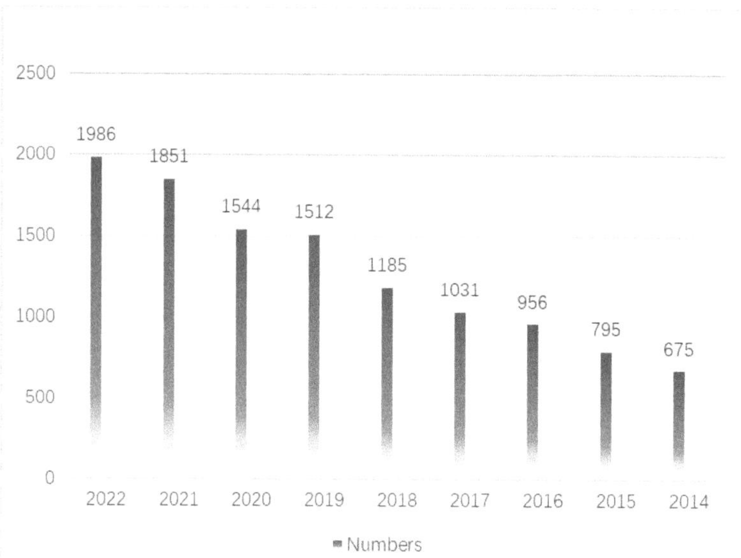

Figure 7.8 Publication numbers related to the keyword "bioinspired" (from the Web of Science, July 19, 2023).

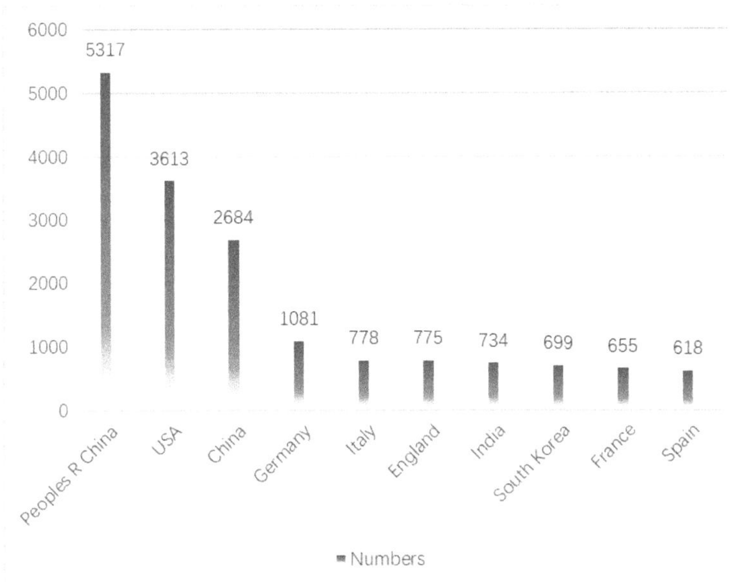

Figure 7.9 Publications related to the keyword "bioinspired" in different states or countries (from the Web of Science, July 19, 2023).

References

Balu B., V. Breedveld, and D. W. Hess. Fabrication of "roll-off" and "sticky" superhydrophobic cellulose surfaces via plasma processing. *Langmuir* **24**, 4785–4790 (2008).

Bhushan B. Biomimetics: Lessons from nature—an overview. *Philos. Trans. R. Soc.* A **367**, 1445 (2009).

Blossey R. Self-cleaning surfaces: Virtual realities. *Nat. Mater.* **2**, 301–306 (2003).

Boduroglu S., M. Cetinkaya, W. J. Dressick, A. Singh, and M. C. Demirel. Controlling the wettability and adhesion of nanostructured poly-(p-xylylene) films. *Langmuir* **23**, 11391–11395 (2007).

Bormashenko E., T. Stein, G. Whyman, Y. Bormashenko, and R. Pogreb. Wetting properties of the multiscaled nanostructured polymer and metallic superhydrophobic surfaces. *Langmuir* **22**, 9982 (2006).

Chen H., P. Zhang, L. Zhang, H. Liu, Y. Jiang, D. Zhang, Z. Han, and L. Jiang. Continuous directional water transport on the peristome surface of *Nepenthes alata*. *Nature* **532**, 85–89 (2016).

Cheng Z. J., L. Feng, and L. Jiang. Tunable adhesive superhydrophobic surfaces for superparamagnetic microdroplets. *Adv. Funct. Mater.* **18**, 3219–3225 (2008).

Cho W. K. and I. S. Choi. Fabrication of hairy polymeric films inspired by geckos: Wetting and high adhesion properties. *Adv. Funct. Mater.* **18**, 1089–1096 (2008).

Clark O. G., R. Kok, and R. Lacroix. Mind and autonomy in engineered biosystems. *Engineering Applications of Artificial Intelligence* **12**, 389–399 (1999).

Esomba S. *Twenty-First Century's Fuel Sufficiency Roadmap*, Published 2012 "bionics." Online Etymology Dictionary.

Feng L., S. H. Li, Y. S. Li, H. J. Li, L. J. Zhang, J. Zhai, Y. L. Song, B. Q. Liu, L. Jiang, and D. B. Zhu. Super-hydrophobic surfaces: from natural to artificial. *Adv. Mater.* **14**, 1857–1860 (2002).

Feng L., Y. A. Zhang, J. M. Xi, Y. Zhu, N. Wang, F. Xia, and L. Jiang. Petal effect: A superhydrophobic state with high adhesive force. *Langmuir* **24**, 4114–4119 (2008).

Feng X. J. and L. Jiang. Design and creation of superwetting/antiwetting surfaces. *Adv. Mater.* **18**, 3063–3078 (2006).

Furstner R., W. Barthlott, C. Neinhuis, and P. Walzel. Wetting and self-cleaning properties of artificial superhydrophobic surfaces. *Langmuir* **21**, 956 (2005).

Gao X. and L. Jiang. Water-repellent legs of water striders. *Nature* **432**, 36 (2014).

Hong X., X. F. Gao, and L. Jiang. Application of superhydrophobic surface with high adhesive force in no lost transport of superparamagnetic microdroplet. *J. Am. Chem. Soc.* **129**, 1478–1479 (2007).

Huang X., D. H. Kim, M. Im, J. H. Lee, J. B. Yoon, and Y. K. Choi. "Lock-and-key" geometry effect of patterned surfaces: Wettability and switching of adhesive force. *Small* **5**, 90–94 (2009).

Jiang L. Bio-inspired, smart, multiscale interfacial materials. *J. Biol. Inorg. Chem.* **19**, S153 (2014).

Jin M. H., X. J. Feng, L. Feng, T. L. Sun, J. Zhai, T. J. Li, and L. Jiang. Superhydrophobic aligned polystyrene nanotube films with high adhesive force. *Adv. Mater.* **17**, 1977–1981 (2005).

Ju J., H. Bai, Y. Zheng, T. Zhao, R. Fang, and L. Jiang. Multi-structure and multi-function integrated fog collection system of cactus. *Nat. Commun.* **3**, 1247 (2012).

Kastrup C. J., M. Nahrendorf, J. L. Figueiredo, H. Lee, S. Kambhampati, T. Lee, S.-W. Cho, R. Gorbatov, Y. Iwamoto, T. T. Dang, P. Dutta, J. H. Yeon, H. Cheng, C. D. Pritchard, A. J. Vegas, C. D. Siegel, S. MacDougall, M. Okonkwo, A. Thai, J. R. Stone, A. J. Coury, R. Weissleder, R. Langer, and D. G. Anderson. Painting blood vessels and atherosclerotic plaques with an adhesive drug depot. *PNAS* **109**(52), 21444–21449 (2012).

Kreder M. J., J. Alvarenga, P. Kim, and J. Aizenberg. Design of anti-icing surfaces: Smooth, textured or slippery? *Nature Rev. Mater.* **1**, 15003 (2016).

Lai Y. K., C. J. Lin, J. Y. Huang, H. F. Zhuang, L. Sun, and T. Nguyen. Markedly controllable adhesion of superhydrophobic spongelike nanostructure TiO_2 films. *Langmuir* **24**, 3867–3873 (2008).

Li C., R. W. Guo, X. Jiang, S. X. Hu, L. Li, X. Y. Cao, H. Yang, Y. L. Song, Y. M. Ma, and L. Jiang. Reversible switch of water droplet mobility on superhydrophobic surface based on phase transition of SCLPC. *Adv. Mater.* **21**, 4254–4258 (2009).

Li H. J., X. B. Wang, Y. L. Song, Y. Q. Liu, Q. S. Li, L. Jiang, and D. B. Zhu. Super-"amphiphobic" aligned carbon nanotube films. *Angew. Chem. Int. Ed.* **40**, 1743–1746 (2001).

Li Y. Q., T. Yu, T. Y. Yang, L. X. Zheng, and K. Liao. Bio-inspired nacre-like composite films based on graphene with superior mechanical, electrical, and biocompatible properties. *Adv. Mater.* **24**, 3426 (2012).

Liu K. S., Y. Tian, and L. Jiang. Bio-inspired superoleophobic and smart materials: Design, fabrication, and application. *Prog. Mater. Sci.* **58**, 503 (2013).

Liu K., X. Yao, and L. Jiang. Recent developments in bio-inspired special wettability. *Chem. Soc. Rev.* **39**, 3240–3255 (2010a).

Liu M. J., S. T. Wang, Z. X. Wei, Y. L. Song, and L. Jiang. Bioinspired design of a superoleophobic and low adhesive water/solid interface. *Adv. Mater.* **21**, 665–669 (2009).

Liu M., Y. Zheng, J. Zhai, and L. Jiang. Bioinspired super-antiwetting interfaces with special liquid-solid adhesion. *Acc. Chem. Res.* **43**, 368–377 (2010b).

Odum H. T. *Ecological and General Systems: An Introduction to Systems Ecology*, University Press of Colorado, 2003.

Ortiz C. and M. C. Boyce. Bioinspired structural materials. *Science* **319**, 1053 (2008).

Parker A. R. and C. R. Lawrence. Water capture by a desert beetle. *Nature* **414**, 33–34 (2001).

Parker A. R. and H. E. Townley. Biomimetics of photonic nanostructures. *Nat. Nanotechnol.* **2**, 347 (2007).

Pokroy B., A. K. Epstein, M. C. M. Persson-Gulda, and J. Aizenberg. Fabrication of bio-inspired actuated nanostructures with arbitrary geometry and stiffness. *Adv. Mater.* **21**, 463 (2009).

Song X. Y., J. Zhai, Y. L. Wang, and L. Jiang. Fabrication of superhydrophobic surfaces by self-assembly and their water-adhesion properties. *J. Phys. Chem.* B **109**, 4048–4052 (2005).

Su B., Y. Tian, and L. Jiang. Bioinspired interfaces with superwettability: From materials to chemistry. *J. Am. Chem. Soc.* **138**, 1727–1748 (2016).

Sun T. L., L. Feng, X. F. Gao, and L. Jiang. Bioinspired surfaces with special wettability. *Acc. Chem. Res.* **38**, 644–652 (2005).

Videler J. J. *Avian Flight*. Oxford University Press.

Videler J. J., E. J. Stamhuis, and G. D. Povel. Leading-edge vortex lifts swifts. *Science* **306**, 1960–1962 (2004).

Vincent J. F. V., O. A. Bogatyreva, N. R. Bogatyrev, A. Bowyer, and A.-K. Pahl. Biomimetics: Its practice and theory. *J. Royal Soc. Interface* **3**(9), 471–482 (2006).

Winkleman A., G. Gotesman, A. Yoffe, and R. Naaman. Immobilizing a drop of water: Fabricating highly hydrophobic surfaces that pin water droplets. *Nano Lett.* **8**, 1241–1245 (2008).

Xi J. M., L. Feng, and L. Jiang. A general approach for fabrication of superhydrophobic and superamphiphobic surfaces. *Appl. Phys. Lett.* **92**, 053102 (2008).

Xia F. and L. Jiang. Bio-inspired, smart, multiscale interfacial materials. *Adv. Mater.* **20**, 2842 (2008).

Yao H. B., H. Y. Fang, Z. H. Tan, L. H. Wu, and S. H. Yu. Biologically inspired, strong, transparent, and functional layered organic–inorganic hybrid films. *Angew. Chem. Int. Ed.* **49**, 2140 (2010).

Zhang S., J. Huang, Z. Chen, and Y. Lai. Bioinspired special wettability surfaces: From fundamental research to water harvesting applications. *Small* **13**, 1602992 (2017).

Zhao N., Q. D. Xie, X. Kuang, S. Q. Wang, Y. F. Li, X. Y. Lu, S. X. Tan, J. Shen, X. L. Zhang, Y. Zhang, J. Xu, and C. C. Han. A novel ultra-hydrophobic surface: Statically non-wetting but dynamically non-sliding. *Adv. Funct. Mater.* **17**, 2739–2745 (2007).

Zhao Y. J., Z. Y. Xie, H. C. Gu, C. Zhu, and Z. Z. Gu. Bio-inspired variable structural color materials. *Chem. Soc. Rev.* **41**, 3297 (2012).

Zhao Y., Q. H. Lu, D. S. Chen, and Y. Wei. Superhydrophobic modification of polyimide films based on gold-coated porous silver nanostructures and self-assembled monolayers. *J. Mater. Chem.* **16**, 4504–4509 (2006).

Zheng Y., D. Han, J. Zhai, and L. Jiang. In situ investigation on dynamic suspending of microdroplet on lotus leaf and gradient of wettable micro- and nanostructure from water condensation. *Appl. Phys. Lett.* **92**, 084106 (2008).

Zheng Y., H. Bai, Z. Huang, X. Tian, F.-Q. Nie, Y. Zhao, J. Zhai, and L. Jiang. Directional water collection on wetted spider silk. *Nature* **463**, 640–643 (2010).

Zheng Y., X. Gao, and L. Jiang. Directional adhesion of superhydrophobic butterfly wings. *Soft Matter* **3**, 178–182 (2007).

Index